CELL BIOLOGY ASSAYS

ESSENTIAL METHODS

CELL BIOLOGY ASSAYS

ESSENTIAL METHODS

Edited by

GERI KREITZER
Weill Medical College of Cornell University

FANNY JAULIN
Weill Medical College of Cornell University

CEDRIC ESPENEL
Weill Medical College of Cornell University

AMSTERDAM • BOSTON • HEIDELBERG • LONDON • NEW YORK • OXFORD
PARIS • SAN DIEGO • SAN FRANCISCO • SINGAPORE • SYDNEY • TOKYO
Academic Press is an imprint of Elsevier

Academic Press is an imprint of Elsevier
32 Jamestown Road, London NW1 7BY, UK
30 Corporate Drive, Suite 400, Burlington, MA 01803, USA
525 B Street, Suite 1900, San Diego, CA 92101-4495, USA

First edition 2010

Material in this work originally appeared in *Cell Biology, Third Edition*, edited by Julio E. Celis, (Elsevier, Inc 2006).

British Library Cataloguing-in-Publication Data
A catalogue record for this book is available from the British Library

Library of Congress Cataloging-in-Publication Data
A catalog record for this book is available from the Library of Congress

ISBN : 978-0-12-375152-2

Typeset by Macmillan Publishing Solutions
www.macmillansolutions.com

Printed and bound in United States of America

10 11 12 13 14 15 10 9 8 7 6 5 4 3 2 1

Contents

IV

MEASURING INTRACELLULAR SIGNALING EVENTS USING FLUORESCENCE MICROSCOPY

V

IN SITU CELL ELECTROPORATION TO STUDY OF SIGNALING PATHWAYS

VI

CYTOSKELETON ASSAYS

Preface

One of the primary aims of biology is to understand how expression, activation and inactivation of genes and proteins is orchestrated to generate the wide variety of cells, organs and tissues that comprise a living organism. Ultimately, elucidation of regulatory networks controlling these processes will provide key mechanistic and molecular insight into normal development and have the potential to lead to the discovery of cellular factors that can be targeted in attempts to combat numerous diseases that affect cell and tissue architecture and function. Regardless of where the work starts, at some point the need to study specific genes and proteins within the cellular milieu, becomes central toward achieving this goal.

The chapters included in this volume, taken from *Cell Biology: A Laboratory Handbook, 3rd edition*, describe an array of cell biological assays that can be used to define and characterize experimentally the machinery that regulates events critical to the function of diverse cell types. Broadly defined, these encompass methods to analyze translocation of proteins between sub-cellular compartments, intracellular signaling and cytoskeletal organization. Each of these processes is key to the definition of cell shape, function and responsiveness to both intracellular and extracellular cues.

The methods described in these chapters can be separated into three general categories. The first group of assays lays out the rationale and techniques to measure protein transport between sub-cellular compartments or the plasma membrane *in vitro*, in semi-intact cells and in intact cells. Each of these assays have been used to gain fundamental insight into the mechanisms regulating exocytosis and endocytosis, nuclear-cytoplasmic shuttling and translocation of proteins (including co-translational transport into the endoplasmic reticulum) between intracellular compartments. In combination, these techniques provide researchers with a full arsenal of approaches with which to identify, characterize and determine the minimal essential machinery for protein transport in cells. The second group of assays provides in depth methods to study intracellular signaling and localized protein activity within single, living cells. These assays take advantage of vital fluorescent sensors to measure local changes in intracellular calcium, protein conformation, protein phosphorylation and nucleotide triphosphate occupancy, all read-outs of protein activity. In addition, these assays incorporate methods to control and analyze signaling events with tremendous spatial and temporal resolution in cells. The third group of methods detail *in vitro* and cell-based assays to evaluate the regulation and organization of actin and microtubule networks in response to individual proteins, defined cytosol, signaling cascades and environmental insult by bacteria. These assays also provide researchers with the ability to identify molecular effectors of cytoskeletal dynamics and organization that play key roles in how cells respond physically to a variety of physiological stimuli. Results obtained in any one groups of assays can be fruitfully incorporated into experiments using the others to gain more complete insight into regulation individual cellular processes. Indeed, *in vitro* experiments showing that motor proteins move along actin and microtubules have been combined successfully with protein transport and signaling assays to identify cytoskeletal motors responsible for numerous protein and organelle transport events in cells.

Using detailed experimental examples, the chapters in this volume examine a wide range of techniques and assays that can be applied in most laboratories. While several of these methods are technically demanding, most are straight-forward, utilize equipment, tools and supplies commonly available at research institutions, and can be adapted to suit specific questions of individual scientists. In this era of cell biology, the ability to apply diverse technical approaches enables researchers to address, at a molecular level, the many questions associated with complex biological events. Together, the step-by-step instructions with detailed discussion of each technique make this laboratory handbook an essential resource.

List of Contributors

Celia Antonio Department of Biochemistry & Molecular Biophysics, College of Physicians & Surgeons, Columbia University, 701 W 168ST HHSC 724, New York, NY 69117

Nathalie Q. Balaban Department of Physics, The Hebrew University-Givat Ram, Racah Institute, Jerusalem, 91904, ISRAEL

William E. Balch Department of Cell and Molecular Biology, The Scripps Research Institute, 10550 North Torrey Pines Road, La Jolla, CA 92037

Stefanie Benesch Department of Cell Biology, Gesellschaft fur Biotechnoogische Forschung, Mascheroder Weg 1, Braunschweig, D-38124, GERMANY

Heather L. Brownell Office of Technology Licensing and Industry Sponsored Research, Harvard Medical School, 25 Shattuck Street, Gordon Hall of Medicine, Room 414, Boston, MA 02115

Nigel P. Carter The Wellcome Trust, Sanger Institute, The Wellcome Trust, Genome Campus, Hinxton, Cambridge, CB10 1SA, UNITED KINGDOM

Maria Carmo-Fonseca Institute of Molecular Medicine, Faculty of Medicine, University of Lisbon, Av. Prof. Egas Moniz, Lisbon, 1649-028, PORTUGAL

Samit Chatterjee Margaret M. Dyson Vision Research Institute, Department of Ophthalmology, Weill Medical College of Cornell University, 1300 York Avenue, New York, NY 10021

Mark S. F. Clarke Department of Health and Human Performance, University of Houston, 3855 Holman Street, Garrison—Rm 104D, Houston, TX 77204-6015

Pascale Cossart Unite des Interactions Bacteries-Cellules/Unité INSERM 604, Institut Pasteur, 28, rue du Docteur Roux, Paris Cedex 15, F-75724, FRANCE

Robert A. Cross Molecular Motors Group, Marie Curie Research Institute, The Chart, Oxted, Surrey, RH8 0TE, UNITED KINGDOM

Ami Deora Margaret M. Dyson Vision Research Institute, Department of Ophthalmology, Weill Medical College of Cornell University, 1300 York Avenue, New York, NY 10021

Bernhard Dobberstein Zentrum fur Molekulare Biologie, Universitat Heidelberg, Im Neuenheimer Feld 282, Heidelberg, D-69120, GERMANY

Daniel L. Feeback Space and Life Sciences Directorate, NASA-Johnson Space Center, 3600 Bay Area Blvd, Houston, TX 77058

Kevin L. Firth ASK Science Products Inc., 487 Victoria St, Kingston, Ontario, K7L 3Z8, CANADA

Margarida Gama-Carvalho Faculty of Medicine, Institute of Molecular Medicine, University of Lisbon, AV. Prof. Egas Moniz, Lisbon, 1649-028, PORTUGAL

Susan M. Gasser Friedrich Miescher Institute fr Biomedical Research, Maulbeerstrasse 66, Basel, CH-1211, SWITZERLAND

Benjamin Geiger Department of Molecular Cell Biology, Weizman Institute of Science, Wolfson Building, Rm 617, Rehovot, 76100, ISRAEL

Cemal Gurkan Department of Electron Microscopy. The Cyprus Institute of Neurology and Genetics, Nicosia 1683, Cyprus

Jean Gruenberg Department of Biochemistry, University of Geneva, 30, quai Ernest Ansermet, Geneva 4, CH-1211, SWITZERLAND

Gerald Hammond Molecular Neuropathobiology Laboratory, Cancer Reasearch UK London Research Institute, 44 Lincoln's Inn Fields, London, WC2A 3PX, UNITED KINGDOM

Rebecca Heald Molecular and Cell Biology Department, University of California, Berkeley, Berkeley, CA 94720-3200

Leda Helen Raptis Department of Microbiology and Immunology, Queen's University, Room 716 Botterell Hall, Kingston, Ontario, K7L3N6, CANADA

Florence Hediger Department of Molecular Biology, University of Geneva, 30, Quai Ernest Ansermet, Geneva, CH-1211, SWITZERLAND

Klaus P. Hoeflich Division of Molecular and Structural Biology, Ontario Cancer Institute, Department of Medical Biophysics, University of Toronto, 610 University Avenue, 7-707A, Toronto, Ontario, M5G 2M9, CANADA

Elina Ikonen The LIPID Cell Biology Group, Department of Biochemistry, The Finnish National Public Health Institute, Mannerheimintie 166, Helsinki, FIN-00300, FINLAND

Mitsuhiko Ikura Division of Molecular and Structural Biology, Ontario Cancer Institute, Department of Medical Biophysics, University of Toronto, 610 University Avenue 7-707A, Toronto, Ontario, M5G 2M9, CANADA

Jeff A. Jones Space and Life Sciences Directorate, NASA-Johnson Space Center, TX77058

Ralph H. Kehlenbach Hygiene-Institut-Abteilung Virologie, Universitat Heidelberg, Im Neuenheimer Feld 324, Heidelberg, D-69120, GERAMNY

Geri E. Kreitzer Cell and Developmental Biology, Weill Medical College of Cornell University, LC-300, New York, NY 10021

Anna Koffer Physiology Department, University College London, 21 University Street, London, WC1E 6JJ, UNITED KINGDOM

Frank Lafont Department of Biochemistry, University of Geneva, 30, quai Ernest-Ansermet 1211, Geneva 4, CH-1211, SWITZERLAND

Paul LaPointe Department of Cell and Molecular Biology, The Scripps Research Institute, 10550 North Torrey Pines Road, La Jolla, CA 92037

Andre Le Bivic Groupe Morphogenese et Compartimentation Membranaire, UMR 6156, IBDM, Faculte des Sciences de Luminy, case 907, Marseille cedex 09, F-13288, FRANCE

Chuan-PU Lee The Department of Biochemistry and Molecular Biology, Wayne State University School of Medicine, 4374 Scott Hall, 540 E. Canfield, Detroit, MI 48201

Silvia Lommel Department of Cell Biology, German Research Center for Biotechnology (GBF), Mascheroder Weg 1, Braunschweig, D-38124, GERMANY

Patti Lynn Peterson Department of Neurology, Wayne State University School of Medicine, 5L26 Detroit Receiving Hospital, Detroit Medical Center, Detroit, MI 48201

Anne Muesch Developmental and Molecular Biology, Albert Einstein College of Medicine, Bronx, NY10461

Alan D. Marmorstein Cole Eye Institute, Weill Medical College of Cornell Cleveland Clinic, 9500 Euclid Avenue, i31, Cleveland, OH 44195

Bruno Martoglio Institute of Biochemistry, ETH Zentrum, Building CHN, Room L32.3, Zurich, CH-8092, SWITZERLAND

Atsushi Miyawaki Laboratory for Cell Function and Dynamics, Advanced Technology Center, Brain Science Institute, Institute of Physical and Chemical Research (RIKEN), 2-1 Horosawa, Wako, Saitama, 351-0198, JAPAN

Frank R. Neumann Department of Molecular Biology, University of Geneva, 30, Quai Ernest Ansermet, Geneva, CH-1211, SWITZERLAND

Hendrik Otto Institut für Biochemie und Molekularbiologie, Universität Freiburg, Hermann-Herder-Str. 7, Freiburg, D-79104, GERMANY

Bryce M. Paschal Center for Cell Signaling, University of Virginia, 1400 Jefferson Park Avenue, West Complex Room 7021, Charlottesville, VA 22908-0577

Javier Pizarro Cerdá Unite des Interactions Bacteries-Cellules/Unité INSERM 604, Institut Pasteur, 28, rue du Docteur Roux, Paris Cedex 15, F-75724, FRANCE

Helen Plutner Department of Cell and Molecular Biology, The Scripps Research Institute, 10550 North Torrey Pines Road, La Jolla, CA 92037

Linda J. Robinson

Enrique Rodriguez-Boulan Margaret M Dyson Vision Research Institute, Department of Opthalmology, Weill Medical College of Cornell University, New York, NY 10021

Manfred Rohde Department of Microbial Pathogenicity, Gesellschaft fur Biotechnoogische Forschung, Mascheroder Weg 1, Braunschweig, D-38124, GERMANY

Sabine Rospert Institut für Biochemie und Molekularbiologie, Universität Freiburg, Hermann-Herder-Str. 7, Freiburg, D-79104, GERMANY

Ulrich S. Schwarz Theory Division, Max Planck Institute of Colloids and Interfaces, Potsdam, 14476, GERMANY

Antonio S. Sechi Institute for Biomedical Technology-Cell Biology, Universitaetsklinikum Aachen, RWTH, Pauwelsstrasse 30, Aachen, D-52057, GERMANY

James R. Sellers Cellular and Motility Section, Laboratory of Molecular Cardiology, National Heart, Lung and Blood Institute (NHLBI), National Institutes of Health, 10 Center Drive, MSC 1762, Bethesda, MD 20892-1762

Kai Simons Max Planck Institute of Molecular Cell Biology and Genetics, Pfotenhauerstrasse 108, Dresden, D-01307, GERMANY

Evi Tomai Department of Microbiology and Immunology, Queen's University, Room 716 Botterell Hall, Kingston, Ontario, K7L3N6, CANADA

Angela Taddei Department of Molecular Biology, University of Geneva, 30, Quai Ernest Ansermet, Geneva 4, CH-1211, SWITZERLAND

Kevin Truong Division of Molecular and Structural Biology, Ontario Cancer Institute, Department of Medical Biophysics, University of Toronto, 610 University Avenue, 7-707A, Toronto, Ontario, M5G 2M9, CANADA

Isabelle Vernos Cell Biology and Cell Biophysics Programme, European Molecular Biology Laboratory, Meyerhofstrasse 1, Heidelberg, D-69117, GERMANY

Adina Vultur Department of Microbiology and Immunology, Queen's University, Room 716 Botterell Hall, Kingston, Ontario, K7L3N6, CANADA

Xiaodong Wang Southwestern Medical Center, Department of Biochemistry, Dallas, TX 75390

Jürgen Wehland Department of Cell Biology, Gesellschaft fur Biotechnoogische Forschung, Mascheroder Weg 1, Braunschweig, D-38124, GERMANY

Ye Xiong The Department of Biochemistry and Molecular Biology, Wayne State University School of Medicine, 4374 Scott Hall, 540 E. Canfield, Detroit, MI 48201

Charles Yeaman Department of Cell and Developmental Biology, Weill Medical College of Cornell University, New York, NY 10021

Chiara Zurzolo Department of Cell Biology and Infection, Pasteur Institute, 25,28 rue du Docteur Roux, Paris, 75015, FRANCE

SECTION I

MEMBRANE PROTEIN TRANSLOCATION ASSAYS: EXOCYTOSIS, ENDOCYTOSIS, ER-GOLGI, TRANSLOCATION THROUGH THE ER

CHAPTER

1

Permeabilized Epithelial Cells to Study Exocytic Membrane Transport

Frank Lafont, Elina Ikonen, and Kai Simons

OUTLINE

I. INTRODUCTION

Polarized exocytic transport in epithelial cells can be measured by performing *in vitro* assays on filter-grown Madin–Darby canine kidney cells (MDCK strain II cells). Techniques have been developed to establish conditions where cytosol-dependent transfer of a viral marker protein is monitored in MDCK cells permeabilized with the cholesterol-binding and pore-forming toxin *Streptococcus* streptolysin-O (SLO). The transport is assayed either early in the biosynthetic pathway, from the endoplasmic reticulum (ER) to the Golgi complex, or from the *trans*-Golgi network (TGN) to the apical or from the TGN to the basolateral plasma membrane (for a detailed characterization of the assays and their properties, see Lafont *et al.*, 1998).

The assays provide the possibility to analyze the effects of exogeneous molecules added to the cytosol. Cytosolic molecules can be inactivated prior to their addition to permeabilized cells or depleted using antibodies. Membrane-impermeable molecules can gain access to lipids or membrane proteins facing the cytosol. We have successfully used fluorescent lipid analogues, chemicals, antibodies, peptides, purified proteins, and toxins. The specificity of the inhibition of transport after depleting cells for a selected cytosolic molecule may be tested by rescuing transport when the purified molecule is added back to the transport reaction. Efficiency of the transport steps can also be tested after cells have been treated with drugs affecting biosynthetic pathways, e.g., after cholesterol biosynthesis inhibition using methyl-β-cyclodextrin.

II. MATERIALS AND INSTRUMENTATION

The recombinant SLO fused to a maltose-binding protein is obtained from Dr. S. Bhakdi (University of Mainz, Mainz, Germany). SLO is also now available from several commercial sources. (Hersey and Steiner (1985)). PhosphorImager and ImageQuant software are purchased from Molecular Dynamics. Sonifier cell disruptor B 15 is purchased from Branson. Media and reagents for cell culture are purchased from Gibco Biocult and Biochrom.; chemicals are from Merck; ATP (Cat. No. A-6033), antipain (Cat. No. A-6271), $CaCO_3$ (Cat. No. C-5273), chymostatin (Cat. No. C-7268), cytochalasin D (Cat. No. C-5394), leupeptin (Cat. No. L-2884), pepstatin (Cat. No. P-4265), and pyruvate (Cat. No. P2256) are from Sigma; creatine kinase (Cat, No. 127566), creatine phosphate (Cat. No. 621714), dimathyl sulfoxide (DMSO) (Cat. No. D-5879), endoglycosidase H (Cat. No. 1088734), and NADH (Cat No. 107727) are from Boehringer Mannheim; trypsin and soybean trypsin inhibitor are from Worthington Biochemical Corporation; protein A–Sepharose (Cat. No. 17-0780-01) is from Pharmacia; and cell filters (0.4-μm pore size; No. 3401, 12-mm-diameter Transwell polycarbonate filters) are from Costar.

III. PROCEDURES

A. SLO Standardization

Each batch of toxin is standardized for the amount of lactate dehydrogenase (LDH) released from the filter-grown MDCK cells, and the amount of LDH released is determined according to a previously described protocol. Alternatively, it is possible to use antibody cytosol accessibility to monitor SLO permeabilization.

Solutions

1. *Growth medium (GM)*: Eagle's minimal essential medium with Earle's salts (E-MEM) supplemented with 10 m*M* HEPES, pH 7.3, 10% (v/v) fetal calf serum, 2 m*M* glutamine, 100 U/ml penicillin, and 100 μg/ml streptomycin.
2. Either purified SLO or recombinant SLO fused to a maltose-binding protein is obtained from Dr. S. Bhakdi (University of Mainz, Mainz, Germany). The recombinant SLO preparation is strored as lyophilisate at −80°C.
3. *10X KOAc transport buffer*: To prepare 1 liter, dissolve 250 m*M* HEPES (59.6 g); 1150 m*M* KOAc (112.9 g), and 50 m*M* $MgCl_2$ (25 ml from 1 *M* stock) in 800 ml water. Adjust the pH to 7.4 with ~50 ml 1 *M* KOH and make it 1 liter with water.
4. *KOAc+ buffer*: KOAC buffer containing 0.9 m*M* $CaCO_3$ (90 mg/liter) and 0.5 m*M* $MgCl_2$ (0.5 ml/liter from 1 *M* stock)
5. *Assay buffer:* KOAc plus 0.011% Triton X-100 for media and KOAc for filter
6. NADH 14 mg/ml KOAc
7. 60 m*M* pyruvate: 30 mg in 50 ml water

Steps

1. Grow MDCK cells strain II on 1.2-mm-diameter filters for 3.5 days *in vitro* and change the medium every 24 h.
2. Wash cells by dipping in KOAc.
3. Prepare a range of SLO, e.g., from 0.5 to 10 μg of recombinant SLO per filter. Activate the SLO for 30 min at 37°C in 50 μl KOAc plus 10 m*M* dithiothreitol (DTT) per filter. Place the filters on a Parafilm sheet put on a metal plate on an ice bucket.

4. Add 50 μl of activated SLO on the apical side for 15 min at 4°C.
5. Wash excess SLO by dipping the filters twice in KOAc+.
6. Transfer the filters to a 12-well dish containing 0.75 ml TM at 19.5°C. Add 0.75 ml TM to the apical side and incubate at 19.5°C for exactly 30 min in a water bath.
7. Collect apical and basolateral media, cut the filters, and shake the filters with 1 ml KOAc containing 0.1% Triton X-100.
8. Mix in a disposable cuvette:
   a. 800 μl of assay buffer (KOAc or KOAc/Triton X-100)
   b. 200 μl of sample
   c. 10 μl of NADH
   d. 10 μl of pyruvate.

   Turn the cuvette upside down twice using Parafilm to close it.
9. Read immediately the OD at 340 nm for 60 s (the OD should be around 1.5).
10. Calculate the LDH activity as the change in OD in 10 s, taking the average of the first 30 s.
11. Dissolve the recombinant SLO in KOAc buffer and store in aliquots at −80°C with the amount necessary for one set of assays per aliquot.

B. Preparation of Cytosols

Solutions

1. *Growth medium*: For HeLa cells, use Joklik's medium supplemented with 50 ml/liter newborn calf serum (heat inactivated for 30 min at 56°C), 2 m*M* glutamine, 100 U/ml penicillin, 100 μg/ml streptomycin, and 150 μl 10 *M* NaOH. For MDCK cells, use Eagle's minimal essential medium with Earle's salts (E-MEM) supplemented with 10 m*M* HEPES, pH 7.3, 10% (v/v) fetal calf serum, 2 m*M* glutamine, 100 U/ml penicillin, and 100 μg/ml streptomycin.
2. *10X KOAc transport buffer*: To prepare 1 liter, dissolve 250 m*M* HEPES (59.6 g); 1150 m*M* KOAc (112.9 g), and 50 m*M* $MgCl_2$ (25 ml from 1 *M* stock) in 800 ml water. Adjust the pH to 7.4 with ~50 ml 1 *M* KOH and make it 1 liter with water.
3. *PBS+*: PBS containing 0.9 m*M* $CaCl_2$ and 0.5 m*M* $MgCl_2$.
4. *Swelling buffer (SB)*: For 100 ml, prepare 1 m*M* EGTA 0.5 *M* (200 μl), 1 m*M* $MgCl_2$ 1 *M* (100 μl), 1 m*M* DTT 1 *M* (100 μl), and 1 μ*M* cytochalasin D 1 m*M* (100 μl).
5. *Protease inhibitor cocktail (CLAP)*: To prepare a 1000X stock, dissolve antipain, chymostatin, leupeptin, and pepstatin each at 25 μg/ml DMSO and combine.

Steps

HeLa Cytosol

1. To grow 20 liters of HeLa cells in suspension to a density of 6×10^5 cells/ml:
   a. Inoculate 250 ml cell suspension (6×10^5 cells/ml) in 1 liter medium and leave stirring in a 37°C room.
   b. On the third day, split 1:4 by adding 250 ml to 750 ml of fresh medium. Leave stirring in the 37°C room. This will give 4 liters of cell suspension.
   c. On the fifth day, again split 1:4 by adding 1 liter of the cell suspension to 3 liters of fresh medium in a 6-liter round-bottom flask. If the cells seem overgrown, add some fresh medium. Thus, between 16 and 20 liters of cell suspension are obtained.
2. Concentrate the cell suspension to 2 liters either by centrifugation or with any cell-concentrating system.
3. Centrifuge cells at 5000 rpm in a Sorvall GS-3 rotor (400-ml buckets) at 4°C for 20 min.
4. Discard the supernatant and wash cells by resuspending the pellet with 10 ml of ice-cold PBS with a sterile 10-ml pipette (10 ml PBS/bucket). Pool the suspension in 250-ml Falcon tubes.

5. Spin at 3000 rpm for 10 min at 4°C in a minifuge. The pellet will be loose. Discard the supernatant.
6. Fill the tube with cold SB (about 35 ml). Resuspend the cells carefully with a sterile 10-ml pipette in SB. Let the pellet swell for 5 min on ice.
7. Spin in the minifuge at 2000 rpm at 4°C for 10 min. Remove as much as possible of the supernatant. The pellet should be loose.
8. Transfer the pellet to a 30-ml Dounce homogenizer (use a spatula) and, after adding SB, perform five strokes. Add 0.1 volume of 10× KOAc and homogenize further by 15–20 strokes.
9. Spin the homogenate at 4°C at 10,000 rpm for 25 min in SS-34 tubes (Sorvall). Collect the supernatant.
10. Spin at 50,000 rpm in Ti70 tubes (Beckmann) for 90 min at 4°C. Aliquot the supernatant in screw-top 1.5-ml tubes. Avoid including lipids that might be on the top of the supernatant. Freeze the aliquots in liquid N_2 and store at −70°C.
11. Measure the protein concentration; it should be around 5 mg/ml. Lower concentrations do not work.
12. When needed, thaw the aliquots quickly and keep them on ice (up to 6 h) until use. Aliquots can be refrozen at least twice.

MDCK Cytosol

1. Trypsinize thirty 24 × 24-cm dishes of confluent MDCK cells. Resuspend trypsinized cells in cold growth medium (containing 5% FCS to inactivate trypsin) and leave on ice until all dishes are trypsinized and cells are pooled. After this, all handling should be done on ice using ice-cold solutions.
2. Centrifugate at 4°C for 10 min at 2000 rpm in a RF Heraeus centrifuge. Wash medium away with PBS.
3. Wash in PBS containing 2 mg/ml soybean trypsin inhibitor (STI). Wash out STI with PBS and resuspend the cell pellet in the SB (cf. see earlier discussion) and keep 10 min on ice.
4. Centrifugate at 2000 rpm for 10 min at 4°C in a RF Heraeus centrifuge.
5. Sonicate the loose cell pellet (~10 ml) (power 6, 0.5-s pulse, sonifier cell disruptor B 15, Branson) until cells are broken as judged by light microscopy. Add 0.1 volume of 10X KOAc to the sample and spin it for 20 min at 3000 rpm.
6. Spin the supernatant again at 75,000 rpm for 1 h in a TLA 100.2 rotor.

The procedure routinely yields ~6 ml of cytosol at 14 mg/ml. Other cell types, e.g., NIH 3T3 fibroblasts, can also be used as starting material for cytosol preparation using the protocol just described. We have not observed significant differences between the efficiencies of the different cytosols to support transport. However, HeLa cytosol is used routinely because of the ease of preparing large quantities.

C. Transport Assays

Solutions

All solutions are sterilized and kept at 4°C unless indicated.

1. *Growth medium*: E-MEM supplemented with 10 m*M* HEPES (10 ml/liter), pH 7.3, 10% (v/v) fetal calf serum, 2 m*M* glutamine 200 m*M* (10 ml/liter), 100 U/ml penicillin 10^4 U/ml (10 ml/liter), and 100 μg/ml streptomycin 10^4 μg/ml (10 ml/liter).
2. *Infection medium (IM)*: E-MEM supplemented with 10 m*M* HEPES, pH 7.3, 0.2% (w/v) BSA, 100 U/ml penicillin, and 100 μg/ml streptomycin.
3. *Labeling medium (LM)*: Methionine-free E-MEM containing 0.35 g/liter sodium bicarbonate instead of the usual 2.2 g/liter 10 m*M* HEPES, pH 7.3, 0.2% (w/v) BSA supplemented with 16.5 μCi of [^{35}S]methionine/filter.

4. *Chase medium (CM)*: Labeling medium without [^{35}S]methionine and containing 20 μg/ml cycloheximide and 150 μg/ml cold methionine.
5. *10X KOAc transport buffer*: To prepare 1 liter, dissolve 59.6 g HEPES (250 m*M*); 112.9 g KOAc (1150 m*M*), and add 25 ml from 1 *M* stock $MgCl_2$ (50 m*M*) in 800 ml water. Adjust the pH to 7.4 with ~50 ml 1 *M* KOH and make it 1 liter with water.
6. *KOAc+ buffer*: KOAC buffer containing 0.9 m*M* $CaCO_3$ (90 mg/liter) and 0.5 m*M* $MgCl_2$ (0.5 ml/liter from 1 *M* stock).
7. *PBS+*: PBS containing 0.9 m*M* $CaCl_2$ (0.13 g/liter $CaCl_2 \cdot 2H_2O$) and 0.5 m*M* $MgCl_2$ (0.5 ml/liter from 1 *M* stock).
8. *0.5 M EGTA*: To prepare 100 ml, dissolve 19 g in 60 ml water. The solution should be turbid (pH 3.5). Add slowly ~10 ml 10 *M* KOH. When the pH of the solution starts to clear, adjust to 7.4. Make it 100 ml with water.
9. *0.1 M Ca and 0.5 M EGTA*: To prepare 100 ml, stir 1 g $CaCO_3$ and 3.8 g EGTA in ~70 ml water for at least 45 min (degas). After adding 2 ml of 10 *M* KOH, the pH is about 6. Finally, make the pH 7.4 adding few drops of 1 *M* KOH. Make it 1 liter with water.
10. *Transport medium (TM)*: To prepare 50 ml, combine 50 μl 1 *M* DTT, 200 μl 0.5 *M* EGTA, and 1 ml 0.5 *M* EGTA/0.1 *M* $CaCO_3$ in 1X KOAc.
11. *ATP-regenerating system (ARS)*: 100X stock, prepare three solutions (10 ml each):
    - **a.** 100 m*M* ATP (disodium salt, pH 6–7, neutralized with 2 *M* NaOH; 0.605 g/10 ml)
    - **b.** 800 m*M* creatine phosphate (disodium salt, 2.620 g/10 ml)
    - **c.** 800 U/mg (at 37°C) creatine kinase (0.5 mg/10 ml in 50% glycerol).

    Store stocks in aliquots at −20°C. Mix solutions a–c 1:1:1 just before use.
12. *Lysis buffer (LB)*: PBS+ containing 2% NP-40 and 0.2% SDS.
13. *CLAP*: To prepare a 1000X stock, combine 25 μg/ml DMSO of antipain, chymostatin, leupeptin, and pepstatin.

Steps

The basic steps of the assays can be summarized as follows. Grow MDCK cells on a permeable filter support until a tight monolayer is formed. Grow cells on 12-mm-diameter filters and use when they display a transmonolayer electrical resistance of at least 50Ω × cm^2. Infect cell layers with either vesicular stomatitis (VSV) or influenza virus using the G glycoprotein of VSV (VSV G) or the hemagglutinin (HA) of influenza virus N as basolateral or apical markers, respectively. After a short pulse of radioactive methionine, block the newly synthesized viral proteins either in the ER or in the TGN. At this stage, one cell surface is permeabilized with SLO, which allows leakage of cytosolic proteins, whereas membrane constituents, including transport vesicles, are retained inside the cells. After removal of the endogenous cytosol by washing, add cytosol and ATP and raise the temperature to 37°C. Transport of the viral proteins from the ER to the Golgi complex or from the TGN to the apical or the basolateral plasma membrane is reconstituted in a cytosol-, energy-, and temperature-dependent manner. Measure the amount of viral proteins reaching the acceptor compartment by endoglycosidase H treatment (ER to Golgi transport), trypsinization of HA on the apical surface (TGN to apical transport), or immunoprecipitation of the VSV G at the basolateral surface (TGN to basolateral transport). Obtain quantitations of viral polypeptides resolved on SDS–PAGE by PhosphorImager analysis.

Due to the handling and the various steps, the entire procedure for running one assay requires about 9 h for the apical assay and 11 h for the basolateral assay. Carry out immunoprecipitation for the basolateral assay the following day (takes 5 h). The ER-to-Golgi transport takes about 6 h before an overnight enzymatic treatment step. Samples can be frozen at −20°C before running SDS–PAGEs (routinely performed the following day). It is worth noting that for the reproducibility of the results the assays should be performed strictly, obeying the schedule indicated in the protocol.

Grow the N strain influenza virus (A/chick/Germany/49/Hav2Neq1) in 11-day embryonated chick eggs as described. (Bennett et al., 1988) Prepare a stock of phenotypically mixed VSV (Indiana strain) grown in Chinese hamster ovary C15.CF1 cells, which express HA on their plasma membrane as described. Prepare the affinity-purified antibody raised against the luminal domain of the VSV G as described (Matlin et al., 1983).

Apical Transport Assay

1. *Cell culture*: Seed 1:72 of the MDCK cells from one confluent 75-cm^2 flask per filter (6×10^4 cells/filter) with 0.75 ml of growth medium apically and 2 ml basolaterally. Change the growth medium every 24 h and use cells 3.5 days after plating.
2. *Infection*: Prior to viral infection, wash the monolayers in warm PBS+ and then IM and infect with the influenza virus in 50 μl IM (20 pfu/cell; enough to obtain 100% infection as judged by immunofluorescence) on the apical side. After allowing adsorption of the virus to the cells for 1 h at 37°C, remove the inoculum and continue the infection for an additional 3 h after adding 0.75 ml IM on the apical side and 2 ml IM on the basolateral side of the filter.
3. *Metabolic labeling*: Rinse the cell monolayers by dipping in beakers containing warm PBS+ and LM at 37°C. Place a drop of 25 μl of the LM containing 12.5 μCi of [^{35}S]methionine on a Parafilm sheet in a wet chamber at 37°C in a water bath. Add 100 μl of only LM to the apical side of the filters and place them basal side down on the drop. Incubate for 6 min at 37°C.
4. *TGN block*: Terminate metabolic pulse labeling by moving the filters to a new 12-well plate containing 1.5 ml CM already at 20°C. Add 0.75 ml of CM (20°C) on the apical side and incubate for 75 min in a 19.5°C water bath.
5. *SLO permeabilization*: Activate the SLO in KOAc buffer containing 10 m*M* DTT at 37°C for 30 min and keep at 4°C until used for permeabilization. Use the toxin within 1 h of activation. Wash the filters twice in ice-cold KOAc+ by dipping. Carefully remove excess buffer from the apical side and blot the basolateral side with a Kleenex. Place 25-μl drops of the activated SLO (enough to release 60% LDH in 30 min, about 20 mg/ml) on a Parafilm sheet placed on a metal plate on an ice bucket. Leave the apical side without buffer. Place the filters on the drop for 15 min. Wash the basolateral surface twice by dipping in ice-cold KOAc+ buffer. Transfer the filters to a 12-well dish containing 0.75 ml TM at 19.5°C. Add 0.75 ml TM to the apical side and incubate at 19.5°C for exactly 30 min in a water bath.
6. *Transport*: Remove the filters from the water bath, rinse them once with cold TM, blot the basolateral side, and place on a 35-μl drop on either TM (control) or HeLa cytosol (±treatment or molecule to be tested) supplemented each with ARS (3 μl/100 μl TM or HeLa cytosol). Add the ARS to TM or cytosol immediately before dispensing the drops onto which the filters are placed. Layer 100 μl of TM on the apical side and incubate at 4°C for 15 min in a moist chamber. Transfer the chamber to a warm water bath for 60 min at 37°C. Terminate the transport by transferring on ice and washing the filters with cold PBS+ three times.
7. *Trypsinization*: Add to the apical surface 250 μl of 100 μg/ml trypsin freshly prepared in PBS+ and then add 2 ml of PBS+ to the basolateral chamber. Keep on ice for 30 min and stop the reaction by adding 3 μl of soybean trypsin inhibitor (STI; 10 mg/ml) to the apical side and then wash the apical surface three times with PBS+ containing 100 μg/ml STI.

8. *Cell lysis*: Solubilize the monolayers in 100 μl LB containing freshly added CLAP. Scrape the cells and spin for 5 min in a microfuge; discard the pellet. Analyze 20 μl per sample by SDS–PAGE on a 10% acrylamide gel, fix, and dry the gel.
9. *Quantitation*: Scan the gels with a PhophorImager and measure the band intensities using ImageQuant software. Calculate the amount of HA transported as the percentage of HA (68 kDa) transported to the cell surface = [2 × HA2/(HA + 2 × HA2) × 100] with HA2 (32 kDA) being the small trypsin cleavage product of HA.

Basolateral Transport Assay

1. *Cell culture*: Seed 1:72 of MDCK cells from one confluent 75-cm^2 flask per filter (6×10^4 cells/filter) with 0.75 ml of growth medium apically and 2 ml basolaterally. Change the growth medium every 24 h and use cells 3.5 days after plating.
2. *Infection*: Prior to viral infection, wash the monolayers in warm PBS+ and then in IM and infect with the vesicular stomatitis virus in 50 μl IM (20 pfu/cell; enough to obtain 100% infection as judged by immunofluorescence) on the apical side. After allowing adsorption of the virus to the cells for 1 h at 37°C, remove the inoculum and continue the infection for an additional 3 h after adding 0.75 ml IM on the apical side and 2 ml IM on the basolateral side of the filter.
3. *Metabolic labeling and chase*: Rinse the cell monolayers by dipping in beakers containing warm PBS+ and then LM at 37°C. Place a drop of 25 μl of the LM containing 12.5 μCi of [^{35}S]methionine on a Parafilm sheet in a wet chamber at 37°C in a water bath. Add 100 μl of LM to the apical side of the filter and place the filter on the drop. Incubate for 6 min at 37°C and incubate for an additional 6 min at 37°C in CM with 0.75 ml on the apical side and 2 ml on the basolateral side before the 19.5°C block.
4. *TGN block*: Terminate the pulse by moving the filters to a new 12-well plate containing 1.5 ml CM already at 20°C. Add 0.75 ml of CM (20°C) on the apical side and incubate for 60 min in a 19.5°C water bath.
5. *SLO permeabilization*: Activate the SLO in KOAc buffer containing 10 m*M* DTT at 37°C for 30 min and keep at 4°C until used for permeabilization. Use the toxin within 1 h of activation. Wash the filters twice in ice-cold KOAc+ by dipping. Carefully remove excess buffer from the apical side and blot the basolateral side with a Kleenex. Add 50-μl drops of the activated SLO (enough to release 60% LDH in 30 min, about 40 mg/ml) on a Parafilm sheet placed on a metal plate layered on an ice bucket. Leave the basolateral side without buffer. Incubate for 15 min. Wash the apical surface twice with 0.75 ml of ice-cold KOAc+ buffer. Transfer the filters to a 12-well dish containing 0.75 ml TM at 19.5°C. Add 0.75 ml TM to the apical side and incubate at 19.5°C for exactly 30 min in a water bath.
6. *Transport*: Remove the filters from the water bath, rinse them once with cold TM, blot the basolateral side, and add 100 μl of either TM (control) or HeLa cytosol (±treatment or molecule to be tested) supplemented with ARS each (3 μl/100 μl TM or HeLa cytosol) on the apical side. Add the ARS to TM or cytosol prior to dispension onto drops. Incubate at 4°C for 15 min. Put the filters on already dispensed 35-μl drops of TM supplemented with ARS on a Parafilm in a moist chamber in a water bath at 37°C. Incubate for 60 min. Terminate the transport by transferring on ice and washing the filters with cold PBS+ three times.
7. *Anti-VSV G binding*: Wash twice (2–5 min) with 2 ml of CM containing 10% FCS (CM-FCS) on the basolateral side and once with 0.75 ml on the apical side. Dilute anti-VSV-G antibodies in CM-FCS (1:18, i.e., 50 μg/ml; 300 μl) and place the filters on 25-μl

drops of PBS+/ab (1.5 μl ab/filter). Add nothing on the apical side and incubate in a cold room on a metal plate placed onto an ice bucket for 90 min. Remove the filters and wash (three times) on the basolateral side with CM-FCS (2 ml with constant rocking) and once on the apical side. Shake the filters gently for efficient removal of unbound antibodies. Wash the filters once with PBS+ and place each on a 25-μl drop of PBS+ supplemented with cold virus (1:25; 300 μl). Incubate for 10 min in the cold.

8. *Cell lysis*: Solubilize the monolayers in 200 μl LB containing freshly added CLAP and cold virus (1:166). Scrape the cells, spin for 5 min, and discard the pellet. Remove a 10-μl aliquot from each sample (total).
9. *Immunoprecipitation*: Wash protein A–Sepharose powder with PBS (3×), let it swell 10 min in PBS, wash it with LB, and store as 1:1 slurry in LB at 4°C for 3 weeks maximum. Add 30 μl of the 1:1 slurry of protein A–Sepharose to the lysate. Rotate in a cold room for 60 min. Spin the resin down and wash (3×) with 500 μl LB, elute the bound sample with 35 μl 2X Laemmli buffer, and boil for 5 min at 95°C. Load 20 μl of bound material and 10 μl of total (after boiling with 10 μl of 2X Laemmli buffer) on SDS–PAGE gels (10%). Run SDS–PAGE on 10% acrylamide gels, fix, and dry the gels.
10. *Quantitation*: Scan the gels with a PhophorImager and measure band intensities using ImageQuant software. Calculate the amount of VSV G (67 kDa) transported as follows: percentage of VSV G on the cell surface = (surface immunoprecipitated VSV G / total VSV G) × 100.

ER-to-Golgi Transport Assay

In the ER-to-Golgi transport assay, influenza N- or VSV- infected MDCK monolayers can be used. An infection with the influenza virus is used here as an example.

1. *Cell culture*: Seed 1:72 of the MDCK cells from one confluent 75-cm^2 flask per filter with 0.75 ml of growth medium apically and 2 ml basolaterally. Change the growth medium every 24 h and use cells 3.5 days after plating.
2. *Infection*: Prior to viral infection, wash the monolayers in warm IM and then infect with the influenza virus in 50 μl IM (20 pfu/cell; enough to obtain 100% infection as judged by immunofluorescence) on the apical side. After allowing adsorption of the virus to the cells for 1 h at 37°C, remove the inoculum and continue the infection for an additional 3 h after adding 0.75 ml IM on the apical side and 2 ml IM on the basolateral side of the filter.
3. *Metabolic labeling*: Rinse the cell monolayers by dipping in beakers containing warm PBS+ and then LM at 37°C. Place a drop of 25 μl of the LM containing 12.5 μCi of $[^{35}S]$methionine on a Parafilm sheet in a wet chamber at 37°C in a water bath. Add 100 μl of LM to the apical side of the filter and place the filter on the drop. Incubate for 6 min at 37°C.
4. Terminate the pulse by moving the filters to a new 12-well plate containing 1.5 ml CM already at 4°C. Add 0.75 ml of CM (4°C) on the apical side and incubate for 30 min at 4°C.
5. *SLO permeabilization*: Activate the SLO in KOAc buffer containing 10 m*M* DTT at 37°C for 30 min and keep at 4°C until used for permeabilization. Use the toxin within 1 h of activation. Wash the filters twice in ice-cold KOAc+ by dipping. Carefully remove excess buffer from the apical side and blot the basolateral side with a Kleenex. Place 25-μl drops of the activated SLO (enough to release 60% LDH in 30 min, about 20 mg/ml) on a Parafilm sheet placed on a metal plate layered on an ice bucket. Leave the apical side without buffer. Place the filters on the drops for 15 min. Wash the basolateral

surface twice by dipping in ice-cold KOAc+ buffer. Transfer the filters to a 12-well dish containing 0.75 ml TM at 4°C. Add 0.75 ml TM to the apical side and incubate at 37°C for 3 min and 1.5 ml on the basolateral side of the filter (formation of pores), followed by an incubation at 4°C for 20 min in fresh TM on both sides (cytosol depletion).

6. *Transport*: Rinse the filters once with cold TM, blot on the basolateral side, and put on a 35-μl drop of either TM (control) or HeLa cytosol (±treatment or molecule to be tested) supplemented with ARS each (3 μl/100 μl TM or HeLa cytosol). Add the ARS to TM or cytosol prior to dispension onto drops. Layer the apical side with 100 μl of TM and incubate at 4°C for 15 min in a moist chamber. Transfer the chamber to a water bath at 37°C for 45 min. Terminate the transport by transferring on ice and washing the filters with cold PBS+ three times.
7. *Cell lysis*: Solubilize the monolayers in 100 μl LB containing freshly added CLAP. Scrape the cells and spin for 5 min in microfuge; discard the pellet.
8. *Endoglycosidase H treatment*: Remove a 75-μl aliquot and add to 25 μl of 0.2 *M* sodium citrate buffer, pH of 5.0. The resulting 100 μl mixture has a pH of 5.3. Divide it in two 50-μl aliquots. One receives 5 μl of 1 U/ml endoglycosidase H and the other receives only the citrate buffer. After 20 h at 37°C, terminate the reaction by boiling in Laemmli buffer. Analyze the samples by running SDS–PAGE on a 10% acrylamide gel, fix, and dry the gel.
9. *Quantitation*: Scan the gels with a PhophorImager and measure band intensities using ImageQuant software. Calculate the amount of HA transported as the percentage of HA acquiring endoglycosidase H resistance with the following formula: percentage of HA reaching the Golgi complex = (endo H-resistant HA / total HA) × 100.

IV. COMMENTS AND PITFALLS

In all cases the values are expressed as control cytosol-dependent transport being 100% (transport in the presence of cytosol minus transport in the absence of added cytosol). For each manipulation a matched control is used (e.g., antibody or peptide tested vs. control antibody or peptide, respectively). Assays are carried out routinely on duplicate filters and quantifications represent the mean ± SEM obtained in several experiments.

A critical parameter for the successful performance of these transport assays is the quality of SLO. MDCK cells are difficult to permeabilize compared to several other cell types (e.g., BHK, CHO, and L cells), and with the available commercial sources of SLO, the degree of permeabilization, as measured by LDH release, has not been satisfactory. The wild-type toxin purified from *Streptococci* and the recombinant toxin produced in *Escherichia coli* have both worked equally well.

The available amount of reagent often determines whether the preincubation during cytosol depletion is possible, as cytosol depletion must be carried out in an excess volume of buffer. The routinely used volume (750 μl per filter) can be, however, somewhat reduced. By using 500 μl per filter there is not yet a significant effect on transport efficiency, and by using 200 μl per filter cytosol depletion is compromised moderately, which increases the background, cytosol-independent transport, and results in a transport efficiency of about three-fourths of normal.

Because the transport is carried out in a leaky cellular microenvironment with diluted cytosolic components, the increase in transport obtained with exogenous cytosol is usually two- to threefold (three- to fourfold in the ER-to-Golgi assay). This is the window in which the differences in cytosol dependent transport are measured. In assays measuring transport in the late secretory pathway, part of the efficiency is lost due to retention of some viral marker early in the exocytic route. However, because both markers, HA and VSV G, are glycoproteins whose mobility on SDS–PAGE shifts according to the degree of glycosylation, careful examination of their mobilities will reveal, in apical and basolateral assays, if the test condition retarded significantly the processing of the marker to the terminally glycosylated form. This may therefore serve as an internal control for the specificity of inhibition. A more accurate way to test the effect of a reagent in the early secretory pathway is to assay the ER-to-Golgi transport. The real advantage of having established similar procedures for three different transport assays is the possibility of using them as internal controls for each other. This enables the identification of molecules that are specifically involved in either apical or basolateral transport routes and allows the discrimination between compounds that are needed only in the polarized routes versus those that are common to all three transport processes.

References

Bennett, M.K., Wandinger-Ness, A., and Simons, K. (1988). Release of putative exocytic transport vesicles from perforated MDCK cells. *Embo. J.* **7**, 4075–4085.

Hersey, S.J., and Steiner, L. (1985). Acid formation by permeable gastric glands: enhancement by prestimulation. *Am. J. Physiol.* **248**, G561–8.

Lafont, F., Verkade, P., and Simons, K. (1998). Annexin XIIIb associates with lipid microdomains to function in apical delivery. *J. Cell Biol.* **142**, 1413.

Matlin, K., et al. (1983). Trans epithelial transport of a viral membrane glycoprotein implanted into the apical plasma membrane of Madine-Darby canine kidney cells. I. Morphological evidence *J. Cell. Biol.* **97**, 627–637.

2

Studying Exit and Surface Delivery of Post-Golgi Transport Intermediates Using *In Vitro* and Live-Cell Microscopy-Based Approaches

Geri E. Kreitzer, Anne Muesch, Charles Yeaman, and Enrique Rodriguez-Boulan

OUTLINE

I. INTRODUCTION

Study of the mechanisms of polarized protein sorting in epithelial cells has been facilitated greatly by the use of enveloped RNA viruses, such as vesicular stomatitis virus (VSV) and influenza virus, which bud from the basolateral and apical plasma membranes, respectively (Rodriguez-Boulan and Sabatini, 1978). Following infection, a rapid onset of viral protein synthesis occurs, leading to the vectorial transport of envelope glycoproteins to either the apical or the basolateral surface. This model continues to provide information on the mechanisms of protein sorting (Musch *et al.*, 1996) and the basic protocols are included here. However, because cells cannot be coinfected efficiently with both types of viruses due to reciprocal inhibition of protein synthesis, a major drawback of this paradigm is the inability to study the segregation of apical and basolateral proteins from one another in the same cell.

Two alternative approaches have been developed that improve upon and greatly facilitate studying the molecular effectors of protein sorting in the *trans*-Golgi network (TGN) and polarized transport routes to the plasma membrane in epithelial cells. These approaches utilize either recombinant adenovirus vectors or intranuclear microinjection of cDNAs to introduce exogenous biosynthetic markers into cells. Both methodologies advance previous techniques in numerous ways: (i) they allow for high-level, simultaneous expression of two markers (Marmorstein *et al.*, 2000); (ii) they are amenable to the use of temperature blocks, which allow for accumulation in and synchronous release of newly synthesized proteins from the TGN; (iii) neither method interferes with the ability of cells to synthesize and transport endogenous proteins, permitting the study of marker proteins in a normal cellular environment; (iv) adenoviral infection generally results in transduction of all cells in the culture and is thus ideal for metabolic labeling studies and biochemical analysis of biosynthetic events; (v) microinjection results in the rapid expression of cDNAs, providing a means by which to study anterograde membrane trafficking events selectively and dynamically in individual cells; and (vi) cDNAs or adenoviral particles can be introduced easily into a wide variety of cultured cells, making it relatively simple to compare secretory sorting pathways in different multiple cell types.

The most important advance provided by cDNA microinjection and adenoviral-mediated gene transfer in studying protein-sorting events is the ability to cointroduce two or more genes into cells and express simultaneously multiple secretory cargoes that follow divergent routes out of the TGN. This allows one to evaluate the role(s) of potential molecular effectors of protein sorting and targeting to different cellular domains. Furthermore, the level of expression of exogenous genes can be manipulated [by changing the amount of DNA introduced and the expression time allowed in microinjection-based assays or by varying either the multiplicity of infection (moi) or the time in culture following adenoviral infection] to allow study of the ability to saturate the various sorting pathways available to the cell (Marmorstein *et al.*, 2000). Adenoviral infection also results in high-level expression of reporter proteins in large cell populations, a factor essential in obtaining sufficient incorporation of radioactive amino acids for pulse–chase studies, as well as for immunoisolation of transport vesicles. (Once the adenovirus has been titrated and the infection conditions optimized, 100% of the cells express the desired proteins and remarkably consistent pools of cells are produced from experiment to experiment.) Procedures for adenoviral infection can be modified and adapted easily to a variety of different cell lines.

While cDNA microinjection is limited with respect to the number of cells that can be evaluated, it is exquisitely suited to live cell imaging studies aimed at evaluating highly dynamic membrane trafficking events occurring at specific points throughout the biosynthetic pathway (e.g., ER-to-Golgi events and Golgi-to-plasma membrane events, such as budding of Golgi membranes, transport of post-Golgi carriers, and exocytosis of post-Golgi carriers). Momentum in this area is due primarily to the advent of fluorescent tags, such as green fluorescent protein [GFP(Chalfie *et al.*, 1994)] from the jellyfish *Aequorea* and dsRed from the coral *Discosoma* (Matz *et al.*, 1999; Baird *et al.*, 2000), which can be genetically appended to any DNA of choice. These tags serve as vital fluorescent indicators that facilitate direct observation and analysis of membrane-trafficking events in living cells.

This article describes an assay that monitors post-Golgi vesicle budding from semi-intact MDCK cells following infection either with enveloped RNA viruses or with recombinant adenovirus vectors. The adenoviruses we have found most useful for these applications encode receptors for neurotrophins ($p75^{NTR}$) and low-density lipoprotein (LDLR), which were shown previously to be sorted to the apical and basolateral surfaces of polarized MDCK cells, respectively (Le Bivic *et al.*, 1991; Hunziker *et al.*, 1991; Gridstaff *et al.*, 1998). Each of these proteins, when expressed in MDCK cells, incorporates radiosulfate into carbohydrate moieties during posttranslational processing late in the secretory pathway, providing a convenient method to label markers in the sorting compartment.

In addition, this article describes methods we have developed pertaining to the use of time-lapse fluorescence imaging to study the transport of plasma membrane proteins through the biosynthetic pathway in living cells. While we only discuss this technique specifically with respect to MDCK epithelial cells, with slight modifications we have also successfully employed this approach for similar studies in several different cell lines. Additionally, we have found that these studies can be used with numerous reporter proteins of interest. Thus while we focus on reporters marking the apical or basolateral plasma membrane, the system is not limited to the study of membrane-associated proteins. Finally, this article discusses some quantitative measurements that can be made from this type of data related to protein-trafficking events occurring *in vivo*.

II. MATERIALS AND INSTRUMENTS

Dulbecco's modified Eagle's medium (DMEM, Cat. No. 10-013-CV), MEM nonessential amino acids (Cat. No. 25-025-CI), Hank's balanced salt solution (HBSS, Cat. No.21-023-CV), and l-glutamine (Cat. No. 25-005-CI) are from Cellgro. MEM SelectAmine kits (Cat. No. 19050), MEM vitamins (Cat. No. 11120), penicillin-streptomycin (Cat. No. 15140), 7.5% bovine serum albumin (BSA, Cat. No. 15260), and donor horse serum (Cat. No. 16050) can be purchased from Gibco-BRL (Grand Island, NY). Heat-inactivated fetal bovine serum (FBS, Cat. No. 100-106) is from Gemini Bioproducts. HEPES (Cat. No. H-4034), d-glucose (Cat. No. G-8270), and cycloheximide (Cat. No. C-7698) are from Sigma Chemicals. Polycarbonate filters (Transwells; 0.4 mm pore size, Cat. No. 3412 for 24-mm filters) can be purchased from Corning Costar Corp. (Cambridge, MA). Tissue culture grade plasticware is from Corning Plasticware. Tran^{35}S-label (Cat. No. 51006), [^{35}S]cysteine (Cat. No. 51002), and $H_2{}^{35}SO_4$ (Cat. No. 64040) can be purchased from ICN (Costa Mesa, CA). Reagents for the production of viruses are described elsewhere (Rodriguez-Boulan and Sabatini, 1978; see article by Hitt *et al.*) All other reagents are standard reagent grade and available from several sources. Sterile solutions are autoclaved or sterilized by ultrafiltration (0.2 mm).

There are numerous manufacturers of microscope heating and cooling chambers. We use the Harvard apparatus recording chamber (PDMI-2) and temperature controller (TC-202A). A variety of microinjection apparatuses are available from Narishige Inc. or from Eppendorf. Cooled charged-couple device (CCDs) cameras are also available from several manufacturers. For high-resolution time-lapse imaging, we recommend a camera capable of 12- to 16-bit digitization (4095–65,000 gray levels) with at least a 5- to 10-MHz controller for rapid acquisition rates. The Flaming Brown Micropipet puller (Model P-97) is from Sutter Instruments. Glass capillaries for pulling microinjection needles are available from World Precision Instruments (Cat. No. 1B100F-6). The Sykes-Moore culture chamber, consisting of a 32 × 7-mm chamber (Cat. No. 1943-11111), silicone gaskets (Cat. No. 1943-33315), and the wrench assembly (Cat. No. 1943-44444) can be purchased from Bellco. Number 1 thickness, 25-mm glass coverslips for the chamber (Cat. No. 483800-80) are from VWR.

III. PROCEDURES

A. Adenovirus Transduction

1. *Coinfection of MDCK Cells with Recombinant Adenovirus Vectors*

Solutions

1. *Dulbecco's modified Eagle's medium (DMEM)*: Dissolve DMEM powder in 1 liter H_2O. Sterilize by filtering through a 0.2-μm pore filter.
2. *Complete DMEM (cDMEM)*: Add 50 ml heat-inactivated fetal bovine serum (FBS), 10 ml 0.2 M l-glutamine, 5 ml penicillin–streptomycin (10,000 U/ml and 10 mg/ml, respectively), and 5 ml MEM nonessential amino acids to 430 ml DMEM. Store at 4°C for up to 2 weeks.

Steps

1. For infection with replication-defective viruses, grow cells on semipermeable polycarbonate filter supports. Seed cells at confluency on day 1 and culture for 4 days with medium changed daily. MDCK strain II cells are confluent at a density of ca. 7×10^5 cells/cm^2, so approximately 3.3×10^6 cells should be seeded on each 24-mm filter.
2. Before applying adenovirus vectors, rinse cultures twice with serum-free DMEM (2 ml for the apical chamber and 2.5 ml for the basolateral chamber). Removal of serum proteins from the culture medium results in enhanced adsorption of adenovirus particles to the cell surface and improves infection efficiency. MDCK cells do not appear to be affected adversely by serum deprivation during the infection period, but this should be checked with other cell types before attempting infection.
3. Add recombinant adenoviruses, diluted in serum-free DMEM, at moi ranging from 1 to 1000 pfu/cell. Incubate at 37° for 1 h, gently tilting the plates every 15 min to mix. Two or more adenovirus vectors can be mixed together and applied simultaneously to cells. It is recommended that serial three-fold dilutions of viruses be tested in order to determine the optimal moi required for the quantitative expression of the marker proteins.
4. Vectors must be applied to the *apical* domain of epithelial cells, as infection is markedly more efficient from this surface. The reason for this is unclear, but a preference for adenovirus entry through the apical surface has been observed in every polarized epithelial cell type we have examined, including canine (MDCK), bovine (MDBK), and porcine (LLC-PK_1) kidney; rat thyroid (FRT) and retinal pigment epithelia (RPE-J); and human intestine (Caco-2). Serum-free DMEM, without virus, should be applied to the basolateral chamber. Infection should be performed in a minimum volume of DMEM required to keep the filter submerged. A recommended volume is 0.25 ml apical/0.5 ml basolateral for a 24-mm filter.

5. Following the infection period, add 2 ml cDMEM to both apical and basolateral chambers. It is not necessary to aspirate the virus because the addition of serum effectively terminates the infection. Culture the cells at 37°C for the desired incubation time. Expression of adenovirus-encoded proteins should be detectable by 4–6 h following infection, will rise gradually over the next 18 h and should reach a plateau by 24 h. This level of expression will be maintained for at least 1 week, provided the cells are fed daily. We routinely use cultures of polarized MDCK cells between 20 and 24 h postinfection.
6. Monitor infection after 24 h as follows:
   a. *Use indirect immunofluorescent staining* to determine (i) the percentage of cells in the culture expressing the transfected gene products and (ii) the intracellular distribution of the gene products. Under optimum infection conditions, at least 95% of the cells will express each adenovirus-encoded protein. If the ultimate goal of the experiment is the study of the molecular trafficking of two or more proteins in the same cell, it is essential that all of the cells that express one marker also express the second marker. Methods for fixation, permeabilization, and staining of epithelial cells grown on polycarbonate filters are described elsewhere in this manual (see Chapter 4).
   b. *Use immunoprecipitation or Western blotting* to determine (i) the molecular weight of the adenovirus-encoded proteins and (ii) the level of expression of the proteins in the culture. Ideally, both adenovirus-encoded proteins will be expressed at comparable levels.

Pitfalls

Each batch of adenovirus virus vector must be tested for optimum transduction. We find a batch-to-batch variability in the correspondence of pfu obtained from plaque assays to the moi needed.

B. Infection of MDCK Cells with Enveloped RNA Viruses

Solutions

1. *DEAE-dextran (100X stock)*: Dissolve 100 mg DEAE-dextran (Sigma Cat. No. D1162) in 10 ml H_2O. Filter, sterilize, and store 1-ml aliquots at −20°C.
2. *Infection medium*: Add 13 ml 7.5% bovine serum albumin and 5 ml 1*M* HEPES, pH 7.4, to 482 ml DMEM. Filter, sterilize, and store at 4°C. Immediately before use, add 0.1 ml DEAE-dextran stock to 10 ml medium. DEAE-dextran is only necessary for infection with VSV.
3. *Virus stocks*: Vesicular stomatitis virus, Indiana strain (VSV), and influenza virus A (WSN strain) are grown in MDCK strain II cells, harvested, and plaque assayed as described by Rodriguez-Boulan and Sabatini (1978).

Steps

1. Set up 10-cm dishes of MDCK strain II cells, passages 6–20, and allow them to reach confluency. Cultures are infected with VSV or influenza virus 3 days after becoming confluent.
2. Before infecting cells, rinse cultures twice with infection medium. For viral infection, inoculate MDCK cells with 50 pfu/cell VSV or influenza WSN in 3.5 ml infection medium containing 0.1 mg/ml DEAE-dextran.
3. Incubate cultures for 1 h at 37°C.
4. Aspirate viral medium and rinse cultures twice with fresh infection medium.
5. Return VSV-infected cultures to 37°C and incubate a further 3.5 h before metabolic radiolabeling (see later). Incubate

influenza WSN-infected cultures for 4.5h at 37°C before labeling. Cultures should be examined hourly to monitor cytopathic effects.

C. Metabolic Radiolabeling and Accumulation of Marker Proteins in the Trans-Golgi Network

1. *Radiosulfate Labeling of Glycoproteins at 20°C*

Both $p75^{NTR}$ and LDLR are sulfated when expressed in MDCK cells. Sulfation occurs largely on asparagine-linked carbohydrate moieties on both proteins. Because this posttranslational modification occurs late in the secretory pathway, likely in the *trans*-Golgi or TGN, it provides a convenient method to label markers in the sorting compartment. When labeling is performed at the reduced temperature of 20°C, the labeled markers accumulate in the TGN because post-Golgi vesicular transport is inhibited.

Solutions

1. *Sulfate-free labeling medium*: Essentially, labeling medium is DMEM in which the $MgSO_4$ is replaced by $MgCl_2$. Combine 100ml 10× DME salts (Ca^{2+}, Mg^{2+}-free), 10ml 100× Ca^{2+}, Mg^{2+} stock, 10ml MEM Vitamins, 10ml of 1003 stock MEM amino acid solutions (arginine, glutamine, histidine, isoleucine, leucine, lysine, phenylalanine, threonine, tryptophan, tyrosine, glycine, serine, and valine), 1ml of 100× stock MEM solutions of methionine and cysteine, 10ml MEM nonessential amino acid solution, 20ml 1*M* HEPES, pH 7.4, 27ml 7.5% BSA stock, and H_2O to a final volume of 1000ml. Filter, sterilize, and store at 4°C.
2. *10× DME salts (Ca^{2+}, Mg^{2+}-free)*: Combine 50ml 100 × $Fe(NO_3)_3$, 50ml 100× NaH_2PO_4, 50ml 100 × KCl, 22.5g dextrose, 32g NaCl, and 15ml phenol red solution. Adjust volume to 500ml with H_2O. Filter, sterilize, and store at 4°C.
3. *$Fe(NO_3)_3$ stocks*: Add 0.05g $Fe(NO_3)_3$ to 50ml H_2O to prepare a 100,000× stock solution. Dilute 100µl into 100ml H_2O to prepare 100× stock solution. Filter, sterilize, and store at 4°C.
4. *100× Na H_2PO_4*: Add 1.25g NaH_2PO_4 to 100ml H_2O. Filter, sterilize, and store at 4°C.
5. *100× KCl*: Add 4.0g KCl to 100ml H_2O. Filter, sterilize, and store at 4°C.
6. *100× Ca^{2+}, Mg^{2+}*: Add 2.96g $CaCl_2 \cdot 2H_2O$ and 3.02g $MgCl_2 \cdot 6H_2O$ to 100ml H_2O. Filter, sterilize, and store at 4°C.

Steps

1. Twenty-four to 48h following adenovirus infection, aspirate culture medium and rinse filters three times with sulfate-free labeling medium. Sulfate starve cells for 30min at 37°C in this medium.
2. Label cells for 1h at 20°C in sulfate-free labeling medium containing $H_2{}^{35}SO_4$. To label cells on one 75mm Transwell filter, we use 0.5mCi $H_2{}^{35}SO_4$ in 500µl sulfate-free labeling medium. Place medium, preequilibrated at 20°C, on a sheet of Parafilm in a humid chamber and place the Transwell filter upon this so that label is exposed to the basolateral surface. Apply 2.5ml sulfate-free labeling medium, without label, to the apical chamber to prevent drying.

2. *Pulse–Chase Labeling with [35S]Methionine/Cysteine*

Solutions

1. *Methionine/cysteine-free labeling medium*: Prepare 1000ml of medium following the product specification insert (Gibco SelectAmine kit, Cat. No. 19050), excluding the methionine and cysteine in the kit. Add

10 ml 1*M* HEPES and 27 ml 7.5% BSA stock solution. Filter, sterilize, and store at 4°C. For PC12 cells, replace the BSA will 20 ml dialyzed serum. To prepare the dialyzed serum, under sterile conditions, dialyze a mixture of two-thirds fetal bovine serum and one-third horse serum against PBS in 12,000 molecular weight cutoff dialysis tubing for 12–20 h. Dialyze serum to remove small molecules, which may be used to scavenge sulfate.

2. *Chase medium*: Add 5 ml of 100× MEM methionine and cysteine (left over from the Selectamine kit) solutions to 40 ml complete DMEM (or PC12 growth medium for PC12 cells). Immediately before use, add cyclohexamide to a concentration of 20 μg/ml.

Steps

1. Rinse MDCK cultures three times in methionine/cysteine-free labeling medium before incubation at 37°C for the final 30 min of incubation following viral infection. For PC12 cells, extensive rinsing may not be possible, so rinse once before the 30-min incubation.
2. Label VSV-infected MDCK cells with [^{35}S]methionine/cysteine (Tran^{35}S-label, ICN Cat. No. 51006). Label influenza WSN-infected cells with [^{35}S]cysteine (ICN Cat. No. 51002). Use 0.5 mCi, in a total volume of 3.5 ml labeling medium, to label each 10-cm plate. Pulse–label cells for 10 min at 37°C. Medium can be recycled twice if multiple dishes are to be labeled. Label LDLR- and p75-infected PC12 cells with [^{35}S]cysteine, due to the fact that these proteins are not sulfated as they are in MDCK cells. Use 0.5 mCi, in a total volume of 1 ml labeling medium, to label each 10-cm plate. Pulse–label cells for 15 min at 37°C, with gentle rocking every 5 min.
3. For MDCK cells, aspirate labeling medium and rinse plates three times with chase medium. For PC12 cultures, just aspirate medium.
4. Chase cultures at 20°C for 2 h in chase medium. During the chase at 20°C, roughly 60% of the labeled VSV G/influenza HA proteins are accumulated in the TGN (Musch *et al.*, 1996).

D. Vesicle Budding from the TGN in Semi-intact Cells

Semi-intact MDCK cells are prepared after accumulating marker proteins in the TGN (Musch *et al.*, 1996). Cells are first swollen in a low salt buffer and are subsequently scraped from the substratum, which produces large tears in the plasma membrane. Endogenous cytosol and peripheral membrane proteins are removed by washing with a high salt buffer. Addition of an exogenous source of cytosol, an energy-regenerating system, and incubation at 37°C typically result in the release of 25–65% of the total marker accumulated in the TGN into sealed vesicles. Budded vesicles are separated from the material that remains by a brief, low-speed centrifugation step.

Solutions

1. *Swelling buffer*: Add 7.5 ml 1M HEPES/KOH, pH 7.2, and 7.5 ml 1*M* KCl to 485 ml H_2O. Store at 4°C.
2. *10× transport buffer*: Add 20 ml 1*M* HEPES/KOH, pH 7.2, 2 ml 1*M* $Mg(OAc)_2$, and 18 ml 5*M* KOAc to 60 ml H_2O. Store at 4°C.
3. *1× transport buffer*: Add 1 ml 10× transport buffer to 9 ml H_2O. Immediately before use, bring to 1 m*M* DTT and add protease inhibitors to 1× concentration.
4. *High salt buffer*: Add 10 ml 1*M* HEPES/KOH, pH 7.2, 50 ml 5*M* KOAc, and 1 ml 1*M* $Mg(OAc)_2$ to 4391 ml H_2O. Store at 4°C. Immediately before use, bring to 1 m*M* DTT and add protease inhibitors to 1X concentration.

5. *1M dithiothreitol stock*
6. *500× protease inhibitor stock*: Dissolve 5 mg of each of the following inhibitors individually in 330 μl dimethyl sulfoxide (DMSO) and combine the three solutions.
   a. *Pepstatin A*: (Sigma Cat. No. P-4265)
   b. *Leupeptin*: (Sigma Cat. No. L-8511)
   c. *Antipain*: (Sigma Cat. No. A-6191)
7. *100 mM PMSF stock*
8. *Energy mix*: Pipette in the following order: 3 μl ATP, 2 μl GTP, 4 μl creatine phosphate, and 3 μl creatine kinase.
   a. *0.1M ATP* (Boehringer Mannheim Cat. No. 519 987)
   b. *0.2M GTP* (Boehringer Mannheim Cat. No. 106 399)
   c. *0.6M creatine phosphate* (Boehringer Mannheim Cat. No. 621 722)
   d. *8 mg/ml creatine kinase* (Boehringer Mannheim Cat. No. 127 566)
9. *Bovine brain cytosol*: Prepare gel-filtered bovine brain cytosol in batches exactly as described previously (Malhotra *et al.*, 1989). The protein concentration should be 10–20 mg/ml. Snap freeze 50-μl aliquots in liquid nitrogen and store at −80°C.

Steps

All steps are performed on ice, unless otherwise specified.

1. Following the 20°C incubation, wash the monolayer twice briefly with ice-cold swelling buffer and incubate in the same for 15 min.
2. Scrape cells from the filter (or plastic dish) with a rubber policeman into 2.5 ml transport buffer. DiSPo scrapers (Baxter Scientific, McGaw Park, IL) work well for this purpose. The best way to scrape cells from the Transwell filter is to place the filter inside the lid of the dish so that the bottom lies flat against the plastic. This prevents the scraper from poking through the filter, but care must be exercised to prevent tearing the filter. Scraping does not have to be vigorous, but should be done with long, gentle strokes. Transfer cells to 1.5-ml microfuge tubes and rinse the filter (or dish) with 2.5 ml fresh transport buffer. Combine with cells from first scraping. Discard the filter as radioactive waste.
3. Pellet cells by centrifugation at 800 *g* for 5 min in a refrigerated microfuge.
4. Pool semi-intact cells into one tube and wash with 1.5 ml high salt buffer on ice for 10 min. Pellet cells by centrifugation at 800 *g* for 5 min in a refrigerated microfuge.
5. Resuspend cells in transport buffer. A volume of 250 μl is used to resuspend cells from one filter (or dish).
6. Set up vesicle budding assay.
   a. In standard assays, suspend semi-intact cells (ca. 10 μl = 20–25 μg protein) in an assay volume of 50 μl transport buffer supplemented with 50 μg gel-filtered bovine brain cytosol and an energy-regenerating system (1 m*M* ATP, 1 m*M* GTP, 5 m*M* creatine phosphate, and 0.2 IU creatine kinase). Combine components in the following order: mix 26 μl H_2O, 1 μl 100 m*M* PMSF, 4 μl 10× transport buffer, 4 μl energy mix, 5 μl cytosol (final concentration = 1 mg/ml), and 10 μl semi-intact cells.
   b. Important controls include assays in which either the cytosol or the energy mix or both components are omitted. For the complete depletion of energy from the system, cytosol and semi-intact cells must be preincubated for 10 min on ice with 0.6 U/ml apyrase before assembling the assay. In the absence of either cytosol or energy, vesicular release from semi-intact cells should be negligible. We suggest that serial dilutions of cytosol (i.e., 0.1–10 mg/ml) be tested in order to determine the optimal range of cytosol-dependent vesicle budding in the assay.

7. Incubate assays at 37°C for desired time. In standard assays, we incubate for 30–45 min. However, we suggest that a time course of vesicle budding be performed to optimize the assay for different marker proteins. As additional controls, two complete assays should be assembled and incubated at 0 and 20°C. At these reduced temperatures, vesicular release from semi-intact cells should be insignificant.
8. Pellet semi-intact cells by centrifugation at 800 *g* for 5 min in a refrigerated microfuge. Transfer supernatant fractions to clean microfuge tubes. The pellets (containing nonbudded material remaining in the TGN) and supernatants (containing the vesicles released during the 37°C incubation) can be analyzed further.
   - **a.** To quantify the efficiency of vesicular release of each marker under different conditions, samples can be lysed in SDS–PAGE sample buffer and analyzed directly by PAGE. In cells infected with VSV or influenza WSN, the viral proteins should be the only labeled proteins in the lysates. Alternatively, following adenovirus-mediated transfer of cDNAs encoding $p75^{NTR}$ and LDLR into MDCK cells, these proteins are by far the most heavily labeled proteins when radiosulfate is used as a precursor.
   - **b.** To confirm that markers are present inside sealed vesicles, the supernatant fraction should be treated with either proteinase K or trypsin. In the absence of Triton X-100, only the cytoplasmic domains of the proteins will be cleaved and this can be detected as a relatively small mobility increase during SDS–PAGE. In contrast, protease treatment in the presence of 1% Triton X-100 will result in complete digestion of markers. Use three 50-μl assay samples for this analysis. To tube 1, add nothing. To tube 2, add 2.5 μl protease (10 mg/ml stock → 0.5 mg/ml final). To tube 3, add 2.5 μl protease and 2.5 μl 20% Triton X-100. Incubate on ice for 30 min. Inactivate protease with 1 m*M* PMSF or 1 mg/ml soybean trypsin inhibitor before lysing samples in SDS–PAGE sample buffer. Analyze products by SDS–PAGE.
   - **c.** Immunoisolation of specific classes of transport vesicles is performed using antibodies against the cytoplasmic portions of cargo proteins as well as appropriate negative controls. (i) Use 5 mg protein A-Sepharose (Pharmacia, Piscataway, NJ; Cat. No. 17-0780-01) for each immunoisolation. Swell in transport buffer for 10 min. (ii) If using a murine monoclonal primary antibody, use a bridge. Incubate 5 mg protein A–Sepharose with 50 μg rabbit antimouse IgG (Rockland Labs, supplied by VWR, New York, Cat. No. 610-4102) in 1 ml transport buffer for 60 min at room temperature. Wash twice with transport buffer. (iii) Block nonspecific binding sites in 1 ml transport buffer containing 0.2% BSA for 60 min at room temperature. (iv) Couple primary antibody for 2 h at room temperature in transport buffer. Wash twice with transport buffer. Titrate each antibody to determine amount needed for quantitative recovery of vesicles. (v) Incubate immunoadsorbant with the supernatant fraction from the vesicle budding assay in a total volume of 1 ml transport buffer for 2–18 h at 4°C with end-over-end rotation. In some cases, vesicle coat proteins may mask epitopes on the cytoplasmic tails of cargo. Therefore, it may be necessary to wash the vesicles in high salt buffer prior to immunoisolation to strip coat proteins. Add 33 μl of 1 *M* KOAc to 50-μl vesicles. Incubate on ice for 10 min. Add 333 μl salt-free transport buffer [20 m*M* HEPES/KOH, 2 m*M* $Mg(OAc)_2$]. (vi) Wash immunoprecipitates six times with transport

buffer. Elute bound markers by boiling 5 min in SDS–PAGE sample buffer. Analyze by SDS–PAGE.

E. Vesicle Budding from TGN-Enriched Membranes

A TGN-enriched membrane fraction is prepared from metabolically labeled PC12 cells after marker proteins have been accumulated in the TGN. Initially, a postnuclear supernatant is prepared following the method of Tooze and Huttner (1992). From the postnuclear supernatant a TGN-enriched membrane fraction is prepared (Xu *et al.*, 1995). As with semi-intact MDCK cells, addition of an exogenous source of cytosol and an energy-regenerating system leads to the release of accumulated marker protein from the TGN in sealed vesicles. All steps for the vesicle budding assay are identical to those for intact MDCK cells except that 50-μg aliquots of TGN-enriched membranes are used for each individual assay condition.

F. Expression of cDNA Using Microinjection

Solutions

1. *Preparation of cDNA stocks for microinjection*: Prepare DNA using either a midi or maxi preparation, making sure to suspend the DNA pellet (final concentration of at least 0.2 mg/ml) in sterile water rather than Tris–EDTA. DNA stocks can be stored at either 4 or −20°C.
2. *HEPEs:KCl microinjection buffer*: 10 m*M* HEPES, 140 m*M* KCl, pH 7.4
3. *cDNA for microinjection*: Dilute cDNA stock in HKCl to a final concentration of 5–20 μg/ml (see later for how to choose a concentration)
4. *MDCK cell culture medium*: DMEM prepared as per manufacturer's instructions. Add 50 ml FBS (10% FBS final concentration) and 10 ml of 1*M* HEPES, pH 7.4 (20 m*M* final concentration), and 5 ml of nonessential amino acids to 435 ml DMEM.

Steps

For microinjection, cells must be cultured on sterilized glass coverslips.

1. Sterilize coverslips using either of the following methods.
   a. ***Autoclaving:*** Place coverslips in a glass petri dish and autoclave on the dry cycle for 20 min. If cells adhere well to the glass, this method is most convenient, as many coverslips can be sterilized at once.
   b. ***Acid washing:*** Place coverslips in histology staining racks. Wash coverslips in a beaker of 2*N* hydrochloric acid (2× 5-min washes), rinse in distilled water (2 × 5-min rinses), and wash in 100% ethanol (3 × 2 min). After the final ethanol wash, place the staining racks containing the cleaned coverslips into a dry beaker, cover with heavy-duty foil, and bake at 250°C for 1 h. Acid treatment "etches" the glass, creating a somewhat rough surface that is useful if cells are not sufficiently adherent on the coverslips.
2. Place the sterilized coverslips into a 10-cm tissue culture dish. Seven 25-mm coverslips can be placed in one 10-cm dish.
3. Trypsinize cells and seed onto coverslips placed previously in the 10-cm culture dish(es).
   a. ***For experiments in nonpolarized cells:*** Seed 5×10^5 cells onto coverslips. Use a cell stock that is actively dividing (i.e., sparse cells). Cells should be used between 36 and 48 h after they are seeded onto the coverslips.

b. *For experiments in polarized cells:* Seed cells at confluency onto coverslips. Change the medium 24h after plating and culture the cells another 2–4 days prior to use. Do not change the culture medium again as this can result in changes in cell morphology.

4. On the day of the experiment, transfer coverslips to 3.5-cm culture dishes. Add fresh medium and microinject the cDNA into cell nuclei. As one goal of these experiments is to achieve synchronized exogenous protein expression in a population of cells it is important that you only inject cells on the same coverslip for ~5 min. If a sufficient number of cells was not injected in that time, take another coverslip from the incubator and repeat.
5. After injection, place cells into the incubator and wait for protein to be expressed.
6. Monitor and identify the minimum expression time.[1] We routinely find that 1h is sufficient for the expression of many different cDNAs in both nonpolarized and polarized MDCK cells. However, we have also found that there is heterogeneity in expression time depending on the cDNA being injected, as well as on the cell types being used. Minimum expression time can be determined by merely looking at the injected cells at a series of time points (e.g., 1-h intervals) after microinjection. Fluorescently tagged reporter proteins can be visualized directly without fixation. If your reporter is not fluorescently tagged, then fix the cells at 1-h intervals after microinjection and immunostain with antibodies against the exogenous protein.

G. Synchronizing Transport through the Biosynthetic Pathway

Solutions

1. *Bicarbonate-free MDCK cell culture medium*: DMEM without bicarbonate prepared as per manufacturer instructions. Add 25 ml FBS (5% FBS final concentration), 10 ml of 1*M* HEPES, pH 7.4 (20 m*M* final concentration), and cycloheximide (final concentration 100 μg/ml) to 475 ml DMEM.
2. *Recording medium*: Hank's balanced salt solution with calcium and magnesium (HBSS-CM). Add 5 ml FBS (1% FBS final concentration; serum does autofluoresce so it is important to keep the concentration low), 25 ml of 9% d-glucose (4.5 g/liter glucose final concentration), 5 ml HEPES, pH 7.4 (10 m*M* HEPES final concentration), and cycloheximide (final concentration 100 μg/ml) to 465 ml HBSS-CM.

Steps

Synchronization of protein trafficking can be achieved through the use of a series of temperature shifts. Cycloheximide added during the temperature shifts will effectively create a pulse of newly synthesized protein that can be chased synchronously from ER to Golgi and from the Golgi to the plasma membrane. Newly synthesized protein can be accumulated in the ER or the Golgi when cells are incubated at 15 or 20°C, respectively. These proteins will leave the Golgi when cells are shifted to the permissive temperature for secretion, (30–37°C). Because most laboratories do not maintain a tissue culture incubator set to 15 or 20°C (our laboratory uses a small refrigerator set to the desired temperature), you will need to use a bicarbonate-free

[1] The minimum expression time is the time at which exogenous, newly expressed membrane protein is present in the endoplasmic reticulum, but is not yet found at the plasma membrane. It is critical that no exogenous protein is at the cell surface for studies of Golgi to plasma membrane trafficking.

medium during these incubation periods. Check that this medium remains between pH 7.2 and 7.4 during the course of the temperature blocks.

One hour after microinjection (or the minimum expression time for your protein of interest, see earlier discussion), place cells into bicarbonate-free medium with cycloheximide.

1. To study ER to Golgi events, place cells into recording medium and incubate at 15°C in the thermally regulated recording chamber mounted on a microscope for time-lapse imaging.
   - **a.** Monitor the total cellular fluorescence by acquiring images of your cells at 10-min intervals. When total cellular fluorescence stabilizes for ~10 min, acquire and save both transmitted light and fluorescent images of the cells you will be studying.
   - **b.** Increase the temperature of the recording chamber to 20°C. Wait 5 min for the temperature and focus to stabilize.
   - **c.** Acquire time-lapse images to evaluate ER to Golgi transport events. ER to Golgi transport can be studied at either low or high spatial and temporal resolution depending on the questions being addressed.
2. To study Golgi to plasma membrane events, place cells into bicarbonate-free medium and incubate at 20°C.
   - **a.** Determine the time required to accumulate newly synthesized protein in the Golgi. We routinely find that 1–3 h is sufficient, but this time varies from protein to protein. The extent to which new protein has accumulated in the Golgi can be determined by assessing the degree of colocalization with Golgi markers at a series of time points (e.g., 30-min intervals) after shifting to 20°C.
   - **b.** After accumulating protein in the Golgi in a 20°C incubator, place cells into recording medium and incubate at 20°C in the recording chamber mounted on a microscope for time-lapse imaging. Acquire and save both transmitted light and fluorescent images of the cells you will be studying.
   - **c.** Increase the temperature of the recording chamber to between 33°C and 37°C. Wait 5 min for the temperature and focus to stabilize.
   - **d.** Acquire time-lapse images to evaluate post-Golgi transport events. Post-Golgi transport can be studied at either low or high spatial and temporal resolution depending on the questions being addressed.
   - **e.** At the end of every time-lapse recording, save a transmitted light image of the cells from which you recorded data.

H. Kinetics of Protein Transport through the Secretory Pathway

Steps

To determine the rates at which a fluorescent protein moves from the ER to the Golgi and then to the plasma membrane using time-lapse fluorescence microscopy, it is necessary to introduce a pulse of fluorescence into individual cells. We find that microinjection of cDNA is best for this purpose, as the expression of injected cDNA is generally rapid and can be controlled temporally. The kinetics of ER-to-Golgi and Golgi-to-plasma membrane cargo transport can be determined by measuring the ratio of Golgi-associated fluorescence/total fluorescence over time. It is imperative to be able to distinguish Golgi-associated from non-Golgi-associated fluorescent signals. This can be done in two ways. First, you can coexpress a fluorescently tagged Golgi resident protein to use as a reference during time-lapse recordings. Alternatively, you can make an educated deduction as to whether your reporter is in the Golgi based on its localization and the intensity of its fluorescent signal (supplemental data in Kreitzer *et al.*, 2000). Using this differential in position and intensity, you can define a threshold above, which includes Golgi-associated fluorescence, and below, which includes all other cellular fluorescence. Given that you are trying to account for fluorescence present in the entire cell, these assays are best executed using low-magnification objectives capable of imaging an entire cell in a single focal plane.

One hour after microinjection (or the minimum expression time for your protein of interest, see earlier discussion), place cells into bicarbonate-free medium with cycloheximide and incubate at 20°C to accumulate newly synthesized protein in the Golgi.

1. To evaluate ER to Golgi transport kinetics, mount the coverslip into the precooled (20°C) recording chamber on the microscope in recording medium with cycloheximide.
   a. Acquire time-lapse images at 5- to 30-min intervals until the fluorescently tagged reporter has accumulated in the Golgi.
   b. Determine the minimum time required for maximal transport of protein from the ER to the Golgi by measuring the ratio of Golgi-associated fluorescence/total fluorescence over time.

2. To evaluate kinetics of Golgi emptying, place cells into bicarbonate-free DMEM with cycloheximide and incubate at 20°C. The duration of the 20°C temperature block depends on the time it takes to accumulate maximally your protein of interest in the Golgi (see earlier discussion).
   a. Acquire time-lapse images at 5- to 30-min intervals.
   b. Determine the rate at which fluorescently tagged cargo empties from the Golgi by measuring the ratio of Golgi-associated fluorescence/total fluorescence over time in individual cells.

I. Measuring Delivery of Post-Golgi Carriers to the Plasma Membrane

Solutions and Materials

1. Phosphate-buffered saline with calcium and magnesium (PBS-CM)
2. 2% paraformaldehyde prepared freshly in PBS-CM
3. Antibodies reactive with an extracellular epitope contained in the plasma membrane reporter protein. Surface immunolabeling is dependent on having an antibody that recognizes an extracellular epitope on the plasma membrane protein being studied. If this is not available, it may be desirable to create a cDNA probe that contains an epitope tag (e.g., HA, myc or FLAG tag) that can be immunostained.

Steps

Delivery of newly synthesized proteins to the plasma membrane can be evaluated using single cell assays or by biochemical methods. Evaluation of protein delivery to the plasma membrane in single cells can be analyzed from either fixed or living samples. Analysis in fixed samples involves cell surface selective immunolabeling of the expressed reporter protein. Analysis in living samples requires a relatively robust expression of the GFP-tagged reporter protein and acquisition of time-lapse images at relatively high frame rates using either total internal reflection fluorescence microscopy or spinning disk confocal microscopy. In TIR-FM, membrane-bound transport intermediates containing GFP-tagged fusion proteins are detected only when they move into an evanescent field, which in our experiments was within ~120 nm of the plasma membrane domain in contact with the substratum, i.e., the basal membrane. Schmoranzer and colleagues (2000) have established a quantitative method for detecting *bona fide* fusion of post-Golgi transport intermediates with the plasma membrane. Briefly, exocytic events are defined by a simultaneous rise in both the carrier's total fluorescence intensity and the area occupied by carrier fluorescence as it flattens into the plasma membrane and the cargo diffuses laterally. (For a more extensive description of TIR-FM in studying exocytosis, see Mikhailov). Exocytic events occurring in the lateral membrane of polarized epithelial cells can be identified using similar criteria when time-lapse images are acquired by high-speed confocal microscopy (for a complete description of lateral membrane fusion analysis, see Kreitzer *et al.*, 2003). Biochemical methods for studying delivery to the plasma membrane in a large population of cells, such as pulse–chase, cell surface biotinylation assays, have been described in detail previously (see article by Rodriguez–Boulan and Sabatini, 1978).

1. Measuring the rate of protein delivery to the cell surface in fixed cell, "time-lapse" experiments. If cells are grown on glass (as would be the case if exogenous proteins are expressed by cDNA microinjection), this method is useful in evaluating the delivery of proteins to the apical membrane only. For surface-labeling analysis of delivery to the basolateral membrane, cells must be grown on semipermeable filter supports and exogenous proteins must be introduced by transfection or viral infection methods.
   - **a.** Microinject GFP-tagged cDNA into the cell nuclei.
   - **b.** Accumulate newly synthesized protein in the Golgi at 20°C.
   - **c.** Shift to the permissive temperature for transport out of the Golgi (37°C).
   - **d.** At 15- to 30-min intervals after releasing the Golgi block, fix cells in paraformaldehyde for 5 min at room temperature. Do not permeabilize with detergent. Fixation in a nonpermeabilizing fixative, such as paraformaldehyde, enables selective immunolabeling of surface-associated proteins.
   - **e.** Label, by indirect immunofluorescence, the surface-associated reporter protein. Make sure to use a fluorescently conjugated secondary antibody other than fluorescein (or any dye excitable at 488 nm).
   - **f.** Acquire images of both the surface-associated (immunostained) and the total (GFP) protein expressed at each time point after release of the Golgi block. It is imperative to use identical acquisition settings for the individual fluorophores in each time-lapse sample as this is all that allows you to quantitatively (ratiometrically) evaluate the relative amount of reporter protein delivered to the cell surface.
   - **g.** Calculate the integrated fluorescence intensity of both surface-associated and total fluorescence in each cell expressing the reporter protein. The ratio of surface fluorescence (immunostained) to total fluorescence (GFP) of your reporter reflects the relative amount of protein that has been delivered to the plasma membrane at each time point. Over time, this ratio should increase and will directly reflect the rate of protein delivery from the Golgi to the plasma membrane.

2. Analysis of exocytosis using time-lapse total internal reflection fluorescence microscopy (TIR-FM).
   - **a.** Microinject GFP-tagged cDNA into the cell nuclei.
   - **b.** Accumulate newly synthesized protein in the Golgi at 20°C.
   - **c.** Mount coverslip in the recording chamber on a microscope equipped for TIR-FM. Acquire and save both bright-field and epifluorescent images of the cells you will be studying.
   - **d.** Shift to the permissive temperature for transport out of the Golgi and wait 5 min for the temperature and focus to stabilize.
   - **e.** Acquire time-lapse images (aim for at least four to five frames per second) to visualize

exocytic events occurring in the basal plasma membrane. The typical duration of our recordings is 1–2 min.

f. Analyze images for exocytic events as described in Schmoranzer *et al.* (2000).

3. Analysis of exocytosis using time-lapse, spinning-disk confocal microscopy.

a. Microinject GFP-tagged cDNA into the cell nuclei.

b. Accumulate newly synthesized protein in the Golgi at 20°C.

c. Mount coverslip in the recording chamber on a microscope equipped with a spinning disk confocal head. Acquire and save both bright-field and epifluorescent images of the cells you will be studying.

d. Shift to the permissive temperature for transport out of the Golgi and wait 5 min for the temperature and focus to stabilize.

e. Acquire time-lapse images (aim for four to five frames per second) to visualize exocytic events occurring along the lateral membrane. Photobleaching that occurs during confocal image acquisition typically limits the duration of time-lapse sequences to ~1–2 min.

f. To evaluate the spatial positioning of cargo delivery events, acquire time-lapse sequences as described in **step e** at multiple Z-axis positions throughout individual cells.

g. Analyze images for exocytic events as described in Kreitzer *et al.* (2003).

References

Baird, G. S., Zacharias, D. A., *et al.* (2000). Biochemistry, mutagenesis, and oligomerization of DsRed, a red fluorescent protein from coral. *Proc. Natl. Acad. Sci. USA* **97**(22), 11984–11989.

Chalfie, M., Tu, Y., Euskirchen, G., Ward, W. W., and Prasher, D. C. (1994). Green fluorescent protein as a marker for gene expression. *Science* **263**, 802–805.

Grindstaff, K., Yeaman, C., Anandasabapathy, N, Hsu, S.-C., Rodriguez-Boulan, E., Scheller, R., and Nelson, W. J. (1998). Sec 6/8 complex is recruited to cadherin-medicated cell-cell contacts and specifies transport vesicle delivery to the basal-lateral membrane in polarized epithelial cells. *Cell* **93**, 71–74.

Hunziker, W., Harter, C., *et al.* (1991). Basolateral sorting in MDCK cells requires a distinct cytoplasmic domain determinant. *Cell* **66**, 907–920.

Kreitzer, G., Marmorstein, A., *et al.* (2000). Kinesin and dynamin are required for post-Golgi transport of a plasma-membrane protein. *Nature Cell Biol.* **2**(2), 125–127.

Kreitzer, G., Schmoranzer, J., *et al.* (2003). Three-dimensional analysis of post-Golgi carrier exocytosis in epithelial cells. *Nature Cell Biol.* **5**(2), 126–136.

Le Bivic, A., Sambuy, Y., *et al.* (1991). An internal deletion in the cytoplasmic tail reverses the apical localization of human NGF receptor in transfected MDCK cells. *J. Cell Biol.* **115**, 607–618.

Marmorstein, A. D., Csaky, K. G., Baffi, J., Lam, L., Rahaal, F., and Rodriguez-Boulan, E. (2000). Saturation of, and competition for entry into, the apical secretory pathway. *Proc. Natl. Acad. Sci. USA* **97**, 3248–3253

Matz, M. V., Fradkov, A. F., *et al.* (1999). Fluorescent proteins from nonbioluminescent Anthozoa species. *Nature Biotechnol.* **17**(10), 969–973.

Mikhailov, A. (ed.) "Practical Fluorescence Microscopy: Protein Localization and Function in Mammalian Cells." Humana Press, Clifton, NJ.

Musch, A., Xu, H., *et al.* (1996). Transport of vesicular stomatitis virus G protein to the cell surface is signal mediated in polarized and nonpolarized cells. *J. Cell Biol.* **133**(3), 543–558.

Rodriguez-Boulan, E., and Sabatini, D. D. (1978). Asymmetric budding of viruses in epithelial monlayers: A model system for study of epithelial polarity. *Proc. Natl. Acad. Sci. USA* **75**, 5071–5075.

Schmoranzer, J., Goulian, M., *et al.* (2000). Imaging constitutive exocytosis with total internal reflection fluorescence microscopy. *J. Cell Biol.* **149**(1), 23–32.

CHAPTER

3

Use of Permeabilized Mast Cells to Analyze Regulated Exocytosis

Gerald Hammond, and Anna Koffer

OUTLINE

I. INTRODUCTION

Cell permeabilization allows manipulation of experimental conditions that is typical for *in vitro* assays while maintaining the integrity of cellular architecture. Permeabilized mast cells, both primary and cultured cell lines, such as rat basophilic leukemia (RBL-2H3) or human mast cells (HMC-1), have been used to investigate the mechanism of exocytosis, endocytosis, phospholipid metabolism, and cytoskeletal responses. After permeabilization with a bacterial exotoxin, streptolysin-O (SL-O), cells lose their cytosolic components but retain intact and functional secretory vesicles. These cells provide a well-controlled system, ideal for reconstitution type experiments. Permeabilization by SL-O is usually irreversible. Reversible permeabilization of rat peritoneal mast cells can be achieved by exposure to the tetrabasic anion of ATP (ATP^{4-}). This creates lesions (Cockcroft and Gomperts, 1979) due to interaction with a specific cell surface receptor (Tatham *et al.*, 1988), the dimensions of which vary with increasing ATP^{4-} concentration

(Tatham and Lindau, 1990). The pores created by ATP^{4-} are, however, always smaller than those due to SL-O. Combination of the two methods allows control of the extent of leakage of cytosolic factors (Koffer and Gomperts, 1989).

The SL-O monomer binds to membrane cholesterol and then oligomerises to form large pores, up to 30 nm in diameter (Bhakdi *et al.*, 1993; Buckingham and Duncan, 1983; Palmer *et al.*, 1998). The concentration of SL-O and the time of exposure needed for the permeabilization vary for different cells. Appropriate conditions can be found using nonpermeant fluorescent dyes, such as ethidium bromide or rhodamine phalloidin. Adherent cells usually require higher concentrations of SL-O and/or longer exposure times than suspended cells.

Holt and Koffer (2000) described the use of various recombinant mutants of small Rho GTPases in permeabilized mast cells. Other reviews of permeabilization techniques have been published previously (Gomperts and Tatham, 1992; Larbi and Gomperts, 1996; Tatham and Gomperts, 1990). This article describes a "SL-O prebind" method for permeabilizing glass-attached primary rat peritoneal mast cells. This method is based on the temperature-independent binding of SL-O to the plasma membrane and the temperature-dependent polymerization of bound SL-O molecules required to form pores (Sekiya *et al.*, 1996).

Hexosaminidase, released from stimulated permeabilized cells into the supernatant, is assayed biochemically, while the remaining cells are processed for imaging. Thus, functional and morphological studies can be done in parallel. Immunostaining of mast cells poses specific problems due to the presence of highly charged secretory granules. Cell permeabilization allows introduction of antibodies into cells before fixation. This is advantageous, as antibodies come into contact with granules still intact rather than with those permeabilized by fixatives. Staining before fixation is also recommended for membranous structures (e.g., staining with anti-PIP_2), as these are often disrupted by fixatives. Moreover, pretreatment of permeabilized cells with antibodies is useful for functional studies investigating the effects of the activity of the blocking antigen on cellular responses. We describe pretreatment of permeabilized cells before triggering, the assay of released hexosaminidase, and subsequent immunostaining either before or after fixation, with steps taken to avoid unspecific staining. The former method is an adaptation of a previously described procedure (Guo *et al.*, 1998).

II. MATERIALS AND INSTRUMENTATION

Percoll (1.13 ± 0.005 g/ml stock) is from Amersham Biosciences (1 liter, Cat. No. 17-0891-01). 10× calcium/magnesium-free phosphate-buffered saline (PBS) with defined salt densities is from Gibco RBL (500 ml, Cat. No. 14200-067). Streptolysin-O (SL-O, Cat. No. 302) and sodium dithionite (Cat. No. 303, needed for reduction of SL-O) are from iTEST plus, Ltd. (Czech Republic). Other preparations of SL-O can be obtained from Sigma (e.g., Cat. No. S 5265), and SL-O can also be obtained from VWR Scientific, US (Cat. No. DF 0482-60, manufactured by Difco), but in our hands, the just-described reagent has given the most reproducible results. Sigma provides EGTA (Cat. No. E-4378), HEPES (Cat. No. H-4034), PIPES (Cat. No. P-1851), Tris (Cat. No. T-1503), glutamic acid, monopotassium salt (Cat. No. G-1501), 4-methylumbelliferyl-*N*-acetyl-β-d-glucosaminide (Cat. No. M2133), dimethyl sulphoxide (DMSO, Cat. No. D-8779), Triton X-100 (Cat. X-100), goat serum (Cat. No. G9023), unconjugated succinyl-concanavalin A (SCA, Cat. No. L 3885), lysophosphatidylcholine (LPC, Cat. No. L-4129), paraformaldehyde (Cat. No. P-6148), and polyethylene glycol (molecular weight 3550, Cat. No. P-4338). Na_2ATP (trihydrate, Cat. No. 519979), 100 m*M* GTPγS solution

(Cat. No. 1110 349), and digitonin (1 g, Cat. No. 1500 643) are from Boehringer. Citric acid (Cat. No. 100813M), glycine (Cat. No. 101196X), solutions of 1*M* $MgCl_2$ (Cat. No. 22093 3M) and 1*M* $CaCl_2$ (Cat. No. 190464 K), and 18-mm^2 coverslips, thickness 1 (Cat. No. 406/0187/23), are from BDH. Mowiol 4–88 is from Harco (Harlow Chemical Company Ltd.).

Pointed bottom tubes (10 ml, Cat. No. B99 783) are from Philip Harris. "Multitest" 8-well slides (Cat. No. 6040805) are from ICN Biomedicals, Inc. Ninety-six V-well clear plates (Cat. No. 651 101) and 96-well flat-bottom black plates (Cat. No. 655 076) are from Greiner Bio-One Ltd. The eight channel pipettes are Transferepettes-8, 20–100 μl, from Brand, Germany. The refrigerated centrifuge, Omnifuge 2.ORS, is from Heraeus. The plate reader, Polarstar Galaxy, is from BMG Labtechnologies.

A moisture box is a small plastic box with a well-fitting lid lined with wet tissue paper. A plastic tray that can take up to five 8-well slides is placed inside the box. Nylon mesh is from net curtain material obtained from John Lewis Department Store (London).

III. PROCEDURES

A. Preparation of Rat Peritoneal Mast Cells

This procedure has been described in detail previously (Gomperts and Tatham, 1992) and only a brief version is given here.

Solutions

1. *Percoll at a final density of 1.114 g/ml*: Prepare using the following formula:

$$\rho_{\text{Final}} \cdot V_{\text{Final}} = \rho_{\text{Percoll stock}} \cdot V_{\text{Percoll stock}} + \rho_{10\times\text{ PBS}} \cdot \left(\frac{V_{\text{Final}}}{10}\right) + \rho_{\text{H}_2\text{O}^{\infty}} \cdot \left(V_{\text{Final}} - V_{\text{Percoll stock}} - \frac{V_{\text{Final}}}{10}\right)$$

ρ is density (g/ml) and V is volume (ml). Use 10×PBS stock with defined density and highly purified sterile water so that the density may be taken as 1.000 g/ml. Prepare 50–100 ml of Percoll solution in a sterile environment and store in 2-ml aliquots at −20°C (may be stored several years). Warm at room temperature, mix thoroughly, and transfer into a pointed bottom tube just before use.

2. *Chloride buffer (CB)*: Final concentrations 20 m*M* HEPES, 137 m*M* NaCl, 2.7 m*M* KCl, 2 m*M* $MgCl_2$, 1.8 m*M* $CaCl_2$, 5.6 m*M* glucose, and 1 mg/ml bovine serum albumin (BSA), pH 7.2. Prepare 10× concentrated stock solution without BSA. To make 500 ml of 10× CB, dissolve 40 g NaCl, 1 g KCl, 23.83 g HEPES, 5 g glucose, 10 ml of 1*M* $MgCl_2$, and 9 ml of 1*M* $CaCl_2$ in distilled water, adjust pH to 7.2 with NaOH, and complete to 500 ml. Store in 50-ml aliquots at −20°C. Thaw, mix, dilute, add BSA, and readjust pH to 7.2 before use.

Steps

1. Use Sprague–Dawley rats. "Retired breeders" of either sex were used but other strains of various ages are also suitable; older rats provide more cells.
2. After peritoneal lavage of rats, pellet the cells present in the washings by centrifugation (5 min at 250 *g*).
3. Resuspend cells in ~7 ml of CB and filter the suspension through a nylon mesh.
4. Overlay the filtered suspension onto a 2-ml cushion of Percoll solution placed in a 10-ml pointed bottom tube.
5. Centrifuge (10 min at 250 *g*, room temperature, slow acceleration). Dense mast cells pellet through the Percoll cushion while contaminating cells (neutrophils, macrophages, red blood cells) remain at the buffer–Percoll interface.
6. Remove the buffer, interface cells, and Percoll, resuspend the pellet in ~1 ml

CB, and transfer into a clean tube (avoid touching the tube wall where some of the contaminating cells may still adhere).

7. Add further ~10 ml CB, centrifuge (5 min at 250 *g*), and resuspend pellet in 1.5 ml CB. About 1×10^6 cells (>95% purity) are obtained from one rat. This is sufficient for at least five 8-well slides.

B. Permeabilization

SL-O binds to the cells on ice, excess SL-O (and any other additives) is removed by washing with a cold buffer, and cells are then permeabilized by the addition of warm buffer. Permeabilized cells are finally washed to remove freely soluble ions, nucleotides, and proteins before the addition of triggering solutions. A small loss of responsiveness occurs due to the "run-down" during the washing (10–20%) but the advantage is that nucleotides, ions, and proteins leaking out from cells do not come into contact with those applied exogenously. If this is of no concern (e.g., when ATP levels have been depleted by metabolic inhibition), triggers can be added at the time of permeabilization.

Solutions

1. *Glutamate buffer (GB)*: Final concentrations 137 m*M* Kglutamate, 20 m*M* PIPES, 2 m*M* $MgCl_2$, and 1 mg/ml BSA, pH 6.8. Prepare 10× concentrated stock solution without BSA. To make 250 ml of 10× GB, dissolve 63.43 g glutamic acid, 15.12 g PIPES, and 5 ml 1*M* $MgCl_2$ in distilled water, adjust pH to 6.8 with glacial acetic acid, and complete to 250 ml. Store 50-ml aliquots at −20°C. Thaw, mix, dilute, add BSA, and readjust pH to 6.8 before use.
2. *100 mM EGTA stock solution*: Store in 5- to 10-ml aliquots at −20°C. To make 100 ml of 100 m*M* EGTA, dissolve 3.804 g EGTA in 100 ml distilled water.
3. *Glutamate buffer–3 mM EGTA (GBE)*: Add 300 μl of 100 m*M* EGTA to 10 ml GB.
4. *Streptolysin-O (SL-O)*: Streptolysin-O from iTEST is supplied as a lyophilized powder in vials of 22 IU and must be reduced before use with sodium dithionite. To prepare a working stock at 20 IU/ml, dissolve one vial of SL-O in 550 μl PBS containing 0.1% BSA and one vial of 20 mg sodium dithionite in 550 μl PBS containing 0.1% BSA. Mix these two solutions and incubate at 37°C for 1 h. SL-O is now reduced and the solution is ready for use. For storing, divide the reduced SL-O solution into 100-μl aliquots, freeze under liquid nitrogen, and store at −80°C. Just before use, warm and mix and use on the same day. For the "prebind" method, use SL-O at the final concentration of 1.6 IU/ml: Add 1.15 ml of GBE to 100 μl of 20 IU SL-O/ml. Thus, one aliquot will provide 1.25 ml of working solution, enough for one experiment.

Steps

1. Clean 8-well slides with distilled water and then ethanol. Leave to dry. Label the slides.
2. Pipette 30 μl cell suspension in CB per well (i.e., ~20,000 cells per well). Allow 2–4 wells for each condition. Keep 150 μl of the cell suspension aside for secretion assay (see solution D3 and step D2).
3. Let cells attach for 1 h at room temperature on a plastic tray in a moisture box[1].
4. Place the slides on a metal plate on ice.
5. Wash the cells once with 30 μl ice-cold GBE.

[1] In some experiments, it is desirable to deplete endogenous ATP from intact cells with metabolic inhibitors. In this case, after step 2, incubate attached intact cells with CB where glucose has been omitted and 10 μ*M* antimycin A with 6 m*M* 2-deoxyglucose included. More than 90% depletion of ATP is achieved in 20 min at 30°C (Koffer and Churcher, 1993).

6. Add 30 μl ice-cold SL-O (1.6 IU/ml GBE) to each well and incubate on ice for 8 min to allow SL-O to bind to the plasma membranes.
7. Wash the cells once with 30 μl ice-cold GBE to remove unbound SL-O and additives.
8. Permeabilize the cells by adding 30 μl warm (37°C) GBE and transferring to a prewarmed moisture box at 37°C for 90s.
9. Chill the cells by placing the slide back on ice.
10. Wash once with 30 μl GB (without EGTA) to remove freely soluble components.

C. Pretreatment and Triggering

Solutions

1. *GB–30 μM EGTA*: Add 3 μl of 100 m*M* EGTA to 10 ml GB.
2. *Agent to be tested, dissolved in GB–30 μM EGTA*. These can include an antibody directed against a specific antigen whose activity is being tested. Dilution of the antibody is 10–20× less than that used for immunostaining.
3. *100 mM MgATP*: To make 16.5 ml of 100 m*M* MgATP, add 6.61 ml of 0.5*M* Tris and 1.65 ml of 1*M* $MgCl_2$ to 1 g Na_2ATP and add H_20 to a final volume of 16.5 ml. Store in 100-μl aliquots at −20°C.
4. *10 mM GTPγS*: Dilute the 100 m*M* stock solution with GB just before use.
5. *Ca^{2+} /EGTA buffers system*: These are obtained as described previously (Gomperts and Tatham, 1992; Tatham and Gomperts, 1990) by mixing solutions of Ca:EGTA and EGTA, both at ~100 m*M* and pH 6.8 at specified ratios. Keep 10-ml aliquots of 100 m*M* stock solutions at −20°C. Free Ca^{2+} concentration is controlled by adding the appropriate Ca^{2+}/EGTA buffer to a final concentration of 3 m*M*.
6. *Triggering solutions*: Various combinations of calcium, MgATP, and GTPγS (or GTP) in GB are used for cell stimulation. The following four conditions (final concentrations) are used most frequently: EA (basal condition, 3 m*M* EGTA–3 m*M* ATP); EGA (as EA with 50 μ*M* GTPγS); CA (3 m*M* Ca:EGTA /to maintain [Ca^{2+}] at pCa5)–3 m*M* ATP; and CGA (as CA with 50 μ*M* GTPγS). Table 3.1 gives volumes for making triggering solutions using the aforementioned stock solutions.

Steps

1. Immediately after step B9 (permeabilized cells are still on the cold plate), add 15 μl of a solution containing the protein or reagent to be tested in GB–30 μ*M* EGTA.
2. Incubate on ice (in a moisture box) for 5–30 min. The time of incubation depends on the size of the protein/reagent to be introduced into the cells. For example, small proteins such as Rho GTPases require 5–10 min but antibodies may require up to 30 min to penetrate the SL-O lesions. The time and temperature for action of the agents should be considered.
3. Add 15 μl of 2× concentrated triggering solutions containing combinations of calcium/EGTA, MgATP, and GTPγS. (If no pretreatment was required, omit steps 1 and 2 and add 30 μl of 1× triggering solutions.)
4. Stimulate by transferring to a prewarmed moisture box and incubate for 30 min at 37°C. Note that a shorter time (10–15 min) may be sufficient when GTPγS is included in the triggering solution.
5. Stop reaction by transferring slides back onto the ice-cold metal plate.

D. Secretion Assay

Solutions

1. *Fluorogenic hexosaminidase substrate*: Final concentrations 1 m*M* 4-methylumbelliferyl-*N*-acetyl-β-d-glucosaminide, 0.1% DMSO,

0.01% Triton X-100, and 200 m*M* citrate, pH 4.5. To make 500 ml, mix 189.7 mg of the reagent with 0.5 ml DMSO and then add 500 ml of 0.2 *M* citrate–0.01% Triton X-100. To make citrate solution, dissolve 21 g of citric acid and 0.5 g of Triton X-100 in H_2O, adjust pH to 4.5 using NaOH, and make volume up to 500 ml. Using Whatman filter paper No. 1, filter and store in 20- to 50-ml aliquots at −20°C. Hexosaminidase, released from mast cells, hydrolyses the substrate, producing fluorescent 4-methylumbelliferone. Triton reduces surface tension and thus artifacts.
2. *0.5M Tris*: Dissolve 60.55 g of Tris in water, total volume 1 liter.
3. *GB*
4. *Solution for assaying total hexosaminidase content*: To 150 µl of the original cell suspension (saved at step B2), add 1 ml of 0.2% Triton X-100 in GB and mix to lyse the cells.

Steps

1. After step C5, remove 15 µl (i.e., one-half) of the triggering solutions and transfer into cold transparent 96 V-well plates (on ice).
2. Add 100 µl of cold GB to each V well using a multichannel pipette.
3. Pipette 115 µl of the solution for assaying the total hexosaminidase content into one column of V wells (i.e., 8 × 115 µl).
4. Pipette 115 µl of GB into one column of V wells (i.e., 8 × 115 µl, to be used for blanks).
5. Balance with another 96 V-well plate.
6. Centrifuge for 5 min at 250 *g* at 4°C to remove any detached cells.
7. Using a multichannel pipette, remove 50 µl of the supernatant from each V well and transfer into a black flat-bottom 96-well plate.
8. Add 50 µl of the fluorogenic hexosaminidase substrate to each well to assay for the release of hexosaminidase.
9. Cover with lid and wrap in aluminium foil.
10. Incubate 1–2 h at 37°C or overnight at room temperature.
11. Quench with 100 µl of 0.5 *M* Tris.
12. Read fluorescence using a fluorescence plate reader (excitation and emission filter at 360 and 405 nm, respectively).
13. Calculate percentage secretion as follows: % secretion = $100 \times (X-B)/(T-B)$, where X, B, and T are fluorescence values of the sample, average blank, and average total hexosaminidase, respectively.

E. Immunostaining Before Fixation

Cells remaining on eight-well slides after removal of the trigger solutions for secretion assays can be stained by fluorescent phalloidin to evaluate the morphology of F-actin (Holt and Koffer, 2000) or be immunostained. Processing of the cells should be done immediately after step D1 and the secretion assay delayed until the cells are in a blocking solution or fixed.

Solutions

1. *Stock solution of succinyl-Con A, SCA, 5 mg/ml*: Dissolve 25 mg in 5 ml GB without BSA. Store 100-µl aliquots at −20°C.
2. *Glutamate buffer–3 mM EGTA (GBE)*: Add 300 µl of 100 m*M* EGTA to 10 ml GB.
3. *Blocking solution*: 32% goat serum–250 µg/ml succinyl-Con A (SCA)–GBE. To make 1 ml, add 320 µl goat serum and 50 µl SCA (5 mg/ml stock) to 630 µl GBE.
4. *Antibody solution*: 16% goat serum in GBE. To make 1 ml, add 160 µl goat serum to 840 µl GBE. Higher concentrations of antibodies are usually required if these are introduced into permeabilized cells before fixation than those used for after-fixation staining. *Note*: After dilution, centrifuge all antibodies for 1 min at ~15,000 *g* to remove any aggregates.
5. *Fixative*: 3% paraformaldehyde in GBE (without BSA)–4% polyethylene glycol

(PEG, MW 3200). To make 100 ml of fixative, first prepare GB (without BSA)–3 m*M* EGTA–4% (w/v) PEG. Add 3 g of paraformaldehyde to ~80 ml of GBE-PEG and gently heat and stir in a fume cupboard, adding 10*M* NaOH in small doses until paraformaldehyde is dissolved and colourless. Adjust pH to 6.8. Filter (in a fume cupboard) using Whatman filter paper No. 1 and store 10-ml aliquots at −20°C.

6. *Mounting solution*: Add 12 ml of water to 12 g glycerol and 4.8 g Mowiol 4-88 in a conical flask and mix for 2 h or overnight at room temperature. Add 24 ml 0.2 *M* Tris, pH 8.5, and 400 μl of 100 m*M* EGTA and stir further at 50°C (~10 min). Centrifuge for 30 min at ~5000 *g*. Discard pellets, stir the supernatants again, and store in 1-ml aliquots at −20°C.

Steps

1. After step C5 and D1 (i.e., after triggering and removal of aliquots for secretion assay), remove the remaining triggering solutions.
2. Wash once in 30 μl GBE to remove residual calcium/GTPγS.
3. Add 30 μl of blocking solution to block unspecific binding.
4. Incubate in a moisture box for 30 min at room temperature.
5. Remove blocking solution and add 30 μl of the primary antibody diluted in antibody solution.
6. Incubate in a moisture box for 30–60 min at room temperature.
7. Wash the cells 4× with 30 μl GBE.
8. Add 30 μl of secondary antibody, conjugated to a fluorescent probe and diluted in antibody solution.
9. Incubate in a moisture box for 30 min at room temperature.
10. Wash the cells 4× with 30 μl GBE
11. Fix the cells by adding 30 μl fixative for 20 min at room temperature in a moisture box.
12. Wash the cells 4× with 30 μl GBE.
13. Aspirate wells until dry and add 4 μl mounting solution and remove any bubbles.
14. Cover eight wells with 18-mm^2 coverslips.

F. Immunostaining After Fixation

To improve access of antibodies into fixed cells, permeabilization with either lysophosphatidyl choline (LPC) or digitonin is required. Blocking can be done with a lower concentration of serum when cells are fixed, but succinyl–concanavalin A should still be included because fixation exposes granule matrices that may bind antibody nonspecifically. Including a 5-min wash step with 0.4 *M* NaCl before blocking also reduces nonspecific binding of antibodies to the granule matrices without affecting cellular morphology in fixed cells.

Solutions

1. *Fixative and mounting solutions as described earlier.*
2. *GBE*: Add 300 μl of 100 m*M* EGTA to 10 ml GB.
3. *LPC-GBE, 80 μg/ml of LPC in GBE*: Used for permeabilization of fixed cells. Stock solution is 40 mg of LPC/ml ethanol. Store at −20°C. Warm to ~40°C and mix before use. Dilute 500×, i.e., 2 μl into 1 ml of GBE to obtain the final concentration.
4. *0.2 μM digitonin–GBE*. Used as an alternative solution for permeabilization of fixed cells. Digitonin is stored as a 1 m*M* stock solution in DMSO at room temperature. Stock is prepared by dissolving 1.23 mg digitonin/ml in DMSO. Dilute 10 μl of stock digitonin in 990 μl GBE and then add 50 μl of this 10 μ*M* solution to 950 μl GBE to prepare the 0.2 μ*M* solution.
5. *50 mM glycine–GBE*: Make 10 ml of stock 1 *M* glycine solution by adding 0.75 g glycine to 10 ml GB (without BSA). Store in 1-ml

aliquots at −20°C. Add 50 μl of 1 *M* glycine stock to 950 μl GBE.

6. *Blocking solution*: 5% goat serum–250 μg/ml succinyl-Con A (SCA)–GBE. To make 1 ml, add 50 μl goat serum and 50 μl stock SCA (5 mg SCA / ml GB without BSA) to 900 μl GBE.
7. *0.4M NaCl*: For 1 ml, add 100 μl 4 *M* NaCl to 900 μl GBE.
8. *Antibody solution*: 5% goat serum in GBE. To make 1 ml, add 50 μl goat serum to 950 μl GBE. After dilution, centrifuge all antibodies for 1 min at ~15,000 *g* to remove any aggregates.

Steps

1. After step C5 and D1 (i.e., after triggering and removal of aliquots for secretion assay), remove the remaining triggers.
2. Fix the cells by adding 30 μl fixative for 20 min at room temperature in a moisture box.
3. Wash the cells 4× with 30 μl 50 m*M* glycine–GBE.
4. Add 30 μl of LPC-GBE (or digitonin-GBE).
5. Permeabilize fixed cells for 20 min with LPC-GBE or for 5 min with digitonin–GBE, both at room temperature.
6. Wash the cells 4× with 30 μl GBE.
7. Remove GBE and add 30 μl of 0.4 *M* NaCl in GBE to each well.
8. Incubate in a moisture box for 5 min at room temperature.
9. Wash the cells 4× with 30 μl GBE.
10. Add 30 μl of blocking solution.
11. Incubate in a moisture box for 30 min at room temperature.
12. Remove blocking solution and add 30 μl of the primary antibody diluted in antibody solution.
13. Incubate in a moisture box for 30–60 min at room temperature.
14. Wash the cells 4× with 30 μl GBE.
15. Add 30 μl of secondary antibody, conjugated to a fluorescent probe, diluted in antibody solution.
16. Incubate in a moisture box for 30 min at room temperature.
17. Wash the cells 4× with 30 μl GBE.
18. Aspirate wells until dry and add 4 μl mounting solution; remove any bubbles.
19. Cover eight wells with 18-mm^2 coverslips.

IV. COMMENTS

We have described a method that allows assessment of mast cell secretory function in parallel with studies of cell morphology. An example is shown in Fig. 3.1. Permeabilized mast vafter

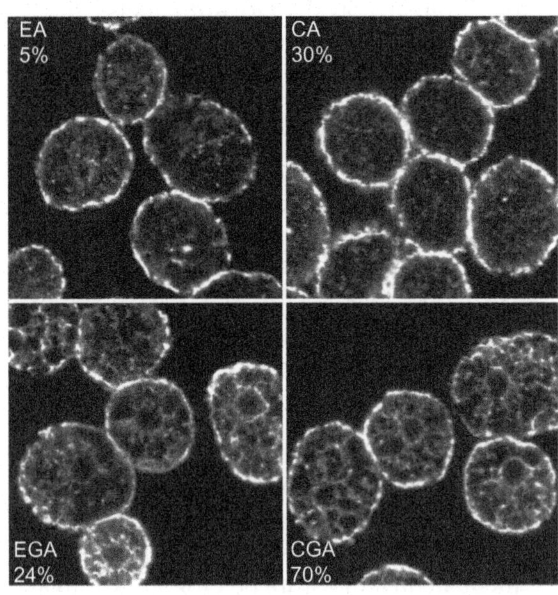

FIGURE 3.1 Glass-attached mast cells were permeabilized and exposed to the triggers described in Table 3.1: EGTA/ATP (EA), calcium (pCa5)/ATP (CA), EGTA/GTP-γ-S/ATP (EGA), or calcium (pCa 5)/ GTPγS/ATP (CGA). After 20 min at 37°C, cells were fixed and stained with anti-β-actin monoclonal antibody (clone AC-15) from Sigma (used at 1/200 dilution). The secondary antibody (at 1/50 dilution in GBE) was goat antimouse IgG biotin from Sigma. Cy2 streptavidin (1/50, from Amersham Biosciences, UK) was the tertiary layer. Confocal micrographs of equatorial slices are shown. Numbers indicate the percentages of released hexosaminidase.

TABLE 3.1 Assembling 1 ml Each of the Four Typical Triggering Solutions

Stock	EA	EGA	CA	CGA
EGTA (100 mM)	30 μl	30 μl	0	0
ATP (100 mM)	30 μl	30 μl	30 μl	30 μl
GTPγS (10 mM)	0	5 μl	0	5 μl
Ca:EGTA (pCa5) (100 mM[a])	0[a]	0[a]	30 μl[a]	30 μl[a]
GB[a]	940 μl	935 μl	940 μl	935 μl

[a] Note that 100 mM is an ideal stock concentration of Ca:EGTA buffer, although the final concentration of stock is calculated empirically when preparing the buffers (Tatham and Gomperts, 1990). Therefore, the volume required to give the final 3 mM concentration of Ca:EGTA final from stock will need to be adjusted.

exposure to the combinations of calcium, ATP, and GTγS described in Table 3.1. Confocal images of equatorial slices are shown together with typical secretory responses shown in parentheses (percentage of released hexosaminidase). The method provides a large potential for investigating the significance of rearrangements of specific molecules to exocytotic function.

V. PITFALLS

Keep cells in the moisture box as much as possible to avoid evaporation. Maintain permeabilized mast cells at low (25–50 μM) concentrations of EGTA before exposure to triggers to avoid spontaneous degranulation. When comparing activities of cells after various pretreatments, keep the time between permeabilization and triggering standard; the length of the "rundown" period affects secretion. Keep 96-well plates clean, washing them with detergent and plenty of water immediately after use.

Note that when staining is performed before fixation, the nature of the endogenous antigen should be taken into account. Because soluble proteins will leak out of the permeabilized cells, use this method for molecules known or expected to be tethered either to the membrane or to the actin cytoskeleton. Determination of the extent of leakage of the antigen under various conditions (by immunoblotting of the supernatants) is always helpful.

References

Bhakdi, S., Weller, U., Walev, I., Martin, E., Jonas, D., and Palmer, M. (1993). A guide to the use of pore-forming toxins for controlled permeabilization of cell membranes. *Med. Microbiol. Immunol. Berl.* **182**, 167–175.

Buckingham, L., and Duncan, J. L. (1983). Approximate dimensions of membrane lesions produced by streptolysin S and streptolysin O. *Biochim. Biophys. Acta* **729**, 115–122.

Cockcroft, S., and Gomperts, B. D. (1979). ATP induces nucleotide permeability in rat mast cells. *Nature* **279**, 541–542.

Gomperts, B. D., and Tatham, P. E. R. (1992). Regulated exocytotic secretion from permeabilized cells. *Methods Enzymol.* **219**, 178–189.

Guo, Z., Turner, C., and Castle, D. (1998). Relocation of the t-SNARE SNAP-23 from lamellipodia-like cell surface projections regulates compound exocytosis in mast cells. *Cell* **94**, 537–548.

Holt, M., and Koffer, A. (2000). Rho GTPases, secretion and actin dynamics in permeabilized mast cells. *Methods Enzymol.* **325**, 356–369.

Koffer, A., and Churcher, Y. (1993). Calcium and GTP-gamma-S as single effectors of secretion from permeabilized rat mast cells: Requirements for ATP. *Biochim. Biophys. Acta* **1176**, 222–230.

Koffer, A., and Gomperts, B. D. (1989). Soluble proteins as modulators of the exocytotic reaction of permeabilized rat mast cells. *J. Cell Sci.* **94**, 585–591.

Larbi, K. Y., and Gomperts, B. D. (1996). Practical considerations regarding the use of streptolysin-O as a permeabilizing agent for cells in the investigation of exocytosis. *Biosci. Rep.* **16**, 11–21.

Palmer, M., Harris. R., Freytag, C., Kehoe, M., Tranum Jensen, J., and Bhakdi, S. (1998). Assembly mechanism of the oligomeric streptolysin O pore: The early membrane lesion is lined by a free edge of the lipid membrane and is extended gradually during oligomerization. *EMBO J.* **17**, 1598–1605.

Sekiya, K., Danbara, H., Yase, K., and Futaesaku, Y. (1996). Electron microscopic evaluation of a two-step theory of pore formation by streptolysin O. *J. Bacteriol.* **178**, 6998–7002.

Tatham, P. E. R., Cusack, N. J., and Gomperts, B. D. (1988). Characterisation of the ATP^{4-} receptor that mediates permeabilization of rat mast cells. *Eur. J. Pharmacol.* **147**, 13–21.

Tatham, P. E. R., and Gomperts, B. D. (1990). Cell permeabilization. *In "Peptide Hormones: A Practical Approach"* (K. Siddle and J. C. Hutton, eds.), pp. 257–269. IRL Press, Oxford.

Tatham, P. E. R., and Lindau, M. (1990). ATP-induced pore formation in the plasma membrane of rat peritoneal mast cells. *J. Gen. Physiol.* **95**, 459–476.

CHAPTER

4

Cell Surface Biotinylation and Other Techniques for Determination of Surface Polarity of Epithelial Monolayers

Ami Deora, Samit Chatterjee, Alan D. Marmorstein, Chiara Zurzolo, Andre Le Bivic, and Enrique Rodriguez-Boulan

OUTLINE

I. INTRODUCTION

A fundamental property of epithelial cells is the polarized distribution of proteins and lipids in the apical and basolateral domains of the plasma membrane. These two domains are physically separated from each other by the tight junction. Many studies have been done over the past 20 years to understand the mechanisms that lead to the establishment and maintenance of the polarized distribution of proteins and lipids in the plasma membrane of epithelial cells (Rodriguez-Boulan *et al.*, 2005; Yeaman *et al.*, 1999; Keller and Simons, 1997).

A major advance in the study of epithelial cell polarity was achieved with the introduction of porous filter supports for the growth of epithelial cell cultures (reviewed in Rodriguez-Boulan *et al.*, 2005). This method differs from classical cell culture in that it allows direct access to the basolateral surface of cultured cells. Epithelial cells and cell lines grown on such filter supports (either nitrocellulose or polycarbonate) attain a more differentiated appearance and become polarized after relatively short times in culture. While most studies on epithelial polarity and trafficking of plasma membrane proteins have been performed using a limited number of cell lines (MDCK, FRT, Caco-2), this culture technique has been gaining in popularity and has been used now for primary cultures as well. The confluency of cells grown on permeable supports can be determined by the measurement of transepithelial electrical resistance or transepithelial [^{3}H]inulin flux (Hanzel *et al.*, 1991).

One of the biggest advantages of this culture system is the accessibility of either the apical or the baso-lateral surface to any reagent added to the medium and the ability to add different reagents to contact either surface. This is the basis of the biotinylation techniques that have been developed to selectively label proteins present on the apical or basolateral domains of the plasma membrane of filter-grown epithelial cells.

The proteins present on the surface of filter-grown monolayers can be selectively modified by the water-soluble cell-impermeable biotin analog sulfo-NHS-biotin. Taking advantage of the access afforded by the filter support, the addition of sulfo-NHS-biotin to only one surface of the cell results in the selective labeling of only the apical or basolateral surface proteins. The biotinylated proteins can then be detected by blotting with [125I]streptavidin or streptavidin conjugated to any number of enzymatic reporters. Furthermore, the cells can be metabolically pulse labeled and the proteins of interest can then be studied using biotinylation, immunoprecipitation and subsequent streptavidin–agarose precipitation. This technique is very versatile and is applicable to the study of diverse aspects of epithelial cell polarity, such as the steady-state distribution of specific antigens, or dynamic processes, such as targeting to the cell surface and transcytosis of membrane proteins. Several biotin analogs are available, including one that contains a disulfide bond. By differentially labeling the surfaces of epithelia in situ with the cleavable NHS-S-S-biotin and a noncleavable biotin, we have been able to study the polarity of a native epithelium *in situ* (Marmorstein *et al.*, 1996).

The first edition of this article described a basic protocol for selective cell surface biotinylation, plus some modifications of the assay to study protein targeting and endocytosis. The second edition included a basic protocol for *in situ* domain-specific biotinylation. In this edition, we have added a protocol of surface immunolabeling as an alternative to biotin labeling for determination of polarity. This technique is critical in cell lines that may exhibit leakiness to biotin. Additionally, we have also included two sections determining the polarity of a protein by means of intranuclear microinjection of its cDNA and quantitative microscopic analysis to determine distribution of the protein relative to known polarized markers.

II. MATERIALS

Sulfo-NHS-biotin (sulfosuccinimidobiotin, Cat. No. 21217); NHS-LC-biotin (sulfosuccinimidyl-6-(biotinamido)-hexanoate, Cat. No. 21335); NHS-SS-biotin (sulfosuccinimidyl-2-(biotinamido)-ethyl-1,3-dithioproprionate, Cat. No. 21331); and immunopure-immobilized streptavidin (Cat. No. 20347) are from Pierce (Rockville, IL). Protein A–Sepharose Cl-4B (Cat. No. 17-0780-01) is from Pharmacia/LKB (Piscataway, NJ). Glutathione (Cat. No. G-6529) and cycloheximide (Cat. No. C-7698) are from Sigma Chemical Co. (St. Louis, MO). *Staphylococcus aureus* cells (Pansorbin, Cat. No. 507858) are from Calbiochem (La Jolla, CA). Cells are grown on polycarbonate filters (Transwell, 12 mm diameter, Cat. No. 3401; 24 mm diameter, Cat. No. 3412) from Corning-Costar (Cambridge, MA); MEM-Select-Amine kits (Cat. No. 19050-012) are from GIBCO BRL Life Technologies (Grand Island, NY); [^{35}S]EXPRE^{35}S ^{35}S (methionine/cysteine) and [^{35}S]cysteine are from Dupont NEN (Boston, MA) (Cat. No. NEG 072 for Express and Cat. No. NEG 022T for [^{35}S]cysteine); [^{125}I]streptavidin can be obtained from Amersham (Arlington Heights, IL) (Cat. No. IM236); streptavidin conjugated to horseradish peroxidase can be obtained from Sigma (Cat. No. S-5512).

III. PROCEDURES

A. Cell Surface Biotinylation

This procedure is used to determine the relative percentage of a plasma membrane protein(s) in the apical versus basolateral plasma membrane of epithelial cells grown on permeable filter supports (modified from Sargiacomo *et al.*, 1989).

Solutions

1. *PBS-CM*: Phosphate-buffered saline containing 1.0 m*M* $MgCl_2$, and 1.3 m*M* $CaCl_2$
2. *Sulfo-NHS-biotin or sulfo-NHS-LC-biotin*: Stock solution is 200 mg/ml in dimethyl sulfoxide (DMSO), which can be stored for up to 2 months at −20°C. Thaw just prior to use and dilute to a final concentration of 0.5 mg/ml in PBS-CM. Use immediately.
3. *50 mM NH_4Cl in PBS-CM or Dulbecco's modified Eagle's medium (DMEM)*: Use to quench the excess biotin at the end of the labeling reaction.

Steps

All steps are carried out on ice and with ice-cold reagents.

1. For all experiments, use confluent monolayers of cells plated at confluency (for most cell lines, 2.5–3.5 × 10^5 cells/cm^2 of filter) 4–5 days prior to biotinylation. Measure transepithelial electrical resistance (TER) and discard monolayers that do not exhibit acceptable resistances (different cell lines exhibit different TER values ranging from tens to thousands of Ω cm^2; monolayers should be used that exhibit TER values in the normal range for your cell line. We have successfully performed this assay on cells with TERs as low as 50Ω cm^2).
2. Wash filters on both sides three times with ice-cold PBS-CM.
3. Add a fresh solution of sulfo-NHS-biotin (0.5 mg/ml in PBS-CM) to the apical or basolateral chamber. Add PBS-CM to the other chamber. We use 0.7 ml apical and 1.4 ml basolateral for 24-mm-diameter filters and 0.4 and 0.8 ml for 12-mm-diameter filters. Incubate with gentle shaking for 20 min at 4°C and then repeat this step.
4. Quench the reaction by removing the solutions from both chambers and replacing

with 1 ml of 50 m*M* NH_4Cl in PBS-CM. Incubate with gentle shaking for 10 min at 4°C.

5. Rinse twice with PBS-CM.
6. Excise filters and either freeze at −80°C (the freeze thaw involved with storage at −80°C appears to inactivate some proteases) or immediately proceed with the extraction of biotinylated proteins as outlined in Section III,F.

Analysis of Results

The amount of protein present on the apical or basolateral surface is determined by a densitometric analysis of the autoradiographs. Multiple exposures are necessary if using the film to ensure that the values obtained are in the linear range of the film. Polarity is expressed as the percentage of total surface protein present on one surface of the monolayer.

B. Biotin Targeting Assay

This procedure is used to determine if proteins are delivered directly, indirectly (transcytotically), or nonpolarly to the apical and/or basolateral surface of an epithelial cell (modified from Le Bivic *et al.*, 1990).

Solutions

1. *Starvation medium*: DMEM without methionine or cysteine. This solution is prepared using a MEM Select-Amine kit by not adding the methionine and cysteine.
2. *[^{35}S]EXPRESS (methionine/cysteine:) or [^{35}S]cysteine*: 1 mCi per multiwell plate (12 × 1.2-cm or 6 × 2.4-cm-diameter filters). In some cases, proteins are effectively labeled with [^{35}S]SO_4, an advantage for studies of post-Golgi sorting, because the addition of sulfate occurs in the *trans*-Golgi network (see Chapter 2 by Kreitzer *et al.*).
3. *Chase medium*: DMEM containing a 10× concentration of methionine and cysteine (made by addition of methionine and cysteine to starving medium) or the normal medium in which the cells grow.
4. *HCO_3-free DMEM containing 20 mM HEPES and 0.2% bovine serum albumin (BSA)*
5. *Sulfo-NHS-biotin (NHS-LC-biotin or NHS-SS-biotin)*: 0.5 mg/ml in PBS-CM + all of the reagents used in the cell surface biotinylation protocol.
6. *Lysis buffer*: 1% Triton X-100 in 20 m*M* Tris, 150 m*M* NaCl, 5 m*M* EDTA, 0.2% BSA, pH 8.0, and protease inhibitors
7. *Immunopure-immobilized streptavidin on agarose beads*
8. *10% SDS*: sodium dodecyl sulfate

Steps

1. Wash cells on filters three times with starvation medium and incubate for 20–40 min in starvation medium. The starving period will depend on your cell type. MDCK cells work well with a 20-min starvation; RPE-J require longer times.
2. Pulse for 20–30 min (again depends on cell line) in starving medium containing [^{35}S]EXPRESS of [^{35}S]cysteine at 37°C. The pulse solution is starving medium plus the ^{35}S label. Minimal volumes are recommended. For MDCK cells we pulse with 20–40 μl from the basolateral surface. A drop of pulse medium is placed on a strip of Parafilm in a humidified chamber (we use a plastic box lined with wet towels), and the insert containing the filter and MDCK monolayer is removed from the multiwell and dropped on top of the pulse medium. Some cell lines (i.e., RPE-J) are better labeled from the apical surface. For apical pulse, starvation medium is removed from both

chambers and pulse medium is applied only to the apical chamber. For 1.2-cm-diameter filters we use a 100-μl volume; for 2.4-cm-diameter filters we use a 350-μl volume of pulse medium.

3. The pulse is terminated by washing with chase medium three times.
4. At different chase times, aspirate chase medium and replace with ice-cold $NaHCO_3$-free DMEM containing 20 m*M* HEPES and 0.2% BSA and store on ice until all chase points have been collected.
5. Proceed to apical or basolateral biotinylation following the protocol described earlier for cell surface biotinylation.
6. Excise filters and either freeze at −80°C or immediately lyse cells and immunoprecipitate specific proteins as described in the section extraction of biotinylated proteins.
7. Remove immunoprecipitated proteins from beads by adding 40 μl of 10% SDS and heating for 5 min at 95°C. Immediately dilute with 460 μl of lysis buffer and pellet for 1 min in a microfuge. Remove 450 μl and place in a new tube. Dilute the remaining 50 μl with 50 μl of 2× Laemmli sample buffer. This sample is used to normalize for differential incorporation of radiolabel from filter to filter. The remaining 450 μl is diluted with a further 1 ml of lysis buffer to which is added an additional 50 μl of streptavidin agarose that has been preblocked for 1–12 h with lysis buffer.
8. Streptavidin precipitation is allowed to proceed for 1 h to overnight at 4°C. Then the beads are washed successively in TPII, TPIII, and TPIV as described in Section III,F. After the final wash the beads are resuspended in Laemmli sample buffer and heated to 95°C for 5 min.
9. Both the sample representing total and surface protein are resolved on SDS–PAGE gels. The gels are dried and exposed for autoradiography.

C. Targeting Assay by Surface Immunolabeling

Epithelial cells such as LLC-PK1 may not form a tight monolayer and hence could be leaky to biotin analogs (MW~ 400–600). To overcome this problem, we have used antibody labeling against the protein of interest (Gan *et al.*, 2002). Antibodies are less likely to traverse through leaky monolayers. Leakiness of antibodies should be directly determined before using this method.

Solutions

1. *PBS-CM*: See Section III,A
2. *DMEM containing 0.2% BSA*

Steps

1. The initial steps are similar to Section III,B (steps 1–4). Label the ice-cold filters from different chase time points with antibody added to either apical or basolateral domains for 1 h on ice in a cold room kept at 4°C. Dilute the antibody in DMEM containing 0.2% BSA at an approximate concentration of 1 μg/ml. After 1 h, wash filters four times in ice-cold PBS-CM containing 0.2% BSA.
2. Excise filters and either freeze at −80°C or immediately lyse cells in lysis buffer.
3. Pull down the antigen–antibody complex with protein A or G beads from nine-tenths of the postnuclear supernatants. Subject one-tenth of the supernatant again to immunoprecipitation to measure total labeled protein.
4. Wash immunoprecipitates on the beads successively in TPII, TPIII, and TPIV as described in Section III,F.
5. After the final wash, resuspend the beads in Laemmli sample buffer and heat to 95°C for 5 min.

6. Resolve both the sample representing total and surface proteins on SDS–PAGE gels. Dry and expose the gels for autoradiography.

Analysis of Results

The polarity of the protein is determined at each time point by densitometric analysis of the autoradiographic data. The values obtained for the surface protein should be normalized against the values obtained form the totals (including precursor forms). This controls for differences in the incorporation of label (specific activity) between monolayers. If the protein is highly polarized from the first time point at which it is detected on the cell surface, then it is delivered directly to that surface. If it is polarized on one surface early in the chase and then switches polarity later in the chase, then it is delivered indirectly. If the protein is nonpolar early in the chase and acquires polarity only after longer chase times, then it is not sorted in the TGN, but its final polarity is acquired by differential stability on the apical and basolateral surfaces.

D. Biotin Assay for Endocytosis

This assay examines the internalization of plasma membrane proteins (from Graeve *et al.*, 1989).

Solutions

1. *PBS-CM.*
2. *Cleavable biotin reagent*: NHS-SS-biotin.
3. *DMEM containing 0.2% BSA.*
4. *Reducing solution*: 310 mg glutathione (free acid) dissolved in 17 ml H_2O (50 m*M*). Add 1 ml of 1.5*M* NaCl, 0.12 ml of 50% NaOH, and 2 ml of serum just before use.
5. *Quenching solution*: 5 mg/ml iodoacetamide in PBS-CM containing 1% BSA.

Steps

1. Wash cells on filters four times, 15 min each time with ice-cold PBS-CM.
2. Add 1 ml of NHS-SS-biotin (0.5 mg/ml in ice-cold PBS-CM) to the chamber being labeled and PBS-CM to the other chamber. Incubate for 20 min at 4°C and repeat with fresh solutions.
3. Wash filters twice with DMEM/0.2% BSA. Keep two filters on ice (one of these will represent the total amount of proteins at the surface before internalization and the other will be treated with the reducing solution and represents your control of efficiency of reduction) and transfer the other filters to 37°C for various times to allow the biotinylated proteins to be internalized.
4. Stop incubation by transferring filters back to 4°C.
5. Wash twice in PBS-CM + 10% serum.
6. Incubate filters for 20 min in reducing solution. Repeat. (Mock treat one filter.)
7. After washing, quench free SH groups in 5 mg/ml iodoacetamide in PBS-CM + 1% BSA for 15 min.
8. Lyse cells and immunoprecipitate as described in the section extraction of biotinylated proteins.
9. Run the samples on SDS–PAGE gels, transfer to nitrocellulose or PVDF, blot with [^{125}I]streptavidin, and expose for autoradiography.

Analysis of Results

Endocytosis of the protein of interest is indicated by protection of the NHS-SS-biotin-labeled surface protein from reduction by glutathione. By chasing the cells for various lengths of time, a rate of endocytosis can be calculated by comparing the percentage of protein protected at each time point. Obviously if none of the protein is protected from reduction, it is

100% at the surface; conversely, if all of the protein is protected, 100% has been internalized.

E. *In Situ* Domain Selective Biotinylation of Retinal Pigment Epithelial Cells

The retinal pigment epithelium is uniquely suited for biochemical studies of polarity in situ. The RPE exists as a natural monolayer with a broad apical surface that is easily exposed after gentle enzymatic treatment to remove the adjacent neural retina. The apical surface of the tissue is labeled with a noncleavable biotin analog such as NHS-LC-biotin for the identification of apical proteins. For identification of basolateral proteins the biotinylatable sites on the apical surface are labeled with the cleavable NHS-SS-biotin. After isolation of RPE cells, the nonlabeled basolateral proteins are labeled in suspension with the noncleavable form. Removal of the cleavable NHS-SS-biotin by reduction with 2-mercaptoethanol results in the presence of biotin only in the population of proteins present on the basolateral surface of the RPE (Marmorstein et al., 1996).

Solutions

1. *Sulfo-NHS-biotin, or NHS-LC-biotin, and NHS-SS-biotin stock solutions in DMSO*
2. *HBSS*: 10 m*M* HEPES buffered Hank's balanced salt solution
3. *PBS-CM*
4. *DMEM containing 10 mM HEPES*
5. *Bovine testicular hyaluronidase*
6. *CMF-PBS*: PBS calcium and magnesium free
7. *PBS-EDTA*: CMF-PBS + 1 m*M* EDTA

Steps

All steps are carried out on ice unless otherwise indicated.

1. Rats are euthanized by CO_2 asphyxiation, and the eyes are enucleated and stored for 3 h to overnight in the dark on ice in HBSS.
2. A circumferential incision is made above the ora serrata, and the cornea, iris, lens, and vitreous are removed.
3. The eyecups are incubated for 10–30 min at 37°C in HBSS containing 290 units/ml bovine testicular hyaluronidase.
4. The ora serrata is removed, and the neural retina is peeled carefully away from the RPE. The optic nerve head is severed and the neural retina is removed. The RPE is inspected under the dissecting microscope. Black spots on the outer surface of the retina or tracts of smooth reflective surface in the eyecup indicate damage. Damaged eyecups are discarded.
5. Soluble components of the interphotoreceptor matrix are removed by incubation in 2-ml microcentrifuge tubes (one eye per tube) on a rotator in ice-cold HBSS for 20 min. This is repeated three times.
6. The apical surface of the RPE in one eyecup is biotinylated with 1 ml of PBS-CM containing 2 mg of sulfo-NHS-biotin. The other eyecup is biotinylated with 1 ml of PBS-CM containing 2 mg of NHS-SS-biotin. This procedure is repeated three times.
7. The reaction is quenched with 1 ml of 10 m*M* HEPES buffered DMEM for 10 min at 4°C.
8. The eyecups are rinsed once in ice-cold CMF-PBS and are then incubated in PBS-EDTA on ice for 30 min.
9. The RPE is gently teased from the inner surface of the eyecup using a 22-gauge needle. The RPE layer is collected in a 1.5-ml microcentrifuge tube and pelleted for 10s in a microfuge. Cells labeled apically with noncleavable sulfo-NHS-biotin or NHS-LC-biotin at this stage are held on ice until the basolateral samples are ready.
10. For cells labeled with cleavable NHS-SS-biotin, the pellet is resuspended in 1 ml of

PBS-CM containing 2 mg/ml sulfo-NHS-biotin or NHS-LC biotin and incubated on a rotator at 4°C. After 20 min the cells are pelleted for 10s in a microfuge and this step is repeated.

11. The reaction is quenched with 50 m*M* NH_4Cl in PBS-CM for 10 min. The cells are then pelleted in the microfuge for 10s.
12. At this point, both apical and basolaterally labeled pellets are frozen dry at −80°C or immediately lysed and specific proteins immunoprecipitated as described in the section extraction of biotinylated proteins.
13. Immunoprecipitated proteins are resuspended in Laemmli sample buffer containing 5% 2mercaptoethanol or 50 m*M* dithiothreitol to release the NHS-SS-biotin from the apical proteins in basolateral samples. After heating to 95°C for 5 min, samples are resolved by SDS–PAGE, transferred to nitrocellulose or PVDF membranes, and blotted with streptavidin.

Analysis of Results

Analysis of the results proceeds as in the section on cell surface biotinylation.

F. Extraction and Immunoprecipitation of Biotinylated Proteins

Solutions

1. *TPI*: 1% Triton X-100, 20 m*M* Tris, 150 m*M* NaCl, 5 m*M* EDTA, pH 8.0, containing 0.2% BSA, and protease inhibitors
2. *TPII*: 0.1% SDS, 20 m*M* Tris, 150 m*M* NaCl, 5 m*M* EDTA, pH 8.0, containing 0.2% BSA
3. *TPIII*: 20 m*M* Tris, 500 m*M* NaCl, 5 m*M* EDTA, pH 8.0, containing 0.2% BSA
4. *TPIV*: 50 m*M* Tris, pH 8.0
5. *Laemmli sample buffer*
6. *Protein A–Sepharose*
7. *Staphlycoccus A cells*: Pansorbin

Steps

1. Excise filters from inserts using a #11 scalpel blade or razor blade. Lyse in 1 ml of TPI at 4°C. In some cases it may be necessary to use more stringent conditions for lysis (i.e., RIPA buffer, which contains 0.5% deoxycholate and 0.1% SDS in addition to 1% Triton X-100).
2. Wash 100 μl/sample of Pansorbin three times with TPI. *Do not omit protease inhibitors.*
3. Centrifuge lysate in a microfuge at 4°C at 13,000 *g* for 10 min. Collect the supernatant and discard the pellet.
4. Add 100 μl of washed Pansorbin to each supernatant. Preclear for 1 h at 4°C.
5. Resuspend 5–10 mg of protein A-Sepharose/lysate in 1 ml TPI /lysate. If you are immunoprecipitating with a mouse IgG, after 10 min, when the Sepharose beads are swollen, add 2 mg of rabbit antimouse IgG. After 1 h wash the beads three times with TPI. On the last wash, pellet the beads in microcentrifuge tubes.
6. Pellet the Pansorbin by centrifugation at 13,000 *g* for 10 min. Collect the supernatant, add an appropriate volume of antibody, and incubate at 4°C for 1 h to overnight.
7. Transfer the immunoprecipitates to the tubes containing the protein A–Sepharose and incubate for 1 h at 4°C.
8. Centrifuge the immunoprecipitates in a microfuge at 13,000 *g* for 30s. Remove the supernatant and resuspend the beads in 1 ml of TPI. Repeat this step three times with TPII, three times with TPIII, and once with TPIV.
9. Resuspend with an appropriate volume of Laemmli sample buffer, heat to 95°C for 5 min, and resolve by SDS–PAGE. For streptavidin blotting, transfer the gel to nitrocellulose or PVDF.

G. Streptavidin Blotting

Solutions

1. *Blocking buffer*: 5% Carnation instant milk, 0.3% BSA, in PBS-CM
2. *Rinse buffer*: 1% BSA, 0.2% Triton X-100 in PBS-CM
3. *[^{125}I]Streptavidin or streptavidin conjugated to horseradish peroxidase or alkaline phosphatase (streptavidin-HRP)*
4. A phosphorimager, Kodak X-OMAT AR film and a cassette with an intensifying screen, or an enhanced chemiluminescence reagent kit (such as those supplied by Amersham) and appropriate film.

Steps

1. After transfer to nitrocellulose or PVDF (PVDF is superior for chemiluminescent detection systems and is used in our laboratory for most streptavidin blots), block the blot for 1 h in blocking buffer.
2. Rinse once with rinse buffer.
3. Incubate for 1 h in 40 ml rinse buffer containing 1–2 × 10^6 cpm/ml of [^{125}I] streptavidin or 0.5–1.0 mg/ml streptavidin-HRP.
4. Wash three to five times for 5–10 min each with rinse buffer.
5. Dry the blot and expose to a phosphorimager screen, autoradiograph it with Kodak X-OMAT AR film, or use any of the many enhanced chemiluminescent kits that are available.

H. Quantitation of Polarized Distribution by Confocal Microscopy Imaging Technique

Modern quantitative optical microscopy offers an alternative method to quantify the relative polarity of fluorescently labeled cell surface proteins in cells developing polarity in settings such as the calcium switch assay (Rajasekaran *et al.*, 1996). This quantitation is based on the relative fluorescent pixel intensities in serial horizontal confocal sections of the target protein to a known polarized marker.

Requirements

1. Confocal microscope
2. Software to measure fluorescence intensity, e.g., Metamorph, Image Space Software.

Steps

1. Samples subjected to immunofluorescence technique (not discussed here) are imaged on a laser confocal microscopic system. Samples are subjected to optical sectioning (*xy*) of the entire thickness (*z*) of the monolayer. We select an interval between sections for optimized collection of fluorescence from a given plane without contribution from the neighboring *z* planes. Choice of interval depends on pinhole dimension, which in turn depends on characteristics of the excitation wavelength. [Refer to a handbook on confocal microscopy, e.g., Pawley (1995), or a confocal microscope manufacturer's manual for optimizing microscopic parameters.]
2. For each optical section, quantify the average per-pixel fluorescence intensity of the labeled proteins using the imaging software. Determine the ratio of intensity obtained for the protein of interest to that of a known marker.

Analysis of Results

The ratio of fluorescent intensity obtained represents the relative distribution of the target protein and is interpreted as follows.

a. A constant pixel intensity ratio for all optical sections of a given monolayer suggests overlapping distribution of the target protein with the known marker.

b. Decreasing pixel intensity ratio in from basolateral to apical domain of the target protein and the apical or tight junction marker would mean that the protein is localized in the basolateral domain.

c. Increasing intensity ratio with a basolateral marker would mean that the target protein is apically targeted (see Rajasekaran *et al.*, 1996).

I. Determination of Plasma Membrane Protein Polarity after Intranuclear Microinjection of Its cDNA

This procedure is used for rapid qualitative determination of the polarity of a newly synthesized protein in polarized cells. Analysis is performed using either a fluorescence widefield or a confocal microscope. We have utilized this technique to study the regulation of polarity of basolateral and apical membrane markers by GTPases and their downstream effectors. Normal polarization of the monolayer needs to be confirmed by studying the localization of known apical or basolateral markers.

Solutions

1. *Microinjection buffer*: H-KCl buffer containing 10 m*M* HEPES, pH 7.4, 140 m*M* KCl. Dissolve 1.04 g of KCl in 99 ml deionized H_2O and then add 1.0 ml HEPES from a 1*M* stock (pH 7.4). Sterilize the buffer by passing through a 0.22-μm filter and store at 4°C (Müsch *et al.*, 2001).
2. *PBS/CM*: See Section III,A
3. *H-DMEM*: DMEM containing HEPES. Dissolve bicarbonate-free powdered DMEM in 900 ml deionized H_2O. Add 20 ml HEPES from a 1*M* stock and adjust the pH to 7.4. Sterilize the medium through a 0.22-μm filter and store at 4°C.
4. *B–DMEM*: DMEM containing sodium bicarbonate

Steps

1. Plate MDCK II cells on sterile glass coverslips at a concentration of 1.6×10^6 cells/ml or at an approximate plating density of 2×10^5 cells per cm^2 in B-DMEM and 10% fetal bovine serum (FBS). Allow the cells to polarize for 4–5 days and change the medium only once on day 2 postplating. Growth conditions required to attain polarity for different cell lines vary and require optimization.
2. Dilute the stock of cDNA (stock prepared in deionized H_2O at a concentration of 0.5 mg/ml) in microinjection buffer to a concentration of around 10 μg/ml. It is highly recommended that cDNA constructs also contain a tag sequence, such as Myc, HA, GFP, and its variants, in-frame with the gene of interest so as to distinguish the newly synthesized proteins from endogenous proteins. For experiments involving more than one cDNA construct, cDNAs can be coinjected. However, the efficiency of expression of a construct may vary in the presence of another. Therefore, proper conditions should be established for good expression of each coinjected construct. A range of concentrations between 1 and 20 μg/ml of DNA should be tested to optimize their expression levels.
3. Prepare microinjection needles by pulling 1-mm-diameter and 6-in.-long borosilicate glass capillaries (1B100F-6, World Precision Instruments, Inc, Sarasota, FL) using a micropipette puller (e.g., Flaming/Brown Micropipette Puller Model P-97, Sutter Instrument Co., Novato, CA).
4. Load the cDNA diluted in microinjection buffer through the blunt end of the needle into the needle holder of the

micromanipulator (Narishige Company, Ltd., SE-TAGAYA-KU, Tokyo, Japan) attached to the inverted microscope (Zeiss-Axiovert 25, Germany).

5. Transfer coverslip into 35-mm-diameter tissue culture dishes. Add 4 ml of H-DMEM containing 5% FBS to each dish and place the dish on the dish holder of the micromanipulator-microscope described earlier. Microinject the nucleus of cells. Avoid microinjecting cells that are right next to each other. This simplifies the analysis of distribution of apical and basolateral markers. In order to avoid unsynchronized protein synthesis, preferably microinject within 10–15 min of transferring the dish to the microscope stage.
6. Incubate the microinjected cells with B-DMEM–10% FBS medium at 37°C. Most of the proteins accumulate in the ER within 60 × 90 min at 37°C postmicroinjection. Different levels of expression should be tested to ensure that the sorting pathways are not saturated. For coinjections, it is necessary to standardize the conditions for sufficient expression of each protein. Adjustment of DNA concentrations (as described in step 2) and time of incubation at 37°C postmicroinjection are two steps that need to be tuned for expression of multiple constructs.
7. After appropriate incubation at 37°C, replace medium with B-DMEM–10% FBS containing 100 μg/ml cycloheximide (concentration may be lowered down to 20 μg/ml if cells detach from coverslip) to inhibit new protein synthesis. The chase time for plasma membrane delivery of protein is initiated at 37°C for 3–4 h.
8. After an appropriate chase period, fix cells with either −20°C chilled methanol for 10 min or 2% paraformaldehyde at room temperature for 15 min. Methanol fixation should be followed by a blocking step at room temperature with 1% BSA prepared in PBS-CM for 30 min. Paraformaldehyde fixation of cells is followed by permeabilization at room temperature for 30 min with either 0.2% Triton-X 100 or 0.075% saponin prepared in PBS-CM containing 1% BSA. Cells can now be processed for immunofluorescence with the appropriate primary and secondary antibodies.

Analysis of Results

Cells processed for immunofluorescence are imaged on either a confocal or a wide-field microscope. The correct orientation of the cells is determined by analyzing the staining of known polarized markers. In case of a wide-field microscope, the entire monolayer is subjected to *z* sectioning at least at 0.5-μm intervals with a 60× 1.4 NA objective, and standard deconvolution software is used to enhance the resolution. Alternatively, we use a confocal microscope in frame-scan mode and collect *xyz* stacks of the monolayer and display the *xyz* stack in the orthogonal plane. The cells can be displayed directly as a *xz* cross section by doing a line scan, i.e., scanning in the *xzy* mode. Localization of the protein is determined depending on the staining pattern, e.g., relative to a tight junctional marker such as ZO-1.

IV. COMMENTS

The methods described here represent examples of applications of the biotinylation technique; other examples of possible applications are (1) a transcytotic assay using a combination of the targeting and endocytosis protocols (Le Bivic *et al.*, 1989; Zurzolo *et al.*, 1992) and (2) detection of GPI-anchored proteins at the cell surface using Triton X-114 phase separation and PI-PLC digestion in place of the standard lysis procedure (Lisanti and Rodriguez-Boulan, 1990).

Another analog of biotin, biotin hydrazide, can be used to label oligosaccharides of surface glycoproteins following periodate oxidation (Lisanti *et al.*, 1989).

V. PITFALLS/RECOMMENDATIONS

1. It has been suggested that the use of pH 9.0 buffer to dilute sulfo-NHS-biotin would enhance the efficiency of labeling of surface proteins (Gottardi and Caplan, 1993). In our experience this is not always true and depends on different proteins and cell lines.
2. Always cut the filters out of the plastic holder before lysis. We have found that the cells can grow along the inside of the plastic ring supporting the filter. Lysis of these cells can result in erroneous results (Zurzolo and Rodriguez-Boulan, 1993).
3. Occasionally, in targeting experiments, intracellular nonbiotinylated forms are recovered on streptavidin beads. Using NHS-LC-biotin and keeping the SDS concentration at 0.4% helps reduce this. Another approach is to use NHS-SS-biotin and remove it from the streptavidin beads by incubation in 50 m*M* dithiothreitol of 5–10% 2-mercaptoethanol in 62.5 m*M* Tris, pH 8.0. Then spin the beads out and dilute the supernatant 1:1 with 2× Laemmli sample buffer.
4. The *in situ* biotinylation assay works best for proteins that are restricted to the RPE cell (i.e., RET-PE2 antigen). Quantification of proteins present in adjacent tissues (particularly the choroid) can contaminate the basolaterally labeled material and yield an incorrectly high level of basolateral labeling.
5. Determination of polarity by confocal microscopy can provide visual validation of the biochemical assay. It is useful for measuring the polarity of steady-state protein and dynamic changes involved in tight junction assembly in the Ca^{2+} switch assay. A thorough analysis with known markers of different domains is recommended before determining polarity of the target protein.
6. For determination of polarity by means of intranuclear microinjection of cDNA, thorough optimization regarding the protein expression level and the incubation time is necessary. Moreover, it is critical to avoid saturating the sorting pathway. Hence, coinjecting and monitoring a known apical or basolateral marker are critical for final evaluation. Because the basolateral domain of the polarized monolayer on the coverslip is not accessible to the antibody without permeabilization, it is difficult to distinguish the pool of basolateral protein that is associated with submembrane structures closely juxtaposed to the cytoplasmic side of plasma membrane and the pool present in the external leaflet of the plasma membrane.

References

Gan, Y., McGraw, T.E., and Rodriguez-Boulan, E. (2002). The epithelial-specific adaptor AP1B mediates post-endocytic recycling to the basolateral membrane. *Nature Cell Biol.* **4**, 605–609.

Gottardi, C., and Caplan, M. (1993). Cell surface biotinylation in the determination of epithelial membrane polarity. *J. Tissue Culture Methods* **14**, 173–180.

Graeve, L., Drickamer, K., and Rodriguez-Boulan, E. (1989). Functional expression of the chicken liver asialoglycoprotein receptor in the basolateral surface of MDCK cells. *J. Cell Biol.* **109**, 2909–2816.

Hanzel, D., Nabi, I. R., Zurolo, C., Powell, S. K., and Rodriguez-Boulan, E. (1991) New techniques lead to advances in epithelial cell polarity. *Semin. Cell Biol.* **2**, 341–353.

Keller P. Simons K. (1997). Post-golgi biosynthetic trafficking. *J. Cell Sci.* 1103001–1103009.

Laemmli, U. K. (1970). Cleavage of structural proteins during the assembly of the head of bacteriophage T4. *Nature* **227**, 680–685.

Le Bivic, A., Real, F. X., and Rodriguez-Boulan, E. (1989). Vectorial targeting of apical and basolateral plasma membrane proteins in a human adenocarcinoma cell line. *Proc. Natl. Acad. Sci. USA* **86**, 9313–9317.

Le Bivic, A., Sambuy, Y., Mostov, K., and Rodriguez-Boulan, E. (1990). Vectorial targeting of an endogenous apical membrane sialoglycoprotein and uvomorulin in MDCK cells. *J. Cell Biol.* **110**, 1533–1539.

Le Gall, A., Yeaman, C., Muesch, A., and Rodriguez-Boulan, E. (1995). Epithelial cell polarity: New perspectives. *Semin. Nephrol.* **15**(4), 272–284.

Lisanti, M., Le Bivic, A., Sargiacomo, M., and Rodriguez-Boulan, E. (1989). Steady state distribution and biogenesis of endogenous MDCK glycoproteins: Evidence for intracellular sorting and polarized surface delivery. *J. Cell Biol.* **109**, 2117–2128.

Lisanti, M., and Rodriguez-Boulan, E. (1990). Glycosphingolipid membrane anchoring provides clues to the mechanism of protein sorting in polarized epithelial cells. *Trends Biochem. Sci.* 113–118.

Marmorstein, A. D., Bonilha, V. L., Chiflet, S., Neill, J. M., and Rodriguez-Boulan E. (1996). The polarity of the plasma membrane protein RET-PE2 in retinal pigment epithelium is developmentally regulated. *J. Cell. Sci.* **109**, 3025–3034.

Müsch A., Cohen D., Kreitzer G., and Rodriguez-Boulan E. (2001). cdc42 regulates the exit of apical and basolateral proteins from the trans-Golgi network. *EMBO J.* **20**, 2171–2179.

Pawley J. B. (ed.) (1995). "Handbook of Biological Confocal Microscopy." Plenum Press, New York.

Rajasekaran, A.K., Hojo, M., Huima, T., and Rodriguez-Boulan, E. (1996). Catenins and zonula occudens-1 form a complex during early stages in the assembly of tight junctions. *J. Cell Biol.* **132**, 451–463.

Rodriguez-Boulan, E., Kreitzer, G., and Muesch, A. (2005). Organization of vesicular trafficking in epithelia. *Nature Rev. Mol. Cell Biol.* **6**, 233–247.

Rodriguez-Boulan, E., and Powell, S. K. (1992). Polarity of epithelial and neuronal cells. *Annu. Rev. Cell Biol.* **8**, 395–427.

Sargiacomo, M., Lisanti, M., Graeve, L., Le Bivic, A., and Rodriguez-Boulan, E. (1989). Integral and peripheral protein compositions of the apical and basolateral plasma membrane domains of MDCK cells. *J. Membr. Biol.* **107**, 277–286.

Yeaman, C., Grindstaff, K.K., and Nelson, W.J. (1999). New perspectives on mechanisms involved in generating epithelial cell polarity. *Physiol. Rev.* **79**,73–98.

Zurzolo, C., Le Bivic, A., Quaroni, A., Nitsch, L., and Rodriguez-Boulan, E. (1992). Modulation of transcytotic and direct targeting pathways in a polarized thyroid cell line. *EMBO J.* **11**, 2337–2344.

Zurzolo, C., and Rodriguez-Boulan, E. (1993). Delivery of Na,K-ATPase in polarized epithelial cells. *Science* **260**, 550–552.

CHAPTER

5

Assays Measuring Membrane Transport in the Endocytic Pathway

Linda J. Robinson, and Jean Gruenberg

OUTLINE

I. INTRODUCTION

Significant progress has been made in understanding mechanisms regulating endocytic membrane traffic using cell-free assays (Braell, 1987; Davey *et al.*, 1985; Diaz *et al.*, 1988; Gruenberg and Howell, 1986; Woodman and Warren, 1988) (see Fig. 5.1). Both early and late endosomes exhibit homotypic fusion properties *in vitro*, as *in vivo*, yet they do not fuse with each other (Aniento *et al.*, 1993). Transport from early to late endosomes is achieved by multivesicular

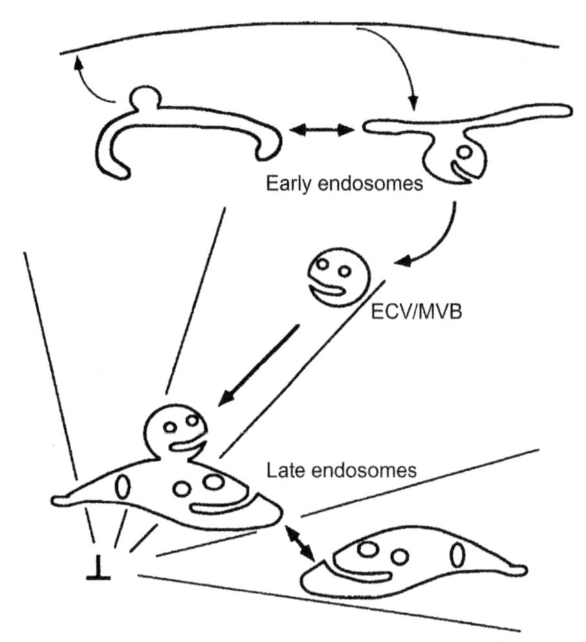

FIGURE 5.1 Membrane trafficking in the endocytic pathway. The reconstituted steps of the endocytic pathway described in this protocol are the fusion of early endosomes with each other, fusion of ECV/MVBs with late endosomes, and fusion of late endosomes with each other. An *in vitro* budding assay for the formation of ECV/MVBs from early endosomes, which are competent to fuse with late endosomes, is described in Aniento *et al.* (1996). As shown, ECV/MVBs are transported along microtubules from early to late endosomes. If microtubules are depolymerized *in vivo*, prior to the loading of cells with an endocytic tracer, this tracer will accumulate in ECW/MVBs. These vesicles will then fuse with late endosomes, loaded with a different marker, *in vitro*.

intermediates termed endosomal carrier vesicles (ECV/MVB), which are presumably translocated on microtubules between the two compartments (Aniento *et al.*, 1996; Bomsel *et al.*, 1990; Gruenberg *et al.*, 1989). The vectorial or heterotypic interactions of ECV/MVBs with late endosomes have also been reconstituted *in vitro*, as has the involvement of microtubules and motor proteins in this process (Aniento *et al.*, 1993; Bomsel *et al.*, 1990). By reducing the components in these *in vitro* assays to a cytosol source, an ATP-regenerating system, salts, and the purified endosomal membranes, the specificity of endosomal fusion events has been addressed, and the molecules and mechanisms involved have been studied. In fact, a number of conserved molecules, as well as molecules specific for different steps of the endocytic pathway, have been identified and/or characterized using cell-free assays such as those described in this protocol (for review, see Gruenberg and Maxfield, 1995).

This assay for endocytic vesicle fusion is based on the formation of a complex resulting from a reaction between two products present in separate populations of endosomes: avidin and biotinylated horseradish peroxidase (bHRP). These reaction products can be internalized into endosomes by fluid phase or receptor-mediated endocytosis *in vivo*. Avidin and a biotinylated compound are used to provide a fusion-specific reaction because of the high binding affinity and low dissociation constant of avidin for biotin. Following internalization, cells are homogenized and purified endosomal fractions are prepared, which are combined in the assay together with cytosol and ATP. If fusion occurs, a complex is formed between avidin and bHRP. At the end of the assay, the reaction mixture is extracted in detergents in the presence of excess biotinylated insulin as a quenching agent. The avidin–bHRP complex is then detected by immunoprecipitation with antiavidin antibodies, and the enzymatic activity of the bHRP associated with the immunoprecipitate is quantified. This article describes techniques for the preparation and partial purification of three different loaded endosomal fractions from BHK cells: early endosomes, endosomal carrier vesicles, and late endosomes. In addition, this article describes the preparation of the cytosol source used, as well as the techniques for the fusion assays themselves.

II. MATERIAL AND INSTRUMENTATION

Standard laboratory rockers for washing cells and a large 37°C water bath, which can fit a metal plate of dimensions of 20 × 33 cm, are used. Large rectangular ice buckets (Cat. No. 1-6030), from NeoLab GmbH, can also accommodate metal plates of the same dimensions. Cell scrapers (flexible rubber policemen) with a silicone rubber piece of about 2 cm, cut at a sharp angle, and attached to a metal bar, are made. A standard low-speed cell centrifuge and Beckman ultracentrifuges and rotors are used. The refractometer (Cat. No. 79729) is from Carl Zeiss Inc., and the pump for collecting sucrose gradients (peristaltic pump P-l) is from Pharmacia Fine Chemicals. A rotating wheel (such as Snijders Model 34528) with a speed of about 10 rotations per minute should be used. All tissue culture reagents, including modified Eagle's medium (MEM), are from either Sigma Chemical Company or GIBCO-BRL/Life Technologies. Peroxidase from horseradish (HRP) (Cat. No. P-8250), ATP (disodium salt, Cat. No. A-5394), and deuterium oxide (D_2O, Cat. No. D-4501) are from Sigma Chemical Company, Ltd. Biotinyl-ε-aminocaproic acid *N*-hydroxysuccinimide ester (biotin-X-NHS, Cat. No. 203188) is from Calbiochem. Avidin (egg white, Cat. No. A-887) is from Molecular Probes. Creatine phosphate (Cat. No. 621714), creatine phosphokinase (Cat. No. 127566), and hexokinase [$(NH_4)_2SO_4$ precipitate of yeast hexokinase, 1400 U/ml, Cat. No. 1426362] are from Boehringer-Mannheim GmbH. Protein A–Sepharose beads (Cat. No. CL-4B) are from Pharmacia. Antiavidin antibodies are generated by injecting purified avidin into rabbits and are affinity purified prior to use. Antiavidin antibodies are also available commercially from several companies. BCA protein assay reagents (Cat. No. 23223) are from Pierce, and Bio-Rad protein assay reagents (Cat. No. 500-0006) are from Bio-Rad Laboratories GmbH.

III. PROCEDURES

A. Internalization of Endocytic Markers into Early Endosomes (EE) from BHK Cells

Solutions

1. *Internalization media (IM)*: MEM containing 10 mM HEPES and 5 m*M* d-glucose, pH 7.4. Filter sterilize and store at 4°C.
2. *Phosphate-buffered saline (PBS)*: 137 mM NaCI, 2.7 m*M* KCL, 1.5 m*M* KH_2PO_4, and 6.5 m*M* Na_2HPO_4; should be pH 7.4. Filter sterilize and store at 4°C.
3. *Biotinylated horseradish peroxidase*: Dissolve 20 mg of HRP in 9.5 ml of 0.1*M* $NaHCO_3$/Na_2CO_3, pH 9.0, buffer (make fresh and check pH carefully) in a small glass Erlenmeyer flask. Dissolve 20 mg of biotin-X-NHS in 0.5 ml dirnethylformamide. Mix by adding the biotin dropwise to the HRP mixture while gently stirring or shaking the Erlenmeyer and incubate at room temperature with gentle stirring for at least 45 min (a 50:1 molar excess of biotin is important). Quench unreacted active groups with 1 ml of 0.2*M* glycine, pH 8.0 (use KOH to pH), by adding dropwise while mixing, and mix for an additional 15 min at room temperature. Transfer to 4°C. Dialyze the mixture extensively against PBS—or IM at 4°C (at least four changes of 200 ml each time). The final dialysis should be in IM. Measure protein concentration (should be about 2 ml/ml) and HRP enzymatic activity (should be unchanged). Aliquot in sterile tubes, freeze in liquid N_2, and store at −20°C until use. Immediately before use, thaw quickly and warm to 37°C.

4. *Avidin*: Avidin powder dissolved in IM at 3 mg/ml. Make fresh immediately before use and warm to 37°C.
5. *PBS/BSA*: 5 mg/ml BSA in PBS. Make fresh before use and cool to 4°C.

Steps

1. *Cell culture*: Maintain monolayers of baby hamster kidney (BHK-21) cells as described in Gruenberg *et al.* (1989). For a fusion assay of 5–10, eight petri dishes (10 cm diameter) should be prepared 16 h before the experiment: four for preparing bHRP-labeled EEs and four for preparing avidin-labeled EEs.
2. *Fluid-phase internalization*: Wash each 10-cm dish of cells twice with 5 ml ice-cold PBS—on ice. This and other washes on ice to follow are performed most easily by placing four dishes onto a metal plate in a large ice bucket on a rocker. After the last wash, remove PBS and place the dish on a metal plate in a 37°C water bath. Add at least 3 ml/dish bHRP or avidin solution prewarmed to 37°C. Incubate for 5 min.
3. *Washes*: From now on, all work should be done at 4°C or on ice. Return the dishes to the metal plate in the ice bucket. Remove the avidin or bHRP solution and wash dishes three times for 5 min with 5 ml ice-cold PBS/BSA followed by 2 × 5 min with 5 ml ice-cold PBS.
4. *Homogenization and fractionation*: Go directly to Section IIIC.

B. Internalization of Endocytic Markers into Endosomal Carrier Vesicles (ECV) and Late Endosomes (LE)

Solutions

1. *Nocodazole stock*: 10 m*M* in dimethyl sulfoxide (DMSO), aliquoted, and stored at −20°C.
2. *IM/BSA*: IM containing 2 mg/ml BSA. Make fresh before use and warm to 37°C.
3. All solutions listed in Section III,A.

Steps

1. *Cell culture*: For a fusion assay of 5–10 points, 10 dishes (10 cm) of BHK cells should be prepared as described in Section III,A, step 1. For ECV–LE fusion assays, use 5 dishes for bHRP-labeled ECVs and 5 dishes for avidin-labeled LEs, For LE–LE fusion assays, use 5 dishes for bHRP-labeled LEs and 5 dishes for avidin-labeled LEs.
2. *Nocodazole pretreatment for ECV preparation*: Intact microtubules are required for the delivery of endocytosed markers to the LE. Therefore, markers accumulate in transport intermediates (ECVs) in the absence of microtubules. Whereas stable microtubules are cold sensitive, dynamic microtubules are depolymerized easily in the presence of nocodazole (Aniento *et al.*, 1993; Bomsel *et al.*, 1990). In BHK cells, microtubules can be depolymerized efficiently in the presence of nocodazole, whereas cold treatment is without effect. For ECV preparation, depolymerize the microtubules immediately before the experiment with 10 μ*M* nocodazole at 37°C for 1–2 h in media used to grow cells in a 5% CO_2 incubator. Following this step, nocodazole (10 μ*M*) should remain present in all solutions up to the homogenization step. For LE preparation, do not treat with nocodazole or include nocodazole in any solutions.
3. *Fluid-phase internalization*: Wash each 10-cm dish of cells twice with 5 ml ice-cold PBS+/− 10 μ*M* nocodazole on ice, as in Section III,A, step 2. After the last wash, remove the PBS and place the dish on a metal plate in a 37°C water bath. Add at least 3 ml bHRP or avidin solution for making LEs or bHRP + 10 μ*M* nocodazole for making ECVs. Incubate for 10 min.

4. *Chase*: Remove bHRP or avidin and wash twice quickly at 37°C with 10 ml PBS/BSA+/− 10 μ*M* nocodazole, prewarmed to 37°C. Remove last wash, and add 8 ml IM/BSA+/− 10 μ*M* nocodazole, prewarmed to 37°C. Incubate at 37°C (in water bath or in a 37°C incubator without CO_2) for 45 min.
5. *Washes*: Remove IM/BSA, move dishes to ice bucket, and wash 2 × 5 min with 5 ml cold PBS/BSA followed by 5 min with 5 ml cold PBS on ice.

C. Homogenization and Fractionation of Cells

Solutions

1. *PBS*: See Section IIIA.
2. *300 mM imidazole stock*: Dissolve imidazole in H_2O and adjust pH to 7.4 with NaOH, filter sterilize, and store at 4°C
3. *Homogenization buffer (HB)*: Add imidazole from 300 m*M* stock to H_2O and dissolve sucrose such that the final concentrations are 250 m*M* sucrose and 3 m*M* imidazole. Filter sterilize and store at 4°C.
4. *62% sucrose solution*: For 100 ml, add 1 ml of imidazole from 300 m*M* stock to 15 ml H_2O. Add 80.4 g sucrose and dissolve by stirring at 37°C. Add H_2O and mix until the refractive index is 1.4464.
5. *10 and 16% sucrose solutions in* D_2O: For 100 ml, add 1 ml imidazole from 300 m*M* stock to 50 ml D_2O. For 10% solution, add 10.4 g sucrose, and for 16% solution, add 17.0 g sucrose. Dissolve sucrose, add D_2O, and mix until the refractive index is 1.3479 for the 10% solution and 1.3573 for the 16% solution.

Steps

1. *Cell scraping*: All of the following steps should be performed on ice or at 4°C. After the last wash, remove all PBS. Add 2 ml/dish PBS and rock the dish so that cells do not dry. Using a flexible rubber policeman, scrape round 10-cm dishes by first scraping in a circular motion around the outside of the dish, followed by a downward motion in the middle of the dish. Scrape gently in order to obtain "sheets" of cells. Using a plastic Pasteur pipette, gently transfer the scraped "sheets" of cells from four or five dishes into a 15-ml tube on ice.
2. Centrifuge at 1200 rpm for 5 min at 4°C. Gently remove supernatant.
3. Add 1 ml HB to pellet, using a plastic Pasteur pipette, gently pipette up and down one time and add an excess of HB (4–5 ml) to change buffer. Centrifuge again at 2500 rpm for 10 min at 4°C. Remove supernatant.
4. *Homogenization*: It is important that cells are homogenized under conditions where endosomes are released from cells, yet where latency is high so that the endosomes are not broken and retain their internalized marker. First add 0.5 ml HB to the cell pellet. Using a 1-ml pipetman, gently pipette up and down until the pellet is resuspended and particles can no longer be seen by eye. Do not introduce air bubbles. Using a 22-gauge needle connected to a narrow 1-ml Tubercutine syringe, prewet the needle and syringe with HB so that no air is introduced. Insert the needle into the cell homogenate, slowly pull up on the syringe until most of the cell homogenate is in the syringe, and gently expel without bubbles. Repeat this procedure until plasma membranes are broken, yet nuclear membranes are not. Monitor homogenization as follows. Take 3 μl of homogenate and place in a 50-μl drop of HB on a glass slide. Mix and cover with a glass coverslip. Observe by phase-contrast microscopy, using a 20× objective. Homogenize until unbroken cells are no longer observed, yet nuclei, which appear as dark round or oblong structures, are not broken. Usually between 3 and 10 up-and-down strokes through the needle

are necessary. Centrifuge homogenate at 2000 rpm for 10 min at 4°C, and carefully collect the postnuclear supernatant (PNS) and nuclear pellet.

5. Save a 50-μl aliquot of each PNS fraction for measuring latency and for calculating the balance sheet as described in Section III,D. Adjust the sucrose concentration of the remaining PNS to 40.6% by adding about 1.1 volume of 62% sucrose solution per volume of PNS. Mix gently but thoroughly, without bubbles. Check sucrose concentration using a refractometer.
6. Place adjusted PNS in the bottom of a SW60 centrifuge tube. On top of the PNS, layer 1.5 ml of 16% sucrose solution in D_2O, followed by 1 ml of 10% sucrose solution in D_2O, and fill tube with HB. Steps should be layered so that interfaces are clearly seen and not disturbed. See Gruenberg and Gorvel (1992) for diagram of gradients.
7. Centrifuge gradients in SW60 rotor at 35,000 rpm for 1 h at 4°C.
8. Carefully remove the interfaces from the gradients after centrifugation by first placing gradients in a test tube rack with a black backdrop. The interfaces should appear white. The layer of white lipids on top of the gradient should be removed carefully. Collect fractions at 4°C using a peristaltic pump at speed 2, with capillary tubes connected to each end. Place the outgoing end into a collection tube and collect the top interface carefully (10%/HB interface = LE + ECV fraction) first. Collect by holding the capillary tube directly in the middle of the wide interface and slowly move in a circular motion until most of the white interface is collected into the smallest possible volume. Wash the pump tubing with water and then collect the EE (16/10%) interface into another tube. Fractions can be frozen and stored in liquid N_2 until use in fusion assays if they are carefully frozen quickly in liquid N_2 and thawed quickly at 37°C immediately before use.

D. Measurement of Latency and Balance Sheet for Gradients

Solutions

1. *HRP stocks*: 1–10 ng HRP in 0.1 ml HB, for standards.
2. *HB*: See Section IIIC.
3. *HRP reagent*: 0.342 m*M* *o*-dianisidine and 0.003% H_2O_2 in 0.05*M* Na-phosphate buffer, pH 5.0, containing 0.3% Triton X-100. To prepare, use very clean glassware or plasticware (as in for tissue culture) and mix 12 ml of 0.5 *M* Na-phosphate buffer, pH 5.0 (filter sterilized), and 6 ml of 2% Triton X-100 (filter sterilized) with 111 ml sterile H_2O. Add 13 mg *o*-dianisidine, dissolve gently, and add 1.2 ml 0.3% H_2O_2 (filter sterilized). Avoid magnetic stirring. Solution should be clear. Store at 4°C in the dark.
4. *1 mM KCN in* H_2O
5. Protein assay system (such as the BCA protein assay reagent or the Bio-Rad protein assay system)

Steps

1. Load a 20-μl aliquot of bHRP PNS into an airfuge tube or a small tabletop ultracentrifuge tube of the Beckman TL-100 type and fill the tube with a known volme of HB. Mix thoroughly by pipetting without air bubbles. Centrifuge at 4°C for 20 min at 20 psi in an airfuge or at 200,000 *g* for 20 min in a tabletop ultracentrifuge rotor (such as Beckman TLA-100.1). Transfer the supernatant to another tube. Resuspend the pellet in 50 μl HB.
2. To measure the latency, adjust samples, blanks, and standards with HB so that the final volume of each is 0.1 ml. Assay both the pellet and the supernatant of the latency measurement. If the supernatant volume is over 0.1 ml, assay only 0.1 ml. Add 0.9 ml of HRP reagent to each tube, mix quickly, and record the time with a stop clock. Allow

color to develop in the dark, as this reagent is light sensitive. When a brown color has begun to develop, read the absorbance at 455 nm and record the time (results expressed as OD units/min or ng HRP/min). Stop the reaction with 10 μl of 1.0 m*M* KCN if necessary.

3. Calculate latency by first adding the value (OD/min) for HRP in the pellet to that of HRP in the supernatant (OD/min after correcting for total supernatant volume). The value for the pellet divided by the total value is the percentage latency. Latency should be over 70% in order to measure endosome fusion.
4. The amount of HRP in each gradient fraction collected from the bHRP gradient can be measured by assaying an aliquot (about 50 μl) of each fraction as described in step 2.
5. Measure the amount of protein in each gradient fraction using a standard protein assay system, as described in the manual.
6. Calculate percentage yield (percentage of HRP in each fraction compared to total amount of HRP in PNS), specific activity (SA) (HRP activity per unit protein), and relative specific activity (RSA) (divide specific activity of each fraction by the specific activity of the PNS). See Gruenberg and Gorvel (1992) for an example of a typical balance sheet.

E. Preparation of BHK Cell Cytosol

Solutions

1. *PBS-*: See Section IIIA.
2. *HB*: See Section IIIC.
3. *HB + protease inhibitors*: HB with the following protease inhibitors added immediately before use: 10 μ*M* leupeptin, 1 μ*M* pepstatin A, 10 ng/ml aprotinin, and, if needed, 1 μm phenylmethylsulfonyl fluoride.

Steps

Two possible cytosol sources for all of the assays described are BHK and rat liver cytosol. For rat liver cytosol preparation, refer to Aniento *et al.* (1993).

1. BHK cells, maintained as described in Section IIIA, should be plated approximately 16 h before the experiment. Large (245 × 245 × 25 mm) square dishes are convenient for large cytosol preparations.
2. All steps should be performed on ice or at 4°C. Wash dishes four times with excess PBS (50 ml per dish for large square dishes).
3. Remove PBS from the last wash, add 12 ml PBS per dish, and rock the dish so that cells do not dry. Scrape cells with a rubber policeman using firm, downward motions, going from top to bottom while holding the plate at an angle, as described in Section IIIC, step 1.
4. Collect scraped cells into 15-ml tubes (one tube per dish). Centrifuge at 1200 rpm for 5 min at 4°C.
5. Remove supernatant and gently add 5 ml HB with a plastic Pasteur pipette and pipette up and down one time.
6. Centrifuge at 2500 rpm for 10 min at 4°C. Remove supernatant and resuspend pellet in 1.2 ml HB + protease inhibitors. Separate into two tubes (about 0.7 ml/tube) for homogenization and homogenize as described in step 4 of Section IIIC.
7. Centrifuge at 2500 rpm for 15 min at 4°C. Add supernatant (PNS) to a centrifuge tube for the TLS-55 rotor (for the Beckman TL-100 tabletop ultracentrifuge) and centrifuge in TLS-55 for 45 min at 55,000 rpm at 4°C. Remove fat from the top using an aspirator. Transfer supernatant (cytosol fraction) to a new tube without disturbing the pellet. Determine the protein concentration of supernatant. Cytosol should be at least 15 mg/ml to give a good signal for fusion assays. Aliquot on ice, freeze quickly, and store in liquid N_2 until use.

F. Preparation of Antiavidin Beads for the *in vitro* Fusion Assay Described in Section IIIG

Solutions

1. *PBS/BSA*: Dissolve 5 mg/ml BSA in PBS. Filter sterilize and store at 4°C.
2. *Sterile PBS*: PBS as described in Section III, A, filter sterilize or autoclave, and store at 4°C.
3. *Antiavidin antibody*: Affinity purify and store aliquoted in 50% glycerol/PBS at −20°C.

Steps

To determine how many antiavidin beads to prepare, first determine the number of fusion assay points. From a typical gradient (see Gruenberg and Gorvel, 1992) about 150 μg of EE and 70 μg of ECV or LE are obtained. Optimal amounts of endosomes to use for fusion assays are 20 μg of each EE fraction and 10 μg of each ECV or LE fraction. Therefore, a typical experiment (one gradient each of avidin and bHRP-labeled fractions) will provide enough endosomes for about seven fusion assay points.

1. Swell 1.5 g of protein A–Sepharose beads in 10 ml sterile H_2O at room temperature overnight.
2. Wash beads three times in 10 ml sterile PBS by centrifuging beads in 15-ml tubes at 3000 rpm for 2 min, resuspending in PBS each time.
3. After the final wash, resuspend beads in an equal volume of sterile PBS per volume of packed beads. Store beads this way up to several months at 4°C.
4. One hundred microliters of this 1:1 slurry is required per fusion assay point. Therefore, for 10 assay points, block 1 ml of beads by washing 3× in 10 ml PBS/BSA, as described in step 2.
5. After final wash, resuspend beads in 10 ml PBS/BSA. For 10 assay points, add 50 μg of antiavidin antibody (5 μg per 100-μl beads). Rotate tube for at least 5 h at 4°C.
6. Wash beads four times in PBS/BSA. After the last wash, for 10 assay points, resuspend beads in 10 ml PBS/BSA.
7. Aliquot 1 ml to each of 10 labeled Eppendorf tubes. Centrifuge in Eppendorf centrifuge at maximum speed for 2 min. Remove supernatant. Beads are now ready for the immunoprecipitation step of the fusion assay (Section III,G, step 10).

G. *In vitro* Assay of Endocytic Vesicle Fusion

Solutions

1. *50 × salts*: 0.625*M* HEPES, 75 m*M* Mg-acetate, 50 mM dithiothreitol, pH 7, with KOH. Filter sterilize, aliquot, and store at −20°C.
2. *K-acetate (KOAc stock)*: 1 *M* in H_2O. Filter sterilize, aliquot, and store at −20°C. *Note:* Depending on the counterion requirement of the experiment, KOAc must be replaced by KCl (see Aniento *et al.*, 1993).
3. *Biotinylated insulin*: 1 mg/ml in H_2O. Store at 4°C.
4. *ATP-regenerating system (ATP-RS)*: Mix 1:1:1 volumes of the following immediately before use.
   - **a.** *100 mM* ATP: Dissolve in ice-cold H_2O, titrate to pH 7.0 with 1*M* NaOH, filter sterilize, aliquot on ice, and store at −20°C.
   - **b.** *800 mM creatine phosphate*: Dissolve in ice-cold H_2O, filter sterilize, aliquot on ice, and store at −20°C.
   - **c.** 4 *mg/ml creatine phosphokinase*: To make 4 ml, add 80 *p*\ of 0.5*M* $NaHPO_4$ buffer, pH 7.0, to 1.6 ml H_2O on ice. When cool, add 16 mg creatine phosphokinase. Vortex until dissolved. Add 2.3 ml ice-cold 87% glycerol. Vortex until well mixed. Aliquot on ice and store at −20°C.
5. *Hexokinase*: Vortex the suspension, pipette the desired amount (e.g., 10 μg–0.1 mg for one assay point), centrifuge for 2 min

in Eppendorf at maximum speed, and aspirate supernatant. Dissolve pellet in the same volume of 0.25 *M* d-glucose. Prepare immediately before use.

6. *TX100 stock*: 10% stock of Triton X-100 in H_2O
7. *PBS/BSA and sterile PBS*: See Section IIIF.
8. *HEP reagent*: See Section IIID.
9. *PBS/BSA/TX100*: PBS/BSA containing 0.2% Triton X-100, make immediately before use

Steps

1. For each fusion assay, at least three points should be included: −ATP, +ATP, and the total. To determine fusion efficiency, determine the total (maximal possible fusion value) by mixing 50 //I of each endosomal fraction in an Eppendorf tube on ice. Add 25 μl TX100 stock and vortex well. Leave on ice at least 30 min and add PBS/BSA and continue as described in step 9.
2. For all other fusion assay points, 3 μl of 50× salts, 8 μl of biotinylated insulin, and 11 μl of KOAc stock are needed for each point. Make a mixture of these three components by multiplying the number of assay points by 3, 8, and 11 and mix the respective amounts of each component together in one tube. Number Eppendorf tubes for the appropriate number of assay points and put them on ice. Add 22 μl of the aforementioned mixture to each tube.
3. Add 50 μl (750 μg–1 mg) of cytosol to each tube and mix.
4. Add either 5 μl of ATP-RS or 10 μl of hexokinase to each tube, as appropriate.
5. Add 50 μl (7–25 μg) of bHRP-labeled endosomes and 50 μl (7–25 μg) of avidin-labeled endosomes to each tube. Endosomal fractions from the gradients can be diluted in HB prior to this step, if desired. Mix gently; avoid introducing air bubbles. Leave tubes on ice for 3 min.
6. Transfer tubes to 37°C for 45 min. Avoid agitation during this time.
7. Return tubes to ice. Add 5 μl of biotinylated insulin to each tube and mix.
8. Add 25 μl of Tx100 stock and vortex well. Leave tubes on ice for 30 min.
9. Add 1 ml of sterile PBS/BSA to each tube and mix well.
10. Centrifuge for 2 min at maximum speed in an Eppendorf centrifuge. Transfer supernatants to numbered tubes containing antiavidin beads, prepared as in Section IIIF.
11. Rotate beads for at least 5 h at 4°C.
12. Centrifuge in Eppendorf centrifuge at maximum speed for 2 min. Remove supernatant and wash four times with PBS/BSA/Tx100. Wash once with sterile PBS.
13. Remove final supernatant and add 900 //I of HRP reagent to each tube. Allow color to develop in the dark at room temperature. Vortex periodically for 2–3 h or put tubes on rotating wheel in the dark at room temperature while color develops.
14. Centrifuge tubes for 2 min in Eppendorf centrifuge. Measure the absorbance of the supernatants at 455 nm.

IV. COMMENTS

Refer to Gruenberg and Gorvel (1992) for an example of a typical balance sheet for the sucrose gradient fractionation step. Typical results for fusion assays are shown in Fig. 5.2.

Highly purified loaded endosomes can be prepared by immunoisolation as described in Howell *et al.* (1989). Immunoisolated endosomes can then be used in the fusion assays described in Section IIIG. See Gruenberg and Gorvel (1992) and Howell *et al.* (1989) for details.

ECV–LE fusion is stimulated by the addition of polymerized microtubules to the fusion assay. Endogenous microtubules can be polymerized by adding 20 μ*M* taxol to the fusion assay. The

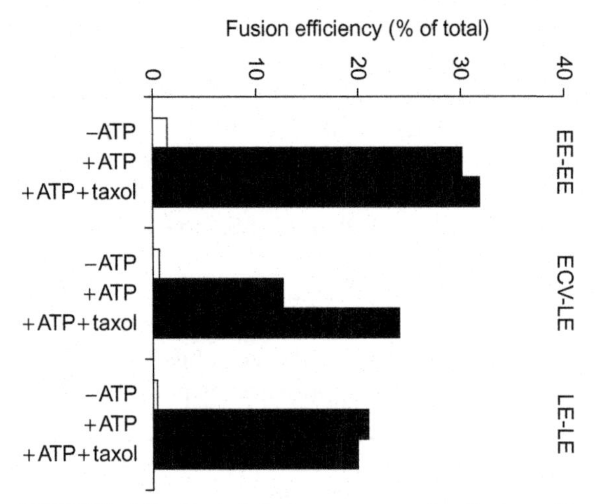

FIGURE 5.2 Typical fusion assay results. Fusion efficiency is expressed as a percentage of total fusion between each set of endosomal membranes. Total, or maximal, endosome fusion is measured by mixing bHRP and avidin-containing endosomal fractions together in the presence of detergent, followed by immunoprecipitation with antiavidin antibodies and HRP determination. Typical "total" values (measured as absorbance at 455 nm) are in the range of 0.6–1.0 ∧455 units for EE fusion assays and 0.3–0.7 ∧455 units for ECV and LE fusion assays. As a control for nonspecific reactions, the assay is typically carried out without ATP (—ATP). As shown, the polymerization of endogenous tubulin present in the cytosol in the presence of taxol is sufficient to facilitate interactions between ECVs and LEs (see Aniento *et al.*, 1993).

preparation of microtubules is described in the article by Ashford and Hyman. See Fig. 5.2 and Aniento *et al.* (1993) for more details on the effects of microtubules and MAPs on ECV–LE fusion.

V. PITFALLS

For ECV and LE preparations, cells should be homogenized until vesicles are no longer seen around the periphery of nuclei. If nuclei begin to aggregate during homogenization, however, this is a sign that some are broken as free DNA causes aggregation. Freezing and thawing of endosomes may cause a partial loss in latency. Use very clean plasticware or glassware for all fusion assay manipulations as HRP contamination can occur easily. Nocodazole, *o*-dianisidine, and KCN are very toxic.

References

Aniento, F., Emans, N., Griffiths, G., and Gruenberg, J. (1993). Cytoplasmic dynein-dependent vesicular transport from early to late endosomes. *J. Cell Biol.* **123**, 1373–1387.

Aniento, F., Gu, F., Parton, R. G., and Gruenberg, J. (1996). An endosomal /3COP is implicated in the pH-dependent formation of transport vesicles destined for late endosomes. *J. Cell Biol.* **133**, 29–41.

Aniento, F., Roche, E., Cuervo, A., and Knecht, E. (1993). Uptake and degradation of glyceraldehyde-3-phosphate dehydrogenase by rat liver lysosomes. *J. Biol Chem.* **268**, 10463–10470.

Bomsel, M., Parton, R., Kuznetsov, S. A., Schroer, T. A., and Gruenberg J. (1990). Microtubule- and motor-dependent fusion *in vitro* between apical and basolateral endocytic vesicles from MDCK cells. *Cell* **62**, 719–731.

Braell, W. A. (1987). Fusion between endocytic vesicles in a cell-free system. *Proc. Natl. Acad. Sci. USA* **84**, 1137–1141.

Davey, J. S., Hurtley, S. M., and Warren, G. (1985). Reconstitution of an endocytic fusion event in a cell-free system. *Cell* **43**, 643–652.

Diaz, R., Mayorga, L., and Stahl, P. D. (1988). *In vitro* fusion of endosomes following receptor-mediated endocytosis. *J. Biol. Chem.* **263**, 6093–6100.

Gruenberg, J., and Gorvel, J.-P. (1992). *In vitro* reconstitution of endocytic vesicle fusion. *In "Protein Targetting, a Practical Approach"* (A. I. Magee and T. Wileman, eds.), pp. 187–216. University Press, Oxford.

Gruenberg, J., Griffiths, G., and Howell, K. E. (1989). Characterization of the early endosome and putative endocytic carrier vesicles *in vivo* with an assay of vesicle fusion *in vitro*. *J. Cell Biol.* **108**, 1301–1316.

Gruenberg, J., and Howell, K. E. (1986). Reconstitution of vesicle fusions occurring in endocytosis with a cell-free system. *EMBO J.* **5**, 3091–3101.

Gruenberg, J., and Maxfield, F. R. (1995). Membrane transport in the endocytic pathway. *Curr. Opin. Cell Biol.* **7**, 552–563.

Howell, K. E., Schmid, R., Ugelstad, J., and Gruenberg J. (1989). Immuno-isolation using magnetic solid supports: Subcellular fractionation for cell-free functional studies. *Methods Cell Biol* **31A**, 264–292.

Woodman, P. G., and Warren, G. (1988). Fusion between vesicles from the pathway of receptor-mediated endocytosis in a cell-free system. *Eur. J. Biochem.* **173**, 101–108.

CHAPTER

6

Microsome-Based Assay for Analysis of Endoplasmic Reticulum to Golgi Transport in Mammalian Cells

Helen Plutner, Cemal Gurkan, Xiaodong Wang, Paul LaPointe, and William E. Balch

OUTLINE

I. INTRODUCTION

The trafficking of proteins along the first stage of the secretory pathway is mediated by small vesicles that bud from the endoplasmic reticulum (ER) and subsequently fuse with the *cis*-Golgi compartment. This article describes a biochemical assay using mammalian microsomes that can be used to measure these events independently. The microsomes are prepared from cells infected at the restrictive temperature (39.5°C) with the ts045 strain of vesicular stomatitis virus (VSV) (Lafay, 1974). As a reporter molecule the assay utilizes ts045 VSV-glycoprotein (VSV-G), which is retained in the ER during infection at 39.5°C due to a thermoreversible folding defect; incubation *in vitro* at the permissive temperature (32°C) results in the synchronous folding and transport of VSV-G to the Golgi complex. To follow vesicle formation, a differential centrifugation procedure is employed to separate the more rapidly sedimenting ER and Golgi membranes from the slowly sedimenting vesicles. Consumption is analyzed using a two-stage assay in which vesicles isolated by differential centrifugation during stage 1 are subsequently added to stage 2 (fusion) reactions containing acceptor Golgi membranes. Transport to the Golgi is measured by following the oligosaccharide processing of VSV-G from the high mannose ER form, which is sensitive to endoglycosidase H (endo H), to the *cis/medial-Golgi* form, which is endo H resistant (Schwaninger *et al.*, 1992). The biochemical characteristics of the overall ER to Golgi transport reaction and the vesicle formation and consumption assays are described elsewhere (Aridor *et al.*, 1995, 1996b, 1998, 1999a,b, 2000, 2001; Rowe *et al.*, 1996a).

II. MATERIALS AND INSTRUMENTATION

Culture medium (α-MEM; Cat. No. 11900-099) is from Life Technologies. The medium is supplemented with penicillin/streptomycin from a 100× stock solution (Cat. No P0781; Sigma). Fetal bovine serum (FBS; Cat. No FB-01) is from Omega Scientific. d-Sorbitol (Cat. No. S-1876), leupeptin (Cat. No L-2884), chymostatin (Cat. No C-7268), pepstatin (Cat. No P-4265), phenylmethylsulfonyl fluoride (PMSF; Cat. No P-7626), actinomycin D (Cat. No A-1410), uridine 5′-diphospho-*N*-acetylglucosamine (UDP-GlcNAc; Cat. No. U-4375), and dimethyl sulfoxide (DMSO; Cat. No. D-2650), are from Sigma. The nitrocellulose membrane (Cat. No. 68260) is from Schleicher & Schuell. Horseradish peroxidase-conjugated goat anti-rabbit IgG (Cat. No. 31460) is from Pierce. Chemiluminescence reagent (Cat. No. NEL-101) and autoradiography film ("Reflection") are from NEN. Polyallomer microfuge tubes (Cat. No. 357448) are supplied by Beckman Instruments Inc. A polyclonal antibody to VSV-G is generated in rabbits immunized with the C-terminal 16 amino acids of VSV-G (Indiana serotype) coupled to KLH (Plutner *et al.*, 1991). Centrifugation at 20,000 or 100,000*g* is performed using an Optima TL ultracentrifuge (Beckman) equipped with a TLA 100.3 rotor. A laser-scanning densitometer (personal densitometer; Molecular Dynamics) is used to quantitate data.

III. PROCEDURES

The following procedures are performed on ice unless otherwise stated.

A. Preparation of Cytosol

The following procedure for the preparation of rat liver cytosol is based on that described by Davidson *et al.* (1992).

Solutions

1. *Phosphate-buffered saline (PBS) (10 × stock)*: 90 mM phosphate and 1.5 *M* NaCl (pH 7.4). To make 1 liter, add 80 g NaCl, 2 g KCl, 2 g KH_2PO_4, and 21.6 g Na_2HPO_4 $7H_2O$ to distilled water. Store at room temperature.
2. *25/125*: 0.125 *M* KOAc, 25 m*M* HEPES (pH 7.4). To make 100 ml, add 3.125 ml of 4 *M* KOAc stock and 2.5 ml of 1 *M* HEPES–KOH (pH 7.4) stock to distilled water. Store at 4°C.
3. *Protease inhibitor cocktail (PIC)*: 10 μg/ml leupeptin, 10 μg/ml chymostatin, 0.5 μg/ml pepstatin A, and 1.0 m*M* PMSF. To supplement 50 ml of buffer (e.g., 25/125) with PIC, add 50 μl of 10 mg/ml leupeptin stock (in H_2O), 50 μl of 10 mg/ml chymostatin stock (in DMSO), 5 μl of 5 mg/ml pepstatin A stock (in DMSO), and 500 μl of 0.1 *M*

PMSF stock (in ethanol). Use PIC buffer immediately or store at −20°C.

Steps

1. Euthanize two anesthetized adult Sprague–Dawley rats (~250 g), remove the livers, and place the tissue in a 250-ml glass beaker. Determine the weight of the tissue (typically ~20 g) and then wash two times with −50 ml of Ix PBS and once with ~50 ml of Ix PBS and once with ~50 ml of 25/125.
2. Finely mince the tissue using a pair of scissors and then homogenize in 2–3 volumes (ml/g of tissue) of 25/125 (PIC) with 20 strokes using a 40-ml Dounce (Wheaton). Use a "loose-fitting" Dounce for the first 10 strokes followed by a "tight-fitting" one for the second 10 strokes.
3. Pour the homogenate into a 38-ml polycarbonate tube (Nalgene; Cat. No. 3117-0380) and centrifuge for 10 min at 12,000 *g* (10,000 rpm) in a Beckman JA20 rotor (Beckman Instruments Inc.). Using a pipette, transfer ~13 ml of the supernatant into each of two 14 × 89-mm ultraclear centrifuge tubes (Beckman Cat. No. 344059) and centrifuge at 150,000 *g* (35,000 rpm) for 90 min in a Beckman SW41 rotor.
4. After centrifugation, remove the overlying lipid layer by aspiration and then withdraw the remaining supernatants (cytosol) from each tube using a pipette.
5. Divide the cytosol into 250-μl aliquots in 0.5-ml microfuge tubes, freeze in liquid N_2, and store at −80°C. The protein concentration of the cytosol is ~25 mg/ml.

B. Preparation of Microsomes

Solutions

1. *Actinomycin D (200× stock)*: Add 10 mg actinomycin D to 10 ml of ethanol. Store at −20°C.
2. *Homogenization buffer*: 0.375 *M* sorbitol and 20 m*M* HEPES (pH 7.4). To make 500 ml, add 34.2 g sorbitol and 10 ml of 1 *M* HEPES–KOH (pH 7.4) stock to distilled water. Store at 4°C.
3. *0.21 M KOAc buffer*: 0.21 M KOAc, 3 m*M* $Mg(OAc)_2$, and 20 m*M* HEPES (pH 7.4). To make 100 ml, add 5.25 ml of 4 *M* KOAc stock, 0.3 ml of 1 *M* $Mg(OAc)_2$ stock, and 2 ml of 1 *M* HEPES–KOH (pH 7.4) stock to distilled water. Store at 4°C.
4. *Transport buffer*: 0.25 *M* sorbitol, 70 m*M* KOAc, 1 m*M* $Mg(OAc)_2$, and 20 m*M* HEPES (pH 7.4). To make 100 ml, add 10 ml of 2.5 *M* sorbitol stock, 1.75 ml of 4 *M* KOAc stock, 0.1 ml of 1 *M* $Mg(OAc)_2$ stock, and 2 ml of 1 *M* HEPES–KOH (pH 7.4) stock to distilled water. Store at −20°C.

Steps

1. Prepare vesicular stomatitis virus (VSV; Indiana serotype) strain ts045 according to Schwaninger *et al.* (1992). Store the virus in 1-ml aliquots in screw-capped tubes at −80°C.
2. Grow normal rat kidney (NRK) cells on 150-mm tissue culture dishes (Cat. No. 3025; Falcon) in α-MEM medium supplemented with 5% FBS at 37°C and 5% CO_2. At confluency, infect the cells with a 5-ml cocktail (per dish) containing 0.1–0.25 ml (2–10 pfu/cell) of ts045 VSV (thawed at 32°C) and 25 μg of actinomycin D in serum-free α-MEM as described by Schwaninger *et al.* (1992). Rock the dishes for 45 min to ensure an even spread of the infection cocktail. After infection, add 20 ml of α-MEM medium supplemented with 5% FBS to each dish and incubate in the presence of 5% CO_2 for 3 h and 40 min to 4 h at the restrictive temperature (39.5°C) (see Comment 1). The method detailed later is based on a typical 12-dish microsome preparation.
3. Following incubation at 39.5°C, transfer each dish to ice, aspirate the medium

immediately, and add 12 ml of ice-cold 1X PBS to cool the cells as quickly as possible.

4. Remove the PBS by aspiration, add 5 ml of homogenization buffer, and scrape the cells from the dishes using a rubber policeman. Use a pipette to transfer the cells to 50-ml plastic tubes (Cat. No. 25325-50: Corning Inc.) and then repeat the scraping procedure to ensure that all the cells are collected. Centrifuge at 720 *g* for 3 min and remove the supernatant by aspiration.
5. Resuspend each cell pellet (from four dishes) in 0.9 ml of homogenization buffer supplemented with PIC and homogenize by three complete passes (three downward strokes with both plungers) through a 1-ml ball-bearing homogenizer (Balch and Rothman, 1985).
6. Combine the cell homogenates and dilute with an equal volume (~3 ml) of homogenization buffer + PIC. Divide the diluted homogenate into six 1.0-ml aliquots in 1.5-ml microfuge tubes and centrifuge at 720 *g* for 5 min.
7. Carefully remove the postnuclear supernatant (PNS) fractions and combine in a plastic 15-ml tube (Cat. No. 25319-15: Corning Inc.). Add 0.5 volume (~2.5 ml) of 0.21 *M* KOAc buffer to the PNS and mix. Divide the mixture into 0.8- to 1.0-ml aliquots in 1.5-ml microfuge tubes and centrifuge at 12,000 *g* (12,200 rpm) for 2 min in an Eppendorf Model 5402 refrigerated microfuge using the soft spin function.
8. Remove the supernatants by aspiration and resuspend the pellets (including any membranes on the sides of the tubes) using a P1000 Gilson tip in a total volume of 1.0 ml of transport buffer + PIC. Dispense 0.5-ml aliquots into two 1.5-ml microfuge tubes and recentrifuge the microsomes at 12,000 *g* for 2 min as described earlier.
9. Resuspend the membrane pellets by repeated trituration using a P1000 Gilson tip in 6–8 volumes (~1 ml per 75-μl membrane pellet) of transport buffer containing PIC. In a typical 12-dish preparation, 1.0–1.5 ml of resuspended microsomes (at a protein concentration of 3–4 mg/ml) is obtained depending on the starting cell density. Pool the membranes, divide into 50- or 100-μl aliquots in 0.5-ml microfuge tubes, freeze in liquid N_2, and store at −80°C. The microsomes can be stored for several months with no loss of transport activity.

C. Preparation of Acceptor Golgi Membranes

The following procedure for the preparation of Golgi membranes by flotation on a sucrose density gradient is a modification of that originally described by Balch *et al.* (1984).

Solution

57.5 mM KOAc buffer: 87.5 m*M* KOAc, 1.25 m*M* $Mg(OAc)_2$, and 20 m*M* HEPES (pH 7.4). To make 100 ml, add 2.18 ml of 4 *M* KOAc stock, 0.125 ml of 1 *M* $Mg(OAc)_2$ stock, and 2 ml of 1 *M* HEPES–KOH (pH 7.4) stock to distilled water. Store at 4°C.

Steps

1. Prepare an enriched Golgi membrane fraction from noninfected wild-type Chinese hamster ovary cells by flotation in sucrose density gradients as described by Beckers and Rothman (1992). Recover the membranes at the 29–35% sucrose interface, mix thoroughly with 4 volumes of 87.5 m*M* KOAc buffer, and divide into 0.5- to 1.0-ml aliquots in microfuge tubes. Centrifuge at 16,000 *g* (soft spin) for 10 min.
2. Remove the supernatants by aspiration, wash the pellets with 2 ml (final volume) of transport buffer, and combine the

membranes into two 1.5-ml microfuge tubes. Centrifuge at 16,000 *g* (soft spin) for 10 min.

3. Resuspend the membranes using a P1000 Gilson tip in transport buffer (total volume of 1.0 ml per 1×10^6 cells). Divide the Golgi membrane fraction into 50- to 100-μl aliquots in 0.5-ml microfuge tubes, freeze in liquid N_2, and store at −80°C.

D. Reconstitution of ER to Golgi Transport

Steps

1. Set up 40-μl transport reactions containing the components indicated in Table 6.1 in 1.5-ml microfuge tubes (see Comment 2). A reaction cocktail consisting of the salts, ATP-regenerating system, and water is added first, followed by the cytosol and finally the microsomes. Mix by pipetting up and down four times using a P20 Gilson tip.
2. Transfer the reactions to a 32°C water bath and incubate for 75–90 min.
3. Terminate the reactions on ice, harvest the membranes by centrifugation at 20,000 *g* (27,000 rpm) in a Beckman TLA 100.3 rotor, and remove the supernatant by aspiration (see Comment 3).
4. Solubilize the membranes in one-third volume of 0.3% SDS in 1 *M* Na acetate, pH5.6, and freshly added 30 m*M* β-mercaptoethanol (BME). Boil 5 min, cool, add two-thirds volume 1 *M* Na acetate containing 3 mU of endo H. Incubate overnight, then terminate the reactions by adding Laemmli sample buffer as described by Schwaninger *et al.* (1992).
5. Separate the endo H-sensitive and -resistant forms of VSV-G on 6.75% (w/v) SDS–polyacrylamide gels as described by Schwaninger *et al.* (1992).
6. Transfer the proteins to a nitrocellulose membrane and perform Western blotting using anti-VSV-G polyclonal primary antibody (1:10,000) and peroxidase-conjugated anti-rabbit IgG secondary antibody (1:10,000) according to Rowe *et al.* (1996).

TABLE 6.1 Reaction Cocktail for Standard ER to Golgi Transport Assay

Solution	Volume (μl)	Final concentration
Microsomes	5	~0.5 mg/ml protein [31.25 m*M* sorbitol, 8.75 m*M* KOAc, 0.125 m*M* $Mg(OAc)_2$, 2.5 m*M* HEPES (pH 7.4)]
Cytosol	8	5 mg/ml protein [25 m*M* KOAc, 5 m*M* HEPES (pH 7.4)]
20× ATP-regenerating system	2	1 m*M* ATP, 5 m*M* creatine phosphate, and 0.2IU creatine phosphate kinase
40 m*M* UDP-GlcNAc	1	1 m*M*
10× Ca^{2+}/EGTA buffer	4	5 m*M* EGTA, 1.8 m*M* $Ca(Cl)_2$ (100 n*M* free Ca^{2+}), 2 m*M* HEPES (pH 7.4)
0.1 *M* $Mg(OAc)_2$	1	2.5 m*M*
1 *M* KOAc	1	40 m*M*
2.5 *M* sorbitol	3.5	218.75 m*M*
1 *M* HEPES (pH 7.4)	1.1	27.5 m*M*
H_2O	To 40 μl (final)	

7. Develop the blots using enhanced chemiluminescence and expose to autoradiography film. Quantitate the relative band intensities of the endo H-sensitive and -resistant forms of VSV-G in each lane by densitometry (see Comments 4 and 5).

E. Vesicle Formation Assay

Solution

Resuspension buffer: 0.25 *M* sucrose and 20 m*M* HEPES (pH 7.4). To make 100 ml, add 8.55 g of sucrose and 2 ml of 1 *M* HEPES–KOH (pH 7.4) stock to distilled water. Store at 4°C.

Steps

1. Set up 40-μl reactions as described in Section III,D, step 1, containing 5–10 μl of microsomes, 5–12 μl of cytosol, salts, and an ATP-regenerating system (see Comment 6).
2. Incubate at 32°C for 0–60 min and harvest the membranes as described in Section III,D, step 3. The membrane pellets can be stored for several hours at this stage prior to the differential centrifugation procedure described later.
3. Add 40 μl of resuspension buffer and disperse the membrane pellets by pipetting up and down 10 times using a P200 Gilson tip (see Comments 3 and 7). Incubate the membranes for 10 min on ice and repeat the trituration procedure to resuspend the membranes completely. Add 8.5 μl of a salt mix [7.3 μl of 1 *M* KOAc + 1.2 μl of 0.1 *M* $Mg(OAc)_2$] to the resupended membranes and mix by pipetting up and down 5 times with a P20 Gilson tip. Perform the differential centrifugation step at 16,000 *g* (14,000 rpm) for 3 min in an Eppendorf Model 5402 refrigerated microfuge using the soft spin function.
4. Using a P200 Gilson tip, carefully take the top 34-μ1 supernatant fraction from the side of the tube opposite the pellet. Transfer to a 1.5-ml polyallomer microfuge tube and centrifuge at 100,000 *g* (60,000 rpm) for 20 min. Carefully aspirate the remaining supernatant fraction from the 16,000-*g* (medium speed) pellet and the entire supernatant from the 100,000-*g* (high speed) pellet (see Comment 8).
5. Add 50 and 35 μl of IX Laemmli sample buffer (Laemmli, 1970) to the medium-speed pellet (MSP) and high-speed pellet (HSP) fractions, respectively, and boil at 95°C for 5 min. Determine the relative amounts of VSV-G in the MSP and HSP from each reaction by SDS–PAGE and quantitative immunoblotting as described by Rowe *et al.* (1996).
6. *A quicker alternative method to step 2 is as follows*: After incubation put the tubes on ice and spin at 16,000 *g* (14,000 rpm) for 3 min. Carefully remove the top 32 μl as described in step 4 and transfer to 1.5-ml polyallomer tubes. Spin at 100,000 *g* (60,000 rpm) for 20 min. Aspirate supernatant as in step 4 and add 40 and 32 μl 1× Laemmli sample buffer to the medium-speed pellet (MSP) and high-speed pellet (HSP) fractions, respectively. Vortex carefully and boil at 95°C for 5 min.

F. Two-Stage Fusion Assay

Solution

0.25 M sorbitol buffer: 0.25 *M* sorbitol and 20 m*M* HEPES (pH 7.4). To make 100 ml, add 10 ml of 2.5 *M* sorbitol stock and 2 ml of 1 *M* HEPES–KOH (pH 7.4) stock to distilled water. Store at 4°C.

Steps

1. For stage 1 incubations, prepare scaled-up (100 μl) vesicle formation reactions

containing 25 μl of microsomes and 30 μl of cytosol as described in Section IIIE, step 1. Incubate for 10 min at 32°C.

2. Terminate the reactions by transfer to ice and sediment the membranes as described in Section IIID, step 3.
3. Resuspend the membranes in 90 μl of resuspension buffer, add 9.8 μl of salt mix [7.3 μl of 2 *M* KOAc + 2.5 μl of 0.1 *M* $Mg(OAc)_2$], and perform the differential centrifugation step as described in Section IIIE, step 3.
4. Withdraw the top 75-ml medium-speed supernatant fraction and recover the vesicles by centrifugation at high speed as described in Section IIIE, step 4.
5. Resuspend the HSPs in 25 μl of 0.25 *M* sorbitol buffer by pipetting up and down 10 times with a P200 Gilson tip and then add 1.9 μl of 1 *M* KOAc (70 m*M* KOAc final concentration) (see Comment 9).
6. Set up 40-μl stage 2 reactions containing 10 μl of resuspended HSP fraction, 4 μl of Golgi membranes, 8 μl of cytosol, and an ATP-regenerating system and salts at the final concentrations indicated in Table 6.1 (see Comment 6).
7. Incubate for 60 min at 32°C and terminate the reactions on ice. Harvest the membranes and quantitate the conversion of VSV-G to the endo H-resistant form as described in Section IIID, steps 3–7.

IV. COMMENTS

1. To reproduce the temperature-sensitive phenotype of ts045 VSV-G transport *in vitro*, it is necessary to supplement the postinfection medium and homogenization buffer with dithiothreitol as described by Aridor *et al.* (1996).
2. Preparations of the ATP-regenerating system and Ca^{2+}/EGTA buffer are described by Schwaninger *et al.* (1992). Transport inhibitors such as antibodies are dialyzed against 25/125 prior to addition to the assay, and the volumes of the salts in the reaction cocktail are adjusted to achieve the final concentrations described in Table 6.1. The reactions can be preincubated on ice for up to 45 min with no loss of transport activity.
3. All of the membrane-bound VSV-G is sedimented at 20,000 *g* following the 32°C incubation. In the vesicle formation assay, membranes are resuspended in sucrose buffer prior to differential centrifugation.
4. The detection system is linear over the range of VSV-G concentrations tested.
5. As an alternative to densitometry, VSV-G bands can be detected by direct fluorescence imaging (e.g., using a Bio-Rad GS-363 molecular imaging system).
6. In the vesicle formation and two-stage assays, the volumes of salts (see Table 6.1) added to the reaction cocktail are adjusted to account for the salts present in the cytosol and membrane preparations in the assay. UDP-GlcNAc is omitted from the vesicle formation assay and from stage 1 of the two-stage assay.
7. Although in the original description of the assay (Rowe *et al.*, 1996) a 0.25 *M* sorbitol buffer was used to resuspend the membranes prior to differential centrifugation, we have subsequently found that resuspension in 0.25 *M* sucrose buffer gives higher yields of vesicles at this step.
8. Because the HSP is small and translucent, care should be taken to avoid losing it during aspiration of the high-speed supernatant.
9. In a typical experiment, multiple stage 1 reactions are performed and the HSPs are resuspended successively in the desired final volume of sorbitol buffer and are then adjusted to 70 m*M* KOAc.

Acknowledgments

This work was supported by postdoctoral fellowships from the Cystic Fibrosis Foundation to XW and grants from the National Institutes of Health (GM 42336; GM33301) to WEB.

References

Aridor, M., and Balch, W. E. (1996). Membrane fusion: Timing is everything. *Nature* **383**, 220–221.

Aridor, M., and Balch, W. E. (1999). Integration of endoplasmic reticulum signaling in health and disease. *Nature Med.* **5**, 745–751.

Aridor, M., and Balch, W. E. (2000). Kinase signaling initiates coat complex II (COPII) recruitment and export from the mammalian endoplasmic reticulum. *J. Biol Chem.* **275**, 35673–35676.

Aridor, M., Bannykh, S. I., Rowe, T., and Balch, W. E. (1995). Sequential coupling between COPII and COPI vesicle coats in endoplasmic reticulum to Golgi transport. *J. Cell Biol.* **131**, 875–893.

Aridor, M., Bannykh, S. I., Rowe, T., and Balch, W. E. (1999). Cargo can modulate COPII vesicle formation from the endoplasmic reticulum. *J. Biol Chem.* **274**, 4389–4399.

Aridor, M., Fish, K. N., Bannykh, G. I., Weissman, J., Roberts, H. R., Lippincott-Schwartz, J., and Balch, W. E. (2001). The Sar1 GTPase coordinates biosynthetic cargo selection with endoplasmic reticulum export site assembly. *J. Cell Biol.* **152**, 213–229.

Aridor, M., Weissman, J., Bannykh, S., Nuoffer, C., and Balch, W. E. (1998). Cargo selection by the COPII budding machinery during export from the ER. *J. Cell Biol.* **141**, 61–70.

Balch, W. E., Dunphy, W. G., Braell, W. A., and Rothman, W. E. (1984). Reconstitution of the transport of protein between successive compartments of the Golgi measured by the coupled incorporation of N-acetylglucosamine. *Cell* **39**, 405–416.

Balch, W. E., and Rothman, J. E. (1985). Characterization of protein transport between successive compartments of the Golgi apparatus: Asymmetric properties of donor and acceptor activities in a cell-free system. *Arch. Biochem. Biophys.* **240**, 413–425.

Beckers, C. J. M., and Rothman, J. E. (1992). Transport between Golgi cisternae. *Methods Enzymology* **219**, 5–12.

Davidson, H. W., McGowan, C. H., and Balch, W. E. (1992). Evidence for the regulation of exocytic transport by protein phosphorylation. *J. Cell Biol.* **116**, 1343–1355.

Laemmli, U. K. (1970). Cleavage of structural proteins during the assembly of the head of bacteriophage T4. *Nature* **227**, 680–685.

Lafay, F. (1974). Envelope viruses of vesicular stomatitis virus: Effect of temperature-sensitive mutations in complementation groups III and V. *J. Virol.* **14**, 1220–1228.

Plutner, H., Cox, A. D., Find, S., Khosravi-Far, R., Bourne, J. R, Schwaninger, R., Der, C. J., and Balch, W. E. (1991). Rab 1b regulates vesicular transport between the endoplasmic reticulum and successive Golgi compartments. *J. Cell Biol* **115**, 31–43.

Rowe, T., Aridor, M., McCaffery, J. M., Plutner, H., Nuoffer, C., and Balch, W. E. (1996). COPH vesicles derived from mammalian ER microsomes recruit COPI. *J. Cell Biol.* **135**, 895–911.

Schwaninger, R., Beckers, C. J. M., and Balch, W. E. (1991). Sequential transport of protein between the endoplasmic reticulum and successive Golgi compartments in semi-intact cells. *J. Biol. Chem.* **266**, 13055–13063.

Schwaninger, R., Plutner, H., Davidson, H. W., Find, S., and Balch, W. E. (1992). Transport of protein between the endoplasmic reticulum and Golgi compartments in semi-intact cells. *Methods Enzymology* **219**, 110–124.

CHAPTER

7

Cotranslational Translocation of Proteins into Microsomes Derived from the Rough Endoplasmic Reticulum of Mammalian Cells

Bruno Martoglio, and Bernhard Dobberstein

OUTLINE

I. INTRODUCTION

In mammalian cells, secretory proteins and membrane proteins of the organelles of the secretory pathway are initially transported across and integrated into the membrane of the rough endoplasmic reticulum (ER), respectively (for reviews, see Rapoport *et al.*, 1996; Johnson and van Weas, 1999). This process can be studied in a cell-free translation/translocation system in which a protein of interest is synthesized and imported into microsomal membranes derived from the rough ER. Components used in such a system are rough microsomes (RM) usually prepared from dog pancreas, a cytosolic extract supporting protein synthesis, mRNA coding for a secretory or a membrane protein, and a radio-labeled amino acid, e.g., [^{35}S]methionine, for detection of the newly synthesized protein.

Upon translocation across the ER membrane, proteins become processed and modified, fold with the help of lumenal chaperones, and can assemble into oligomeric complexes (Lemberg and Martoglio, 2002; Daniels *et al.*, 2003). These functions are maintained in rough microsomes and can be studied. This article describes the basic cell-free *in vitro* translocation system and provides methods to analyze the translocation of proteins across the microsomal membrane and integration into the lipid bilayer.

II. MATERIALS AND INSTRUMENTATION

Ambion: RNase inhibitor (Cat. No. 2684). Amersham Pharmacia: DEAE–Sepharose (Cat. No. 17-0709-01), [35S]methionine (Cat. No. AG 1594), and Sephacryl S-200 (Cat. No. 17-0584-10). Roche Applied Science: ATP (Cat. No. 1140965), CTP (Cat. No. 1140922), GTP (Cat. No. 1140957), UTP (Cat. No. 1140949), creatin kinase (Cat. No. 736988), creatin phosphate (Cat. No. 621714), and SP6 RNA polymerase (Cat. No. 810274). Fluka: EDTA (Cat. No. 03610), octaethylene glycol monododecyl ether (Cat. No. 74680), and SDS (Cat. No. 71725). Merck: Acetic acid (Cat. No. 100063), acetone (Cat. No. 100014), calcium chloride (Cat. No. 102389), HCl (Cat. No. 100317), isopropanol (Cat. No. 109634), magnesium acetate (Cat. No. 105819), magnesium chloride (Cat. No. 105833), 2-mercaptoethanol (Cat. No. 805740), potassium acetate (Cat. No. 104820), proteinase K (Cat. No. 124568), sodium acetate (Cat. No. 106268), sodium carbonate (Cat. No. 106392), sodium citrate (Cat. No. 106448), sucrose (Cat. No. 107654), and trichloroacetic acid (Cat. No. 100810). New England Biolabs: 7mG(59)ppp(59)G (Cat. No. 1404) and endoglycosidase H (Cat. No. 0702). Promega: Amino acid mixture minus methionine (Cat. No. L9961). Sigma: Dithiothreitol (DTT, Cat. No. D 0632), HEPES (Cat. No. H 3375), phenylmethylsulfonyl fluoride (PMSF, Cat. No. P 7626), Sephadex G-25 (Cat. No. G-25-150), Tris base (Cat. No. T 1503), and Triton X-100 (Cat. No. T 9284). Amersham Pharmacia: ÄKTAprime chromatography system (Cat. No. 18-2237-18). Braun Biotech Int.: Potter S homogenizer. Qiagen: Plasmid Maxi Kit (Cat. No. 12162). Millipore: Amicon ultrafiltration unit 8050 (Cat. No. 5122) and YM100 ultrafiltration discs (Cat. No. 14422AM).

III. PROCEDURES

A. Preparation of Components Required for Cell-Free *In Vitro* Protein Translocation across ER-Derived Rough Microsomes

1. Preparation of Rough Microsomes from Dog Pancreas

RMs can be prepared from most tissues or cells in culture. Dog pancreas, however, is the most convenient source for functional RMs because this tissue is specialized in secretion and contains an extended rough ER. Most importantly, pancreas from dog, in contrast to other animals, contains very little ribonuclease that would degrade the mRNA required for the translation reaction. Dogs may be obtained from an experimental surgery department or a pharmaceutical company. Ideally, the dogs are healthy and not operated or treated with drugs, and the pancreas is excised immediately after death.

Solutions

Glass and plasticware are autoclaved, and stock solutions and distilled water are either autoclavedorfilteredthrough0.45-μm-pore-sized filters.

1. *Stock solutions*:

   2 *M* sucrose: To make 500 ml of the solution, dissolve 342.25 g sucrose in water and complete to 500 ml. Sucrose stock solution may be stored in aliquots at 220°C.

   1 *M* HEPES–KOH pH 7.6: To make 200 ml of the buffer, dissolve 47.7 g HEPES in water, adjust pH to 7.6 with KOH solution, and complete to 200 ml.

   4 *M* KOAc: To make 200 ml of the solution, dissolve 78.5 g KOAc in water, neutralize to pH 7 with diluted acetic acid, and complete to 100 ml.

   1 *M* Mg(OAc)2: To make 100 ml of the solution, dissolve 21.5 g Mg(OAc)2 · 4 H_2O in water, neutralize to pH 7 with diluted acetic acid, and complete to 100 ml.

   500 m*M* EDTA: To make 100 ml of the solution, dissolve 14.7 g EDTA in water, adjust to pH 8.0 with NaOH solution and complete to 100 ml.

   1 *M* DTT: To make 10 ml of the solution, dissolve 1.54 g DTT in water and complete to 10 ml. Store in aliquots at 220°C.

   20 mg/ml PMSF: To make 2 ml of the solution, dissolve 40 mg PMSF in 2 ml isopropanol. Prepare just before use.
2. *Homogenization buffer*: 250 m*M* sucrose, 50 m*M* HEPES–KOH, pH 7.6, 50 m*M* KOAc, 6 m*M* Mg(OAc)$_2$, 1 m*M* EDTA, 1 m*M* DTT, and 30 μg/ml PMSF. To make 1 liter of the buffer, add 125 ml 2 *M* sucrose, 50 ml 1 *M* HEPES–KOH, pH 7.6, 12.5 ml 4 *M* KOAc, 6 ml 1 *M* Mg(OAc)$_2$, 2 ml 500 m*M* EDTA, and complete to 1 liter with water. Add 1 ml 1 *M* DTT and 1.5 ml 20 mg/ml PMSF to the cold buffer just before use.
3. *Sucrose cushion*: 1.3 *M* sucrose, 50 m*M* HEPES–KOH, pH 7.6, 50 m*M* KOAc, 6 m*M* Mg(OAc)$_2$, 1 m*M* EDTA, 1 m*M* DTT, and 10 μg/ml PMSF. To make 200 ml of the solution, add 130 ml 2 *M* sucrose, 10 ml 1 *M* HEPES–KOH, pH 7.6, 2.5 ml 4 *M* KOAc, 1.2 ml 1 *M* Mg(OAc)$_2$, 0.4 ml 500 m*M* EDTA, and complete to 200 ml with water. Add 0.2 ml 1 *M* DTT and 0.1 ml 20 mg/ml PMSF to the cold solution just before use.
4. *RM buffer*: 250 m*M* sucrose, 50 m*M* HEPES–KOH, pH 7.6, 50 m*M* KOAc, 2 m*M* Mg(OAc)$_2$, 1 m*M* DTT, and 10 μg/ml PMSF. To make 100 ml of the buffer, add 12.5 ml 2 *M* sucrose, 5 ml 1 *M* HEPES–KOH, pH 7.6, 1.25 ml 4 *M* KOAc, 0.2 ml 1 *M* Mg(OAc)$_2$, and complete to 100 ml with water. Add 0.1 ml 1 *M* DTT and 50 μl 20 mg/ml PMSF to the cold solution just before use.

Steps

This procedure is performed in the cold room. Samples and buffers should be kept on ice.

1. Excise the pancreas (10–15 g) from a dog (e.g., beagle) and rinse in ~500 ml ice-cold homogenization buffer.
2. Remove connective tissue and fat, and cut the pancreas into small pieces. (*Optional*: Shock-freeze the pieces in liquid nitrogen and store the tissue at −80°C.)
3. Place the pieces into ~120 ml homogenization buffer and pass the tissue through a tissue press with a 1-mm mesh steel sieve.
4. Homogenize the pancreas in a 30-ml glass/Teflon potter with eight strokes at full speed (1500 rpm).
5. Transfer homogenate into 30-ml polypropylene tubes. Centrifuge at 3000 rpm (1000 *g*) for 10 min at 4°C in a Sorvall SS34 rotor.
6. Collect the supernatant, avoiding the floating lipid. Extract the pellet once more in 50 ml homogenization buffer as described in step 4 and centrifuge again (see step 5).
7. Transfer the two 1000-*g* supernatants to 30-ml polypropylene tubes. Centrifuge at 9500 rpm (10,000 *g*) for 10 min at 4°C in a Sorvall SS34 rotor.

8. Collect the supernatant, avoiding the floating lipid and repeat step 7 (centrifugation at 10,000 *g*) twice.
9. In the meantime, prepare four 70-ml polycarbonate tubes for a Ti45 rotor and add 25 ml sucrose cushion per tube.
10. After the third 10,000 *g* centrifugation (step 8), collect the supernatant and apply carefully, without mixing, onto the 25-ml sucrose cushions (see step 9).
11. Centrifuge at 35,000 rpm (142,000 *g*) for 1 h at 4°C in a Beckman Ti45 rotor.
12. Discard the supernatant and resuspend the membrane pellet, the rough microsomes (RM), in 20 ml RM buffer using a Dounce homogenizer.
13. Measure the absorption at 260 and 280 nm of a 1:1000 dilution of the RM suspension in 0.5% (w/v) SDS. Usually an absorption of 0.05–0.1 A_{280}/ml and a ratio A_{260}/A_{280} of ~1.7 are obtained. When the pancreas is frozen (see step 2), the A_{260}/A_{280} is 1.5–1.6.
14. Freeze 500-μl aliquots in liquid nitrogen and store at −80°C until use.

Note: RMs prepared by this procedure largely retain SRP on the membrane.

2. *Preparation of Signal Recognition Particle (SRP)*

Purification of components involved in protein translocation revealed that a cytosolic component, the signal recognition particle, is required for targeting nascent polypeptide chains to the ER membrane (Walter and Johnson, 1994). SRP is present in cytosolic extracts, but is also associated with RMs to a variable degree. In order to improve the efficiency of the translocation system, purified SRP may be added to the translation/translocation system. This is particularly useful when wheat germ extract is used for cell-free *in vitro* translation because this extract contains low amounts of functional SRP. SRP is isolated most conveniently from RMs by treatment with high salt. Released SRP is then purified by gel filtration and anion-exchange chromatography.

Solutions

Glass and plasticware are autoclaved, and stock solutions and distilled water are either autoclaved or filtered through 0.45-μm-pore-sized filters.

1. *Stock solutions* (see also preparation of RMs, Section III,A,1)

   1 M Tris-OAc, pH 7.5: To make 1 liter of the buffer, dissolve 121.1 g Tris base in water, adjust to pH 7.5 with acetic acid, and complete to 1 liter.

   10% (w/v) octaethylene glycol monododecyl ether: To make 10 ml of the solution, dissolve 1 g of octaethylene glycol monododecyl ether in water and complete to 10 ml.
2. *RM buffer*: As in solution 4 of Section III,A,1 (preparation of RMs).
3. *High salt solution*: 1.5 *M* KOAc and 15 m*M* $Mg(OAc)_2$. To make 50 ml of the solution, add 18.75 ml 4 *M* KOAc, 0.75 ml 1 *M* $Mg(OAc)_2$, and complete to 50 ml with water.
4. *Sucrose cushion*: 500 m*M* sucrose, 50 m*M* Tris–OAc, pH 7.5, 500 m*M* KOAc, 5 m*M* $Mg(OAc)_2$, and 1 m*M* DTT. To make 100 ml of the solution, add 25 ml 2 *M* sucrose, 5 ml 1 *M* Tris–OAc, pH 7.5, 12.5 ml 4 *M* KOAc, 0.5 ml 1 *M* $Mg(OAc)_2$, and complete to 100 ml with water. Add 0.1 ml 1 *M* DTT to the cold solution just before use.
5. *Gel filtration buffer*: 50 m*M* Tris–OAc, pH 7.5, 250 m*M* KOAc, 2.5 m*M* $Mg(OAc)_2$, 1 m*M* DTT, and 0.01% octaethylene glycol monododecyl ether. To make 1 liter of the buffer, add 50 ml 1 *M* Tris–OAc, pH 7.5, 62.5 ml 4 *M* KOAc, 2.5 ml 1 *M* $Mg(OAc)_2$, 1 ml 10% octaethylene glycol monododecyl ether, and complete to 1 liter with water. Add 1 ml 1 *M* DTT to the cold solution just before use.

6. *Washing buffer*: 50 m*M* Tris–OAc, pH 7.5, 350 m*M* KOAc, 3.5 m*M* $Mg(OAc)_2$, 1 m*M* DTT, and 0.01% octaethylene glycol monododecyl ether. To make 100 ml of the buffer, add 5 ml 1 *M* Tris–OAc, pH 7.5, 8.75 ml 4 *M* KOAc, 0.35 ml 1 *M* $Mg(OAc)_2$, 0.1 ml 10% octaethylene glycol monododecyl ether, and complete to 100 ml with water. Add 0.1 ml 1 *M* DTT to the cold solution just before use.
7. *SRP buffer*: 50 m*M* Tris–OAc, pH 7.5, 650 m*M* KOAc, 6 m*M* $Mg(OAc)_2$, 1 m*M* DTT, and 0.01% octaethylene glycol monododecyl ether. To make 100 ml of the buffer, add 5 ml 1 *M* Tris–OAc, pH 7.5, 16.25 ml 4 *M* KOAc, 0.6 ml 1 *M* $Mg(OAc)_2$, 0.1 ml 10% octaethylene glycol monododecyl ether, and complete to 100 ml with water. Add 0.1 ml 1 *M* DTT to the cold solution just before use.

Steps

This procedure is performed in the cold room and samples and buffers should be kept on ice.

1. Prepare ER-derived rough microsomes from two dog pancreas as described previously and resuspend the final RM pellet (see step 8 in Section III,A,1) in 50 ml RM buffer using a Dounce homogenizer.
2. Add 25 ml high salt solution and incubate for 15 min at 4°C on a turning wheel.
3. Distribute the membrane suspension equally to two 70-ml polycarbonate tubes for a Ti45 rotor onto 25 ml sucrose cushion per tube, avoid mixing.
4. Centrifuge at 32,000 rpm (120,000 *g*) for 1 h at 4°C in a Beckman Ti45 rotor.
5. In the meantime, prepare ten 10.4-ml polycarbonate tubes for a Ti70.1 rotor and add 1 ml sucrose cushion per tube.
6. After centrifugation (step 4), collect the supernatant and apply carefully, without mixing, onto the 1-ml sucrose cushions (see step 5).
7. Centrifuge at 65,000 rpm (388,000 *g*) for 1 h at 4°C in a Beckman Ti70.1 rotor.
8. Collect again the supernatant.
9. Concentrate the supernatant to approximately 10 ml in a 50-ml Amicon ultrafiltration unit equiped with a YM100 membrane.
10. Load concentrated sample onto a Sephacryl S-200 (2.6×20 cm) equilibrated with gel filtration buffer and elute with gel filtration buffer. The flow rate is 1 ml/min. Follow elution with a UV monitor (l = 280 nm) by using, e.g., the ÄKTAprime chromatography system from Amersham Pharmacia.
11. Collect flow through (20–25 ml) and load immediately onto a DEAE–Sepharose column (1×3 cm) equilibrated with gel filtration buffer. The flow rate is again 1 ml/min.
12. Wash column with 20 ml gel filtration buffer and 20 ml washing buffer.
13. Elute SRP with SRP buffer and collect peak fraction (2–3 ml). The SRP eluate has an absorption of 1–4 A_{260}/ml and a ratio A_{260}/A_{280} of approximately 1.4. Freeze 100-μl aliquots in liquid nitrogen and store at −80°C until use.

B. *In Vitro* Translation and Translocation Assay

Solutions

1. *Wheat germ extract*: Wheat germ extract for cell-free *in vitro* translation is prepared as described by Erickson and Blobel (1983) except that we use a gel filtration buffer with lower potassium and magnesium concentrations [40 m*M* HEPES–KOH, pH 7.6, 50 m*M* KOAc, 1 m*M* $Mg(OAc)_2$, 0.1% 2-mercaptoethanol]. Fresh wheat germ may be purchased from a local mill or from General Mills California. Store wheat germ in a desiccator over silica gel beads at 4°C.

Considerable differences in translation efficiency may yield from different batches. Store wheat germ extract in 110-μl aliquots at −80°C and thaw only once.

2. *Capped mRNA*: To obtain mRNA coding for a secretory or a membrane protein, clone a cDNA encoding the protein of interest into a suitable expression vector downstream of a T7 or a SP6 promotor (e.g., pGEM from Promega Biotech). We generally prefer the SP6 promotor as the respective transcripts yield more efficient translation. Prepare plasmid DNA from *Escherichia coli* cultures using the Plasmid Maxi kit from Qiagen. For transcription, linearize the purified plasmid DNA with a suitable restriction enzyme that cuts downstream of the coding sequence. Alternatively, a template for transcription may be generated by polymerase chain reaction. Amplify the coding region of interest using *Pfu* DNA polymerase (Stratagene), an appropriate plasmid DNA as template, a forward primer containing the SP6 promotor, a Kozak sequence (Kozak, 1983) and a ATG for the initiation methionine, and a reverse primer starting with 5′-NNNNNNNNNCTA- to introduce a TAG stop codon at the desired position. Perform runoff *in vitro* transcription according to a standard protocol (e.g., in Sambrook *et al.*, 1989) or by using a commercially available transcription kit (e.g., mMESSAGAmMACHINE kits from Ambion Cat. No. 1340 and 1344). Dissolve the resulting capped messenger RNA in water after extraction with phenol and chloroform and precipitation with sodium acetate and ethanol. Store mRNA at −80°C.
3. *Energy mix*: 50 m*M* HEPES–KOH, pH 7.6, 12.5 m*M* ATP, 0.25 m*M* GTP, 110 m*M* creatine phosphate, 10 mg/ml creatine kinase, and 0.25 m*M* of each amino acid *except* methionine. To make 1 ml of the solution, dissolve 41 mg creatin phosphate (disodium salt · $4H_2O$) and 10 mg creatin kinase in 590 μl water and add 50 μl 1 *M* HEPES–KOH, pH 7.6, 125 μl 100 m*M* ATP solution (Roche Applied Science), 2.5 μl 100 m*M* GTP solution (Roche Applied Science), and 250 μl 19 amino acids mix without methionine (Promega, 1 m*M* of each amino acid). Store the energy mix in 22 μl aliquots at −80°C. Do not refreeze!
4. *Salt mix*: 500 m*M* HEPES–KOH, pH 7.6, 1 *M* KOAc, and 50 m*M* $Mg(OAc)_2$. To make 1 ml of the solution, mix 500 μl 1 *M* HEPES–KOH, pH 7.6, 250 μl 4 *M* KOAc, 70 μl 1 *M* $Mg(OAc)_2$, and 180 μl water.
5. *SRP buffer*: As in solution 7 of Section III,A,2. (preparation of SRP).
6. *20% (w/v) trichloroacetic acid*: To make 100 ml of the solution, dissolve 20 g trichloroacetic acid in water and complete to 100 ml.

Steps

1. Per assay (25 μl), mix on ice 6 μl water, 1 μl salt mix, 2 μl energy mix, 10 μl wheat germ extract, 2 μl SRP buffer or SRP solution (see Section III,A,2, step 12), 2 μl RM suspension (see Section III,A,1, step 8), 1 μl [^{35}S]methionine (>1000 Ci/mmol), and 1 μl capped mRNA.
2. Incubate for 60 min at 25°C.
3. Add 25 μl (1 volume) 20% trichloroacetic acid to precipitate proteins and centrifuge at 14,000 rpm for 3 min at room temperature in an Eppendorf centrifuge.
4. Discard supernatant. Wash pellet with 150 μl cold acetone, and centrifuge as in step 3.
5. Discard supernatant and repeat step 4.
6. Discard supernatant, centrifuge tube shortly, and remove residual acetone completely with a pipette.
7. Add 20–30 μl sample buffer for SDS–polyacrylamide gel electrophoresis and heat for 10 min at 65°C.
8. Load 5–10 μl onto a SDS–polyacrylamide gel and analyze the sample by electrophoresis.

Radiolabeled proteins can be visualized by autoradiography, fluorography, or on a phosphorimager.

Controls: Translation without mRNA (add water instead); translation without membranes (add RM buffer instead, see solution 4 in Section III,A,1).

Note: Translation and translocation assays have to be optimized for each mRNA with respect to magnesium and potassium concentrations as well as to the amount of mRNA, SRP, and membranes. Optimal salt concentrations vary from 1 to 3.5 m*M* magnesium and from 70 to 150 m*M* potassium and are adjusted by using adapted salt mixes. The amount of SRP and membranes may be varied by diluting SRP and RM solutions with the respective buffers.

C. Assays to Characterize the Translocation Products

1. *Protease Protection Assay*

Proteinase K is used to test the translocation of proteins or parts of proteins across microsomal membranes (Blobel and Dobberstein, 1975). Membranes are impermeable to the protease and therefore only proteins or protein domains exposed on the cytoplasmic side of the microsomes are digested. To demonstrate that only intact microsomal vesicles protect lumenal proteins or protein domains, nonionic detergent (e.g., Triton X-100) is added to open the membrane.

Protease treatment is also used to characterize the topology of membrane proteins. In this case, protease treatment is often followed by immunoprecipitations with antibodies directed against defined regions of the protein investigated. Successful immunoprecipitations indicate that the respective domains are exposed on the lumenal side of the microsomes and are protected from the protease.

Solutions

1. *Sucrose cushion*: 500 m*M* sucrose, 50 m*M* HEPES–KOH, pH 7.6, 50 m*M* KOAc, 2 m*M* $Mg(OAc)_2$, and 1 m*M* DTT. To make 10 ml of the buffer, add 1.25 ml 2 *M* sucrose, 0.5 ml 1 *M* HEPES–KOH, pH 7.6, 125 μl 4 *M* KOAc, 20 μl 1 *M* $Mg(OAc)_2$, 10 μl 1 *M* DTT, and complete to 10 ml with water.
2. *3 mg/ml proteinase K*: To make 1 ml of the solution, dissolve 3 mg proteinase K in 1 ml water.
3. *20 mg/ml PMSF solution*: As is solution 1 of Section III,A,1 (preparation of RMs).
4. *10% (w/v) Triton X-100*: To make 10 ml of the solution, dissolve 10 g Triton X-100 in water and complete to 10 ml.
5. *20% (w/v) trichloroacetic acid*: As in solution 6 of Section III,B.

Steps

1. Perform a translation/translocation assay (25 μl) as described in Section III,B (steps 1 and 2).
2. In the meantime, prepare a 200-μl thick-wall polycarbonate tube for a TLA100 rotor and add 100-μl sucrose cushion.
3. After translation/translocation, apply the reaction mixture carefully onto the 100-μl sucrose cushion.
4. Wash microsomes by centrifugation through the sucrose cushion at 48,000 rpm (~100,000 *g*) for 3 min at 4°C in a Beckman TLA100 rotor.
5. Remove the supernatant with a pipette.
6. Resuspend the microsome pellet in 25 μl RM buffer (see solution 4 in Section III,A,1) and split the sample in three 8-μl aliquots.

7a. To the first aliquot (mock treatment) add 2 μl water.

7b. To the second aliquot add 1 μl proteinase K solution (3 mg/ml) and 1 μl water.

7c. To the third aliquot add 1 μl proteinase K solution (3 mg/ml) and 1 μl 10% Triton X-100.

8. Incubate the samples for 10 min at 25°C.

9. Stop proteolysis by adding 1 μl 10 mg/ml PMSF per sample.
10. Add 40 μl water and 50 μl 20% trichloroacetic acid to each sample to precipitate protein.
11. Centrifuge samples, wash protein pellets with acetone, and analyze samples by SDS–polyacrylamide gel electrophoresis as described in Section III,B, steps 3–8.

2. *Sodium Carbonate Extraction*

By alkaline treatment with sodium carbonate at pH 11, microsomal membranes are opened and release their content and peripherally associated proteins. The method is used to separate these proteins from proteins integrated into the lipid bilayer (Fujiki *et al.*, 1982).

Solutions

1. *0.1 M Na_2CO_3*: To make 10 ml of the solution, dissolve 106 mg Na_2CO_3 in water and complete to 10 ml. Prepare just prior to use.
2. *Alkaline sucrose cushion*: 0.1 *M* Na_2CO_3 and 250 m*M* sucrose. To make 10 ml of the solution, dissolve 106 mg Na_2CO_3 in water, add 1.25 ml 2 *M* sucrose (see solutions 1 of Section III,A,1), and complete to 10 ml with water.
3. *20% ($^w/_v$) trichloroacetic acid*: As in solution 6 of Section III,B.

Steps

1. Perform a translation/translocation assay (25 μl) as described in Section III,B (steps 1 and 2).
2. Wash microsomes by centrifugation through a sucrose cushion as described in Section III,C,1, steps 2–5.
3. Resuspend microsome pellet in 25 μl carbonate solution and incubate for 15 min on ice.
4. In the meantime, prepare a 200-μl thick-wall polycarbonate tube for a TLA 100 rotor and add 100-μl alkaline sucrose cushion.
5. For centrifugation, apply the sample (step 3) carefully onto a sucrose cushion in the polycarbonate tube.
6. Centrifuge at 55,000 rpm (130,000 *g*) for 10 min at 4°C in a Beckman TLA 100 rotor.
7. Recover the supernatant and pellet.

8a. To the supernatant add 150 μl 20% trichloroacetic acid to precipitate proteins. Centrifuge the sample, wash the protein pellet with acetone, and prepare the sample for SDS–polyacrylamide gel electrophoresis as described in Section III,B, steps 3–8.

8b. To the pellet add directly 20–30 μl sample buffer for SDS–polyacrylamide gel electrophoresis (see Section III,B, steps 7 and 8).

9. Analyze samples by SDS–polyacrylamide gel electrophoresis and autoradiography or phosphorimaging.

3. *Inhibition of N-Glycosylation with Glycosylation Acceptor Tripeptide*

The recognition sites for N-glycosylation in the ER are Asn-X-Ser and Asn-X-Thr. In the cell-free *in vitro* translation/translocation system described herein, the tripeptide *N*-benzoyl-Asn-Leu-Thr-methylamide efficiently competes with newly synthesized proteins for N-glycosylation (Lau *et al.*, 1983). The translocated protein is therefore not glycosylated in the presence of acceptor tripeptide, and the effects of oligosaccharides, e.g., on protein folding and assembly, may be investigated.

Solution

Acceptor tripeptide solution: Synthesize *N*-Benzoyl-Asn-Leu-Thr-methylamide on a peptide synthesizer. Dissolve the tripeptide in methanol at a concentration of 0.5 m*M* (0.23 mg/ml) and store the stock solution at −80°C.

Steps

1. Evaporate per translation/translocation assay (25 μl) 1.5 μl acceptor tripeptide solution in a test tube using a Speed-Vac centrifuge.
2. Add the components for the translation/ translocation assay to the tripeptide, vortex gently, and perform the assay as described in Section III,B (steps 1 and 2).
3. Precipitate the sample with trichloroacetic acid and analyze by SDS–polyacrylamide gel electrophoresis and autoradiography (see Section III,B, steps 3–8).

4. *Endoglycosidase H Treatment*

Treatment with endoglycosidase H is used to test N-glycosylation of proteins translocated into microsomal membranes. The glycosidase cleaves oligosaccharides of the high mannose type from glycoproteins, leaving an *N*-acetylglucosamine residue attached to the polypeptide (Tarentino *et al.*, 1974).

Solutions

1. *Denaturing solution*: 0.5% (w/v) SDS and 1% (v/v) 2-mercaptoethanol. To make 10 ml of the solution, dissolve 50 mg SDS in water, add 100 μl 2mercaptoethanol, and complete to 10 ml with water.
2. *0.5 M Na-citrate, pH 5.5*: To make 10 ml of the buffer, dissolve 1.47 g Na_3-citrate · 2 H_2O in water, adjust to pH 5.5 with HCl solution, and complete to 10 ml.

Steps

1. Perform a translation/translocation assay (25 μl) as described in Section III,B (steps 1 and 2).
2. Wash microsomes by centrifugation through a sucrose cushion as described in Section III,C,1, steps 2–5.
3. Resuspend microsome pellet in 25 μl denaturing solution and incubate for 10 min at 95°C.
4. Add 2.8 μl reaction buffer and 1 μl endoglycosidase H (1000 U/μl). Incubate for 1 h at 37°C.
5. Precipitate the sample with trichloroacetic acid and analyze by SDS–polyacrylamide gel electrophoresis and autoradiography or phorphorimaging (see Section III,B, steps 3–8).

IV. COMMENTS

As an alternative to wheat germ extract, commercially available rabbit reticulocyte lysate (e.g., from Promega) may be used for cell-free *in vitro* translation. Reticulocyte lysate contains sufficient SRP and therefore no additional SRP is usually required for optimization of translocation. When reticulocyte lysate is used, however, RMs should be treated with micrococcal nuclease to digest endogenous mRNA (Garoff *et al.*, 1978). The reticulocyte translation machinery will otherwise promote completion of nascent pancreatic secretory proteins.

References

Blobel, G., and Dobberstein, B. (1975). Transfer of proteins across membranes. II. Reconstitution of functional rough microsomes from heterologous components. *J. Cell Biol.* **67**, 852–862.

Daniels, R., Kurowski, B., Johnson, A. E., and Hebert, D. N. (2003). N-linked glycans direct the cotranslational folding pathway of influenza hemagglutinin. *Mol. Cell* **11**, 79–90.

Erickson, A. H., and Blobel, G. (1983). Cell-free translation of messenger RNA in a wheat germ system. *Methods Enzymol.* **96**, 38–50.

Fujiki, Y., Hubbard, A. L., Fowler, S., and Lazarow, P. B. (1982). Isolation of intracellular membranes by means of sodium carbonate treatment: Application to endoplasmic reticulum. *J. Cell Biol.* **93**, 97–102.

Garoff, H., Simons, K., and Dobberstein, B. (1978). Assembly of the semliki forest virus membrane glycoproteins in the membrane of the endoplasmic reticulum *in vitro*. *J. Mol. Biol.* **124**, 587–600.

Johnson, A. E., and van Waes, M. A. (1999). The translocon: A dynamic gateway at the ER membrane. *Annu. Rev. Cell Dev. Biol.* **15**, 799–842.

Kozak, M. (1983). Comparison of initiation of protein synthesis in procaryotes, eukaryotes and organelles. *Microbiol. Rev.* **47**, 1–45.

Lau, J. T. Y., Welply, J. K., Shenbagamurthi, P., Naider, F., and Lennarz, W. J. (1983). Substrate recognition by oligosaccharyl transferase: Inhibition of translational glycosylation by acceptor peptides. *J. Biol. Chem.* **258**, 15255–15260.

Lemberg, M. K., and Martoglio, B. (2002). Requirements for signal peptide peptidase-catalyzed intramembrane proteolysis. *Mol. Cell* **10**, 735–744.

Rapoport, T. A., Jungnickel, B., and Kutay, U. (1996). Protein transport across the eukaryotic endoplasmic reticulum and bacterial inner membranes. *Annu. Rev. Biochem.* **65**, 271–303.

Sambrook, J., Fritsch, E. F., and Maniatis, T. (1989). Molecular Cloning: A Laboratory Manual. Cold Spring Harbor, NY.

Tarentino, A. L., Trimble, R. B., and Maley, F. (1978). Endo-β-*N*-acetylglucosaminidase from *Streptomyces plicatus*. *Methods Enzymol.* **50**, 574–580.

Walter, P., and Johnson, A. E. (1994). Signal sequence recognition and protein targeting to the endoplasmic reticulum membrane. *Annu. Rev. Cell Biol.* **10**, 87–119.

SECTION II

LOADING PROTEIN INTO CELLS TO ASSESS PM FUNCTION

CHAPTER

8

Syringe Loading: A Method for Assessing Plasma Membrane Function as a Reflection of Mechanically Induced Cell Loading

Mark S. F. Clarke, Jeff A. Jones, and Daniel L. Feeback

OUTLINE

I. INTRODUCTION

In the past, we have described a method for loading large macromolecules into the cytoplasm of cultured cells via the production of transient plasma membrane wounds inflicted by defined amounts of fluid shear stress (Clarke and McNeil, 1992, 1994). This technique is referred to as "syringe loading" and has been utilized to load a variety of macromolecules (i.e., fluorescent dextrans, proteins, immunoglobulins, calcium indicator dyes, plasmid DNA, and antisense oligonucleotides) into various different cell types (i.e., endothelial cells, fibroblasts, epithelial cells, lymphocytes, and amoeba). This technique is very simple and straightforward, relying on the capacity of the plasma membrane to reseal after the infliction of a *transient* plasma membrane disruption. During the syringe loading procedure, the mechanical force applied to the cells to produce plasma membrane wounding is fluid shear stress generated as a consequence of the cell suspension being forced through a narrow orifice in the form of a 30-gauge hypodermic needle. The macromolecule to be loaded is dissolved in the suspension medium and enters the cell cytoplasm across a diffusion gradient during the time the plasma membrane wound is open. As such, this technique for macromolecular loading of cells in suspension is simple, reproducible, and inexpensive.

The series of highly coordinated, multicomponent responses that occur both within the cell cytoplasm and at the plasma membrane in response to mechanical perturbation/rupture of the external plasma membrane is known collectively as the membrane wound response (McNeil and Steinhardt, 1997). Apart from its use as a simple cell loading technique, syringe loading has also been used to investigate the underlying cellular mechanisms involved in this phenomenon (Miyake and McNeil, 1995; Clarke *et al.*, 1995b). It is increasingly evident that the wound response is a fundamental, highly conserved, and normal response to mechanical loading in a wide variety of cell types (McNeil and Terasaki, 2001). In addition, inappropriate levels of membrane wounding may be important in the etiology of a number of pathological conditions, such as atherosclerosis (Reidy and Lindner, 1991; Yu and McNeil, 1992; Clarke *et al.*, 1995b), unloading induced muscle atrophy (Clarke *et al.*, 1998), and left ventricular hypertrophy (Clarke *et al.*, 1995a). The susceptibility of the plasma membrane to mechanically induced membrane disruption and the ability of the membrane to reseal itself after disruption has occurred (*disruption followed by resealing being the definition of membrane wounding*) are both dependent on the biophysical properties of the membrane, including fluidity, elasticity, compressibility, and overall membrane order. These factors have been used to enhance macromolecular loading efficiency by the inclusion of membrane active agents, such as pluronic F68, during the syringe loading procedure in order to enhance the cell membrane resealing process (Clarke and McNeil, 1992).

To date, the mechanical properties of biological membranes have been tested by various means, including the direct measurement of membrane mechanical properties (i.e., elastic area compressibility, tensile strength, membrane toughness), using micropipette aspiration techniques (Needham and Nunn, 1990; Song and Waugh, 1993; Zhelev and Needham, 1993). In addition, indirect measurement of membrane fluidity using steady-state fluorescent anisotropic measurements, nuclear magnetic resonance, and fluorescent probe diffusion techniques have proven useful in determining changes in the physical properties of biological membranes (Tanii *et al.*, 1994; Kuroda *et al.*, 1996; Gimpl *et al.*, 1997) associated with such physiologically relevant alterations as membrane cholesterol content (Pritchard *et al.*, 1991; Clarke *et al.*, 1995b; Whiting *et al.*, 2000). However, all of these techniques have the disadvantage of describing the physical properties of the cell membrane in purely mechanical terms without taking into consideration the complex "biological" nature of the cell membrane system being examined.

Although complex in nature, the membrane wound response *in toto* can be quantified using direct end point measures that describe the final outcome of the process. Such measures include cell survival at a given level of mechanical perturbation, the number of wounded cells present in the surviving cell population, the amount of membrane wound marker that enters the wounded cell, and the relative size of the membrane wound created based on membrane wound marker size. If membrane wounds are produced in a defined and reproducible manner, the effects of various environmental conditions on plasma membrane function, as reflected by alterations in the membrane wound response, can be quantified by the end point measures described earlier. We have used this approach to probe the effects of different gravitational conditions on the biophysical properties of the plasma membrane in cultured ***adherent*** cells using a technique known as impact mediated loading (Clarke *et al.*, 2001). This article describes the use of syringe loading as a means of investigating the effects of environmental conditions on the plasma membrane wound response in ***suspension*** cells.

Syringe loading in its simplest configuration can be carried out utilizing a manual protocol, a single 1-ml hypodermic syringe and a 30-gauge hypodermic needle (Clarke and McNeil, 1992). The second-generation approach utilized a mechanized syringe pump apparatus that produced a defined expulsion pressure for a predetermined period of time (Clarke and McNeil, 1994). This article describes the development of a third-generation syringe loader device that is capable of processing a total of 10 separate samples in an identical fashion at the same time (Fig. 8.1). The multisample syringe loading technology essentially performs syringe loading under the same conditions as described previously except that many replicate samples can be loaded simultaneously. This device was designed to take into consideration the effects of temperature on membrane resealing dynamics, an experimental parameter that has not been fully controlled in previous syringe loading protocols. In the new configuration, individual sample vials containing identical cell suspensions are housed in a heating block maintained at 37°C throughout the syringe loading protocol (Fig. 8.1). As such, the technology produces a greater total yield of loaded cells with less variability between individual samples than using either the manual or the second-generation mechanized protocol. However, it does not exhibit a significant increase in cell loading efficiency relative to earlier syringe loading protocols. Rather, the goal for this technology was to develop a simple and rapid means of testing the biophysical properties of the plasma membrane of cells (as reflected by alterations in the membrane wounding response) exposed to different environmental conditions.

With this concept in mind and utilizing the multisample syringe loader detailed in Fig. 8.1, this article describes a series of experiments that illustrate the utility of the syringe loading technology as an experimental tool to probe the effects of environmental conditions on membrane function. In the following example, the effects of radiation exposure on cell membrane function as it impacts the membrane wounding response are investigated. The experiment described here was designed to determine whether gamma irradiation resulted in acute (i.e., within 2h of radiation exposure) membrane modification that resulted in an increase in susceptibility to mechanical shear force-induced membrane wounding. Membrane wounding is defined as a survivable disruption of the plasma membrane and is detected experimentally using a normally plasma membrane-impermeant fluorescent tracer such as FDx(fluorescent dextran) (M_r 10kDa). Wounded cells trap the wound marker in their cytoplasm by virtue of resealing the plasma membrane disruption. However, immediately after syringe loading there are cells in the loaded sample that are positive for the wound marker but will die within a matter of hours due to irreparable membrane damage. In the case of adherent cells, these cells detach from the culture substratum and hence can be washed away after a minimum period of culture (i.e., 4h) and are not included in any further analysis. Unfortunately, this approach cannot be employed when using suspension cells. However, if the fluorescent cell viability marker, propidium iodide (PI), is introduced to the sample immediately after syringe loading, those cells that are dead or dying are positive for PI regardless if they are also positive for the membrane wound marker. As such, this approach allows rapid discrimination between those cells that are

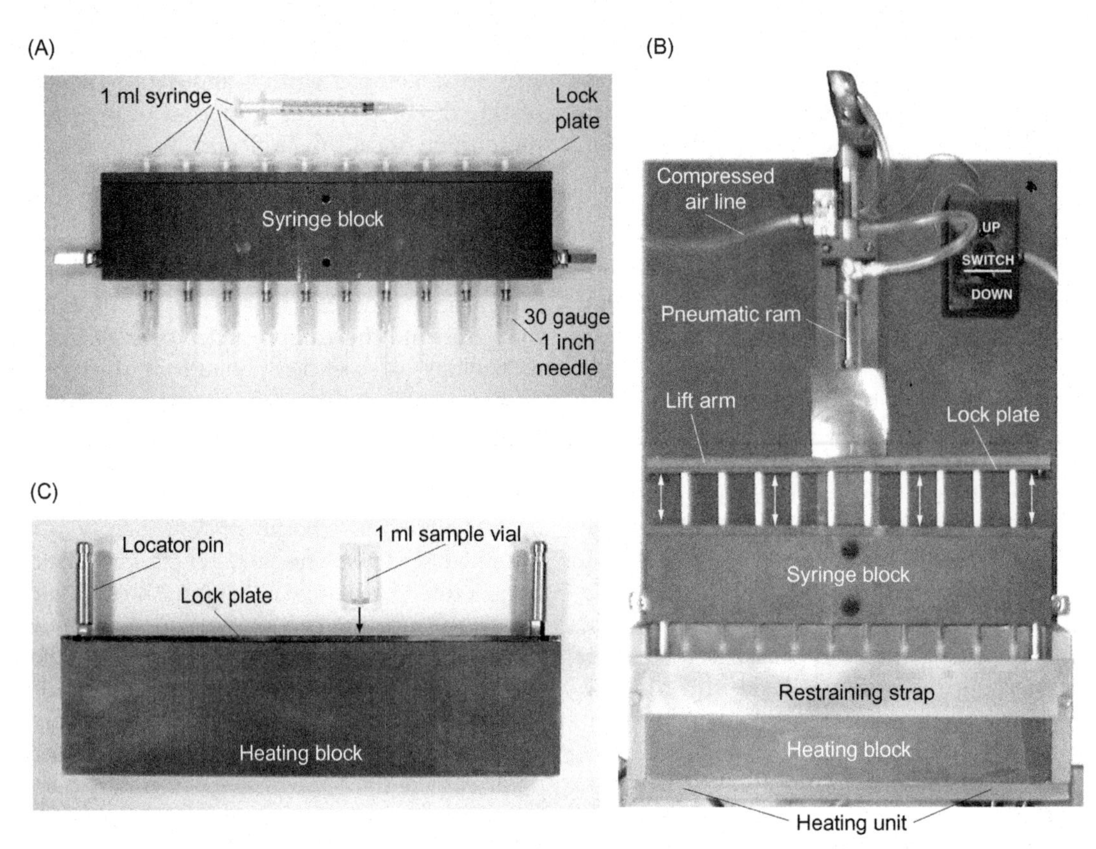

FIGURE 8.1 Multisample syringe loader. Ten, sterile, 1-ml disposable syringes are secured in the syringe block using a lock plate. A sterile, 30-gauge, 1-in. hypodermic needle is affixed to the tip of each syringe, making sure that the needle is seated firmly (Pl A). The syringe block is then attached to the pneumatic ram assembly of the main body of the device with the syringe plungers being secured directly to the lift arm using a lock plate (Pl C). Ten 1-ml sample vials, loaded with 0.5 ml of cells suspended in FDx loading solution, are loaded into a heating block (preequilibrated to 37°C) and secured in place using a lock plate (Pl B). The heating block containing the sample vials is then returned to its heating unit (set at 37°C) in the main body of the device and is secured in place with its restraining strap (Pl C). The pneumatic ram is placed in the "down" position so that the syringes are closed, and the whole syringe block is lowered and locked in place via the locator pins present on the top of the heating block. This arrangement ensures accurate registering of the each hypodermic needle into the center of its corresponding sample vial. The cell suspension in loading solution is pulled up into the syringe barrel by activating the pneumatic ram in the "up" direction and is expelled from the syringe by activating the pneumatic ram in the "down" direction. Expulsion pressure is controlled by a pressure valve attached to the compressed air line that feeds the ram assembly (Pl C).

truly membrane wounded (i.e., FDx positive, PI negative) from those cells that are dead and dying (i.e., PI positive, FDx negative or positive) (Fig. 8.2). By utilizing our multisample syringe loading device and protocol, coupled with subsequent analysis by two-channel fluorescent flow cytometry, the experiment described here provides an example of how the syringe loading technique can be used to quantify the effects of radiation exposure on membrane function in suspension cells. We chose the lymphoblastic cell line Jurkat, as it is a widely used *in vitro* model for studying immune function, an area of specific concern with regard to radiation exposure.

II. MATERIALS AND INSTRUMENTATION

Dulbecco's modified Eagle's medium (DMEM, Cat. No. 320-1885AG), bovine calf serum (CS) (Cat. No. 200-6170AG), and penicillin–streptomycin solution (Cat. No. 600-5140AG) are obtained ready to use from Gibco BRL (Grand Island, NY). Fluorescein isothocyanate-labeled dextran (M_r 10 kDa) (FDx) (Cat. No. D-1821) and propidium iodide (Cat. No. P21493) are obtained from Molecular Probes (Eugene, OR). Tissue culture flasks (T75 and T25) (Cat. Nos. 10-126-41 and 10-126-26) and sterile polypropylene conical centrifuge tubes (50-ml capacity) (Cat. No. 05-538-55A) are obtained from Fisher Scientific (Pittsburgh, PA). Sterile polypropylene conical sample vials (1-ml capacity) (Cat. No. 05-538-55A) are obtained from National Scientific Company (Quakertown, PA). Disposable 1-ml syringes (Cat. No. 9602) and 30-gauge hypodermic needles (Cat. No. 305128) are from Becton-Dickinson (Rutherford, NJ). The multisample syringe loader described in Fig. 8.1 was built by a local machine shop to specifications provided by the authors. As such, sample vial dimensions can be varied depending on experiment requirements, including the use of sterile, septum-sealed sample vials.

III. PROCEDURES

A. Preparation of Tissue-Cultured Cells for Syringe Loading

Stock Solutions and Media Preparation

1. *Stock FDx solution*: Add 200 mg of dry FDx powder to 1 ml of serum-free DMEM and vortex periodically over a period of 30 min to achieve complete solubilization. Centrifuge this solution at 10,000 *g* for 10 min at room temperature to remove any undissolved FDx and sterilize, if desired, by ultrafiltration. Stock FDx (200 mg/ml) can be stored at 4°C in the dark for up to a month or aliquoted and frozen in the dark at −80°C for storage up to a year.
2. *Stock propidium iodide solution*: Add 100 mg of dry PI powder to 1 ml of serum-free DMEM and vortex periodically over a period of 30 min to achieve complete solubilization. Centrifuge this solution at 10,000 *g* for 10 min at room temperature to remove any undissolved PI and collect the supernatant as a stock solution. Stock PI (100 mg/ml) can be stored at 4°C in the dark for up to a month or aliquoted and frozen in the dark at −80°C for storage up to a year.
3. *FDx loading solution*: Add 250 μl of stock FDx solution to 4.75 ml of serum-free DMEM to obtain final concentrations of 10 mg/ml FDx. Use the solution immediately.
4. *5% CS.DMEM*: Add 5 ml of sterile penicillin/streptomycin solution and 50 ml of sterile CS to 445 ml of sterile (1X) DMEM solution to obtain DMEM culture medium containing 5% CS, 100 IU/ml penicillin, and 100 μg/ml streptomycin (5% CS.DMEM). Store at 4°C for up to 21 days.

Steps

1. Grow Jurkat cells to confluence in T75 (75 cm^2) culture flasks using 15 ml of 5% CS.DMEM maintained at 37°C in a 5% CO_2 humidified atmosphere with subculture every third day. Carry out subculture by removing the cell suspension from the T75 flask and placing it in a sterile 50-ml centrifuge tube followed by centrifugation at 100 *g* (~550 rpm) for 5 min at room temperature. Decant the spent medium, resuspend the cell pellet gently in 30 ml of fresh 5% CS.DMEM, and dispense equal volumes into two fresh T75 flasks.

2. On the day of the experiment, collect cells by centrifugation as in step 1. Gently resuspend cells in 10 ml of 5% CS.DMEM and determine cell number using a hemacytometer or electronic cell counting device (i.e., a Coulter particle counter) and adjust the cell density to 2×10^6 cells/ml. Aliquot 5 ml of this cell suspension into T25 flasks in preparation for radiation exposure.
3. Expose cells to different doses of gamma irradiation using a Gammacell 1000 (Cs^{137} source). Our experiments were performed at Baylor College of Medicine (Houston, Texas).
4. Incubate cells at 37°C in a 5% CO_2 humidified atmosphere for a period of 2 h.
5. Remove cells from T25 flasks and collect each sample by centrifugation in sterile 50-ml centrifuge tubes at 100 *g* (~550 rpm) for 5 min at room temperature.
6. Carefully remove medium from the cells and filter through a 0.2-μm cellulose acetate filter to remove any remaining cell debris from the supernatant. Aliquot the medium in 1-ml aliquots and store at −80°C in the dark for subsequent biochemical testing, such as determination of lipid peroxidation marker production.
7. Resuspend each cell pellet in 2 ml of warm serum-free DMEM and determine the cell number by counting in a hemacytometer/Coulter counter.
8. Adjust the cell density of the cell suspension to 1×10^6 cells/ml by the addition of an appropriate amount of warm serum–free DMEM.
9. Dispense 0.5-ml samples of this cell suspension into a minimum of 13 sterile conical polypropylene 1-ml sample vials.
10. Add 25 μl of prewarmed stock FDx solution to all sample vials, mix by gentle vortexing, and load 10 of the 13 sample vials into the heating block (prewarmed to 37°C) of the multisample syringe loader (Fig. 8.1B).
11. Immediately perform syringe loading on the samples from **step 10** while the remaining three vials are incubated at 37°C in ambient air as control samples for FDx uptake by pinocytosis and cell loss due to processing.

B. Syringe Loading Protocol

1. Place heating block containing 10 sample vials containing identical aliquots of the cell suspension into the multisample syringe loader heating unit and attach the restraint strap as shown (Fig. 8.1C).
2. Insert the 10 separate 1-ml syringes with attached 1-in.-long, 30-gauge needles (which have been loaded previously into the syringe block, Fig. 8.1A) into the vials and lock in place using the guide posts (Fig. 8.1C).
3. Draw the cell suspension up into the barrel of the sterile syringes through the 30-gauge hypodermic needles by activating the pneumatic ram attached to the syringe plungers using the "up" switch (Fig. 8.1C). Set the pneumatic ram so that the syringe plungers do not move any further up than the 0.5-ml mark on the syringe barrels.
4. Once the barrels of the syringes are filled (which takes approximately 2s), expel the cell suspensions through the 30-gauge needles back into their respective sample vials at a constant pressure of 35 psi by reversing the direction of the pneumatic ram using the "down" switch. This procedure is defined as two strokes.
5. Repeat this procedure three more times so that each cell suspension has been subjected to eight strokes under identical expulsion pressure conditions.
6. Remove the test sample vials from the multisample syringe loader.
7. Take all the samples (including the control samples, which have been incubated in FDx loading solution at 37°C but not syringe

loaded), add 0.5 ml of 5% CS.DMEM to each vial, collect the cells by centrifugation at 100 *g* for 5 min at room temperature, and remove the supernatant (i.e., FDx loading solution).

8. Add 0.5 ml of warm 5% CS.DMEM to each cell pellet and resuspend the cells by gently vortexing the sample and store at 37°C in a 5% CO_2 humidified atmosphere in preparation for analysis. Analysis should be started as soon as possible.

C. Dual-Label Fluorescent Flow Cytometry Analysis

1. Immediately prior to fluorescent flow cytometry analysis, add a 10-μl aliquot of stock PI solution (final PI concentration of 1 mg/ml) to each sample and mix by gentle vortexing. Add PI to each individual sample immediately before analysis.
2. Analyze control samples first by dual-channel fluorescent flow cytometry using a Becton-Dickinson FACSCalibur system (or similar fluorescent flow cytometer) and plot a two-axis scatter plot of cell fluorescent intensity at 585 nm (FDx signal) and 670 nm (PI signal) using the associated software. These control samples are used to define "quadrant" gating thresholds for the control cell population with regard to FDx and PI background signals (Fig. 8.2). Collect a total of 10,000 events for analysis of each sample.
3. Analyze syringe loaded samples by dual-channel fluorescent flow cytometry using the quadrant gates defined earlier (Fig. 8.2). Collect a total of 10,000 events for analysis of each sample.
4. Calculate the number of dead or dying cells in each sample (expressed as a percentage of the total cells analyzed, Fig. 8.3A) by counting the number of cells in the sample that are positive for PI staining, including

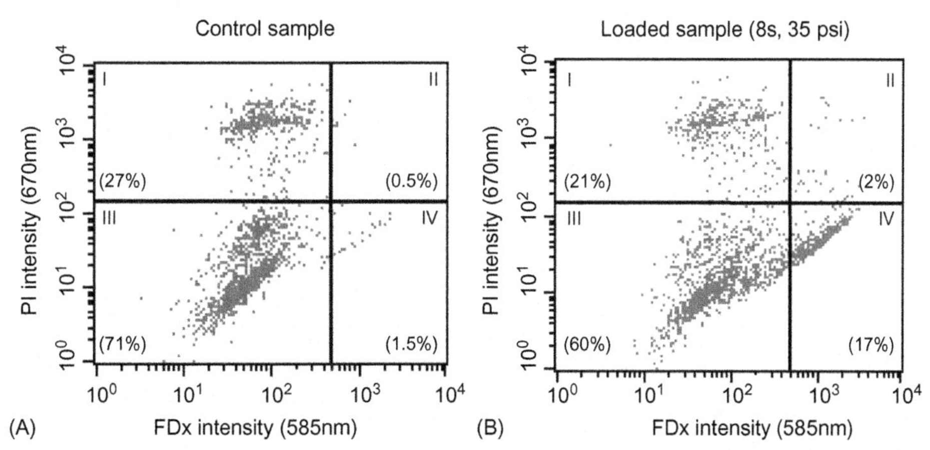

FIGURE 8.2 Dual-channel fluorescent flow cytometry analysis for the simultaneous determination of membrane wounding and irreparable membrane damage in suspension cells. Control samples, consisting of cell suspensions processed in an identical fashion to experimental samples other than they are not syringe loaded, are used to determine quadrant analysis parameters. Dead or dying cells, present even in the control, nonsyringe loaded sample, stain positively for PI (i.e., quadrants I and II) (Pl A). After syringe loading at 35 psi expulsion for a total of eight strokes (Pl B), loaded cells stain positively for FDx (i.e., quadrants II and IV). Truly wounded cells (i.e., FDx positive and PI negative), which have completely resealed their plasma membrane disruptions before the addition of PI to the suspension, are positive only for FDx (i.e., quadrant IV). Numbers in parentheses are values (%) for the cells falling in each quadrant.

those cells that are also positive for FDx (i.e., Fig. 8.2, quadrants I and II, respectively).

5. Calculate the number of wounded cells (expressed as a percentage of the total cells analyzed, Fig. 8.3B) and the mean fluorescent value (MFV; Fig. 8.3C) of the wounded population by determining the number and staining intensity of cells that are positive for FDx staining only (i.e., Fig. 8.2, quadrant IV).
6. Calculate a wound index (which reflects cell survival of mechanically induced plasma

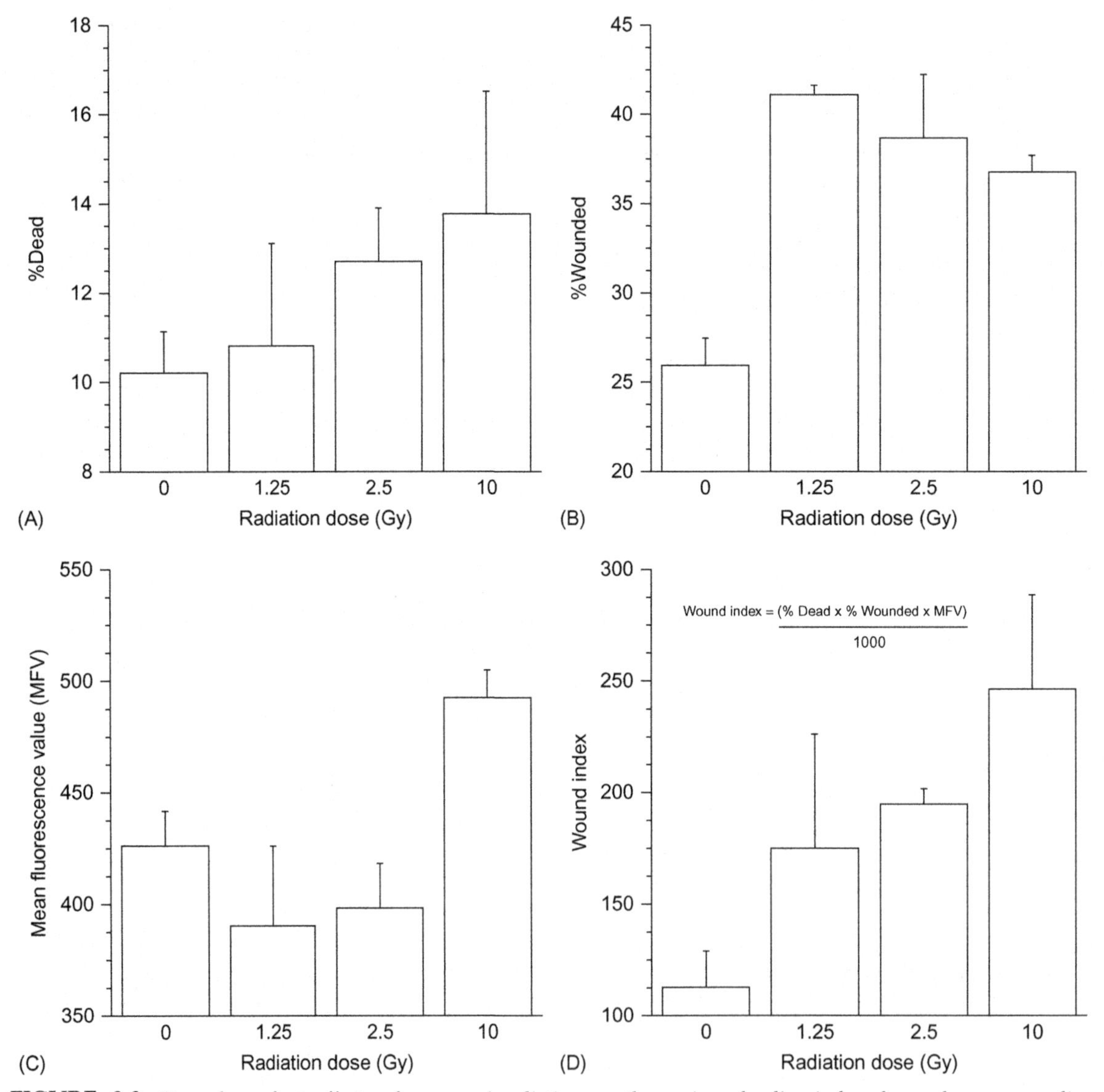

FIGURE 8.3 Dose-dependent effects of gamma irradiation on the syringe loading-induced membrane wounding response of the lymphoblastic cell line Jurkat.

membrane damage, the number, and the MFV of the wounded cells) as described previously (Clarke and McNeil, 1992) for each radiation dose (Fig. 8.3D), where the wound index = % dead cells × % wounded cells × mean fluorescent value/1000.

IV. COMMENTS

The technology described herein allows the effects of a wide range of environmental conditions and their potential countermeasures (e.g., radiation exposure and radio-protectants) to be assessed in a variety of cell suspension models, including peripheral blood lymphocytes from human subjects. A less refined version of this approach has been used previously to study the effects of increasing membrane cholesterol content on plasma membrane susceptibility to mechanical shear-induced membrane wounding (Clarke *et al.*, 1995a). The apparatus and protocol described earlier have taken what was essentially a macromolecular loading technique and adapted it for use as a means of assessing the effects of radiation exposure on the membrane function of lymphoblastic cells in suspension. The technique also has the advantage of being able to immediately discriminate between dead and dying cells and truly wounded cells in suspension after the application of mechanical shear force using two-channel fluorescent flow cytometry (Fig. 8.2).

The apparent dose response in membrane wound susceptibility observed relative to radiation exposure in our model indicates that gamma irradiation induces alterations in the plasma membrane components of Jurkat cells, if not immediately, then most certainly within 2h of exposure (Fig. 8.3). This time course of events suggests that these effects are not associated with genomic damage. The effects of radiation exposure on membrane wounding observed in this study are similar to that described previously after an increase in plasma membrane order. One possible reason for such an increase in membrane order after radiation exposure is the production of reactive oxygen species (ROS), such as superoxide, hydrogen peroxide, hydroxyl, peroxyl, and alkoxyl radicals, which may lead to membrane damage and consequent cross-linking of membrane components (Clarke *et al.*, 2003). This concept is supported by the experimental observation that Jurkat cells exposed to gamma irradiation not only exhibit an increase in susceptibility to membrane wounding within 2h, but also produce significant amounts of lipid peroxidation markers (data not shown).

We have observed previously that in order to obtain optimal levels of membrane wounding experimentally (i.e., maximizing cell loading while minimizing cell death caused by irreparable membrane disruption), a mechanical shear dose–response curve must be constructed. In general, most mammalian cells appear to wound best at expulsion pressures between 30 and 45psi. However, this parameter varies with each individual cell type. Expulsion of the cell suspension under the conditions used in the example given earlier (i.e., 0.5ml volume of cell suspension passing through a 1-in.-long, 30-gauge needle at a constant 35psi expulsion pressure in 1.5s) results in a theoretical average fluid shear stress of approximately 7340 dynes/cm^2. This value is derived from the equation wall shear rate $= 8 \times V_{mean}/D = 32Q/\pi D^3$, where D is the luminal diameter of the hypodermic needle (cm) and Q is the flow rate in cm^3/s, assuming that the viscosity of the loading solution is one centipoise (Clarke *et al.*, 1995b). Due to the non-Newtonian fluid physics of the cell suspension, this calculated value for the shear stress inflicted on the cells syringe loading under these conditions could be artificially low. However, regardless of the actual shear stress value inflicted on the cells while passing through the needle, the multisample syringe loader induces a constant and reproducible level of mechanical shear stress from sample to sample. This translates into the infliction of well-defined and reproducible amounts of membrane wounding in cells subjected to syringe loading. The incorporation of a heating block into the device in order to maintain a constant temperature during the syringe loading procedure has also removed a potential confounding variable from the experimental protocol.

V. PITFALLS

1. Jurkat cells are relatively fragile in tissue culture. Cells should be handled gently with a minimum of agitation, being especially careful when performing manipulations that require pipetting or vortexing. This fragility is evidenced by the relatively large number of PI-positive cells (i.e., dead or dying cells) present in control samples that have not been subjected to syringe loading (see Fig. 8.2A).
2. Jurkat cells should be used for syringe loading experiments 1 day after subculture to ensure the optimal number of healthy cells in the culture.
3. When calculating the theoretical shear stress imposed during syringe loading, it is important to determine the flow rate through the needle experimentally. This will change primarily depending on the expulsion pressure and the number of cells in the suspension; both parameters are under the control of the investigator. These wounding parameters need to be optimized for each particular cell type.
4. It is important to prevent excessive pH changes caused by exposure of the bicarbonate-buffered culture medium to atmospheric air during the procedure as these are potentially harmful to the cells.

References

Clarke, C. H., Weinberger, S. R., and Clarke, M. S. F. (2003). Application of ProteinChip array technology for detection of protein biomarkers of oxidative stress. *Crit. Rev. Oxid. Stress Aging* **1**, 366–379.

Clarke, M. S., Bamman, M. M., and Feeback, D. L. (1998). Bed rest decreases mechanically induced myofiber wounding and consequent wound-mediated FGF release. *J. Appl. Physiol* **85**(2), 593–600.

Clarke, M. S., Caldwell, R. W., Chiao, H., Miyake, K., and McNeil, P. L. (1995a). Contraction-induced cell wounding and release of fibroblast growth factor in heart. *Circ. Res.* **76**(6), 927–934.

Clarke, M. S., and McNeil, P. L. (1992). Syringe loading introduces macromolecules into living mammalian cell cytosol. *J. Cell Sci.* **102**(Pt. 3), 533–541.

Clarke, M. S. F., and McNeil, P. L. (1994). Syringe loading: a method for inserting macromolecules into cells in suspension. *In "Cell Biology: A Laboratory Handbook"* (J. E. Celis, ed.), pp. 30–36, Academic Press, San Diego.

Clarke, M. S. F., Pritchard, K. A., Medows, M. S., and McNeil, P. L. (1995b). An atherogenic level of native LDL increases endothelial cell vulnerability to shear-induced plasma membrane wounding and consequent FGF release. *Endothelium* **4,** 127–139.

Clarke, M. S. F., Vanderburg, C. R., and Feeback, D. L. (2001). The effect of acute microgravity on mechanically-induced membrane damage and membrane-membrane fusion events. *J. Grav. Physiol.* **8**, 37–47.

Gimpl, G., Burger, K., and Fahrenholz, F. (1997). Cholesterol as modulator of receptor function. *Biochemistry* **36**(36), 10959–10974.

Kuroda, Y., Ogawa, M., Nasu, H., Terashima, M., Kasahara, M., Kiyama, Y., Wakita, M., Fujiwara, Y., Fujii, N., and Nakagawa, T. (1996). Locations of local anesthetic dibucaine in model membranes and the interaction between dibucaine and a Na+ channel inactivation gate peptide as studied by 2H- and 1H-NMR spectroscopies. *Biophys. J.* **71**(3), 1191–1207.

McNeil, P., and Terasaki, M. (2001). Coping with the inevitable: How cells repair a torn surface membrane. *Nature Cell Biol.* **3**, E124–E129.

McNeil, P. L., and Steinhardt, R. A. (1997). Loss, restoration, and maintenance of plasma membrane integrity. *J. Cell Biol.* **137**(1), 1–4.

Miyake, K., and McNeil, P. L. (1995). Vesicle accumulation and exocytosis at sites of plasma membrane disruption. *J. Cell Biol.* **131**(6 Pt 2), 1737–1745.

Needham, D., and Nunn, R. S. (1990). Elastic deformation and failure of lipid bilayer membranes containing cholesterol. *Biophys. J.* **58**, 997–1009.

Pritchard, K. A., Schwarz, S. M., Medow, M. S. and Stemerman, M. B. (1991). Effect of low-density lipoprotein on endothelial cell membrane fluidity and mononuclear cell attachment. *Am. J. Physiol.* **260**, C43–C49.

Reidy, M. A., and Lindner, V. (1991). Basic FGF and growth of arterial cells. *Ann. N. Y. Acad. Sci.* **638,** 290–299.

Song, J., and Waugh, R. E. (1993). Bending rigidity of SOPC membranes containing cholesterol. *Biophys. J.* **64**(6), 1967–1970.

Tanii, H., Huang, J., Ohyashiki, T., and Hashimoto, K. (1994). Physical-chemical-activity relationship of organic solvents: Effects on Na(+)-K(+)-ATPase activity and membrane fluidity in mouse synaptosomes. *Neurotoxicol. Teratol.* **16**(6), 575–582.

Whiting, K. P., Restall, C. J., and Brain, P. F. (2000). Steroid hormone-induced effects on membrane fluidity and their potential roles in non-genomic mechanisms. *Life Sci.* **67**(7), 743–757.

Yu, Q. C., and McNeil, P. L. (1992). Transient disruptions of endothelial cells of the rat aorta. *Am. J. Pathol.* **141**, 1349–1360.

Zhelev, D. V., and Needham, D. (1993). Tension-stabilized pores in giant vesicles: Determination of pore size and pore line tension. *Biochim. Biophys. Acta* **1147**(1), 89–104.

SECTION III

PROTEIN TRANSLOCATION BETWEEN CYTOPLASM AND COMPARTMENTS OTHER THAN PLASMA MEMBRANE

CHAPTER

9

Protein Translocation into Mitochondria

Sabine Rospert, and Hendrik Otto

OUTLINE

I. INTRODUCTION

Mitochondria from different sources such as rat liver, rabbit brain, or yeast can be isolated as intact organelles. Isolated mitochondria are able to respire, maintain a membrane potential across their inner membrane, possess an active ATP synthase, and shuttle nucleotides across their membranes. In addition, even a process as complicated as import of mitochondrial precursor proteins can be studied outside the living cell. For this purpose, radiolabeled precursor proteins, synthesized in an *in vitro* transcription/translation system, are mixed with isolated mitochondria (Glick, 1991; Melton *et al.*, 1984). In the presence of ATP, precursor proteins will cross the membranes, become processed to their mature form, and fold to their native state. Building on this basic "import assay," sophisticated experiments have been developed and the results of these experiments provide most of what we know about mitochondrial import today (Neupert, 1997; Pfanner and Geissler, 2001).

This article describes a standard protocol for the *in vitro* synthesis of a radiolabeled precursor protein and the subsequent import of this precursor into isolated yeast mitochondria. As an example, we have selected the precursor protein yeast malate dehydrogenase (Dubaquié *et al.*, 1998). The N-terminal presequence of the yeast malate dehydrogenase precursor, like most mitochondrial precursor proteins, is removed by a protease localized in the mitochondrial matrix (Jensen and Yaffe, 1988). The mRNA of the precursor protein is transcribed with SP6 RNA polymerase (Melton *et al.*, 1984).

II. MATERIALS AND INSTRUMENTATION

SP6 RNA polymerase (Cat. No. 810 274); RNase inhibitor from human placenta (Cat. No. 799 017); set of ATP, CTP, GTP, UTP, lithium salts, 100 m*M* solutions (Cat. No. 1 277 057); creatine kinase from rabbit muscle (Cat. No. 127 566); creatine phosphate, disodium salt (Cat. No. 127 574); and proteinase K (Cat. No. 1092766) are from Roche. Tris (Cat. No. 108382); KCl (Cat. No. 104936); KOH (Cat. No. 105021); $MgCl_2$ (Cat. No. 105833); NaN_3 (Cat. No. 822335); 25% NH_3 solution (Cat. No. 105432); ethanol (Cat. No. 100983); and sodium salicylate (Cat. No. 106602) are from Merck. Spermidine (Cat. No. S 0266); bovine serum albumin, essentially fatty acid free (BSA) (Cat. No. A-7511); dithiothreitol (DTT) (Cat. No. D 5545); HEPES (Cat. No. H 7523); trypsin (Cat. No. T 1426); trypsin inhibitor, from soybean (Cat. No. T 9003); α-nicotinamide adenine dinucleotide disodium salt, reduced form (NADH) (Cat. No. N 6879); EDTA (Cat. No. E 9884); $(NH_4)_2SO_4$ (Cat. No. A 2939); $CaCl_2$, dihydrate (Cat. No. C 5080); magnesium acetate tetrahydrate (Cat. No. M 2545); valinomycin (Cat. No. V 0627); ATP, disodium salt (Cat. No. A 7699); glycerol (Cat. No. G 6279); potassium acetate (Cat. No. P 5708); KH_2PO_4 (Cat. No. P 5379); L-methionine (Cat. No. M 9625); urea (Cat. No. U 5128); phenylmethylsulfonyl fluoride (PMSF) (Cat. No. P 7626); tRNA, from bovine liver (Cat. No. R 4752); and Triton X-100 (Cat. No. T 9284) are from Sigma. Rabbit reticulocyte lysate (Cat. No. L 4960) and amino acid mixture, minus methionine (Cat. No. L 4960), are from Promega. NaCl (Cat. No. 9265.1) and trichloroacetic acid (TCA) (Cat. No. 8789.1) are from Roth. Sorbitol (Cat. No. 2039) is from Baker. l-[^{35}S]methionine, >1000 Ci/mmol (Cat. No. SJ 235), $m^7G(5Ç)ppp(5Ç)G$ (G-cap) (Cat. No. 27 4635 02), and Kodak X-OMAT X-ray film (Cat. No. V1651496) are from Amersham Biosciences. Sorvall centrifuge RC M120 GX, Kendro. Sorvall Rotor S100 AT3-204, Kendro. Eppendorf centrifuge 5417 R, "microfuge" Eppendorf. Greiner PP-tubes 15 ml (Cat. No. 188261), Greiner. X-ray cassettes (Cat. No Rö 13), GLW. Plasmid is pSP65mdh1 (Dubaquié *et al.*, 1998). Highly purified mitochondria (25 mg/ml) are prepared after Glick and Pon (1995).

III. PROCEDURES

Solutions used directly as obtained from the supplier are only listed in Section II. Protocols for the preparation of solutions used throughout the procedure are only given once.

A. Transcription Using SP6 Polymerase

Solutions

1. *1M Tris–HCl stock solution, pH 7.5*: Dissolve 12.1 g Tris in 80 ml H_2O and adjust pH to 7.5 with 5*M* HCl. Add H_2O to 100 ml. Autoclave and store at room temperature.

2. *1M HEPES–KOH, pH 7.4*: Dissolve 23.8 g HEPES in 80 ml H_2O and adjust pH to 7.4 using 4*M* KOH. Add H_2O to 100 ml. Autoclave and store at room temperature.
3. *1M spermidine*: Dissolve 145 mg spermidine in 1 ml H_2O. Store at −20°C.
4. *100 mg/ml BSA*: Dissolve 500 mg BSA in 5 ml H_2O. Store at −20°C.
5. *2.5M $MgCl_2$*: Dissolve 50.8 g $MgCl_2$ in 100 ml H_2O. Autoclave and store at 4°C.
6. *2.5M KCl*: Dissolve 18.6 g KCl in 100 ml H_2O. Autoclave and store at 4°C.
7. *100 mM DTT*: Dissolve 15.4 mg DTT in 1 ml H_2O. Store at −20°C. Make a fresh solution about every 4 weeks.
8. *5× SP6 reaction buffer*: 200 m*M* Tris–HCl, pH 7.5, 30 m*M* $MgCl_2$, 10 m*M* spermidine, and 0.5 mg/ml BSA. To obtain 10 ml of a 5x reaction buffer, mix 2 ml 1*M* Tris–HCl, pH 7.5, 120 μl 2.5*M* $MgCl_2$, 100 μl 1*M* spermidine and 50 μl 100 mg/ml BSA. If necessary, readjust the pH to 7.5. Store in 1-ml aliquots at −20°C.
9. *G-cap ($m^7G(5')ppp(5')G$)*: Dissolve 25 A_{250} units in 242 μl H_2O. Freeze 10-μl aliquots in liquid nitrogen. Store at −70°C.
10. 5 m*M* NTP–GTP: To make a 500 μl stock, add 25 μl 100 m*M* ATP, 25 μl 100 m*M* UTP, and 25 μl 100 m*M* CTP to 425 μl 20 m*M* HEPES–KOH, pH 7.4. Store in 100-μl aliquots at −70°C.
11. 5 m*M GTP*: Mix 475 μl 20 m*M* HEPES–KOH, pH 7.4, with 25 μl 100 m*M* GTP solution.
12. *RNase inhibitor buffer*: 20 m*M* HEPES–KOH, pH 7.4, 50 m*M* KCl, 10 m*M* DTT, and 50% glycerol. Make 10 ml of the buffer by mixing 200 μl 1*M* HEPES–KOH, pH 7.4, 200 μl 2.5 *M* KCl, 1 ml 100 m*M* DTT, and 5 ml glycerol. Add H_2O to 10 ml and store at −20°C.
13. *4 units/μl RNase inhibitor*: Add 500 μl RNase inhibitor buffer to 2000 units of RNase inhibitor. Store at −20°C for up to 6 months.
14. *1 μg/μl linearized plasmid DNA*: Prepare the linearized plasmid (pSP65mdh1) according to standard molecular biology procedures.

Steps

1. Mix the following solutions carefully, avoiding the formation of air bubbles. Follow the indicated order of addition because the DNA might precipitate in 5× SP6 buffer. Precipitation of DNA can also occur if the mixture is placed on ice. Incubate the mixture at 40°C for 15 min.

H_2O	12 μl
4 units/μl RNase inhibitor	1 μl
1 μg/μl linear plasmid (pSP65 mdh1)	5 μl
5 m*M* rNTPs minus GTP	5 μl
5 m*M* G-cap	5 μl
100 m*M* DTT	5 μl
5× SP6 buffer	10 μl
SP6 polymerase	2 μl

2. Start transcription by adding 5 μl 5 m*M* GTP solution and incubate for 90 min at 40°C.
3. Extract the mRNA with phenol/chloroform and then with chloroform/isoamylalcohol, precipitate with 100% ethanol, and wash with 70% ethanol. Resuspend the dried pellet in 125 μl H_2O.
4. The mRNA obtained by this procedure is used directly in the translation protocol. mRNA can be stored in 10-μl aliquots at −70°C. If frozen mRNA is used for translation, thaw rapidly and keep at room temperature before adding the mRNA to the translation reaction.

B. Translation Using Reticulocyte Lysate

Solutions

1. *1M DTT*: Dissolve 154 mg DTT in 1 ml H_2O. Store at −20°C. Make a fresh solution about every 4 weeks.
2. *8 mg/ml creatine kinase*: Dissolve 8 mg creatine kinase in 475 μl H_2O. Add 20 μl 1 *M* HEPES–KOH, pH 7.4, 5 μl 1 *M* DTT, and 500 μl glycerol. Freeze in 10-μl aliquots in liquid nitrogen and store at −70°C.

3. *5 mg/ml tRNA*: Dissolve 10 mg tRNA from bovine liver in 2 ml H_2O. Store in 100-μl aliquots at −20°C.
4. *400 mM HEPES–KOH, pH 7.4*: Mix 6 ml H_2O with 4 ml 1*M* HEPES–KOH, pH 7.4. If necessary, readjust pH.
5. *10 mM GTP*: Mix 450 μl 20 m*M* HEPES–KOH, pH 7.4, with 50 μl 100 m*M* GTP. Store at −20°C.
6. *100 mM ATP*: Dissolve 55.1 mg ATP in 900 μl H_2O. Adjust to pH ~7 using 4 *M* NaOH and pH indicator paper. Adjust volume to 1 ml and store at −20°C.
7. *600 mM creatine phosphate*: Dissolve 153.06 mg creatine phosphate in 1 ml H_2O. Store at −20°C.
8. *4M potassium acetate*: Dissolve 3.92 g potassium acetate in 10 ml H_2O. Do not adjust the pH. Store at −20°C.
9. *50 mM magnesium acetate*: Dissolve 10.7 mg magnesium acetate tetrahydrate in 1 ml H_2O. Store at −20°C.

Steps

1. Prepare the reticulocyte lysate mix and the tRNA mix fresh. To obtain a 100-μl translation reaction, mix the following solutions.

Reticulocyte lysate mix		tRNA mix	
0.4 *M* HEPES–KOH	5 μl	5 mg/ml tRNA	6 μl
10 m*M* GTP	0.6 μl	4 units/μl RNase inhibitor	6 μl
100 m*M* ATP	0.5 μl	4 *M* potassium acetate	3 μl
1 m*M* amino acid mix	3.75 μl		
0.6 *M* creatine phosphate	2 μl		
reticulocyte lysate	50 μl		
8 mg/ml creatine kinase	2 μl		

2. Use the mRNA obtained in Section III, A. It is possible to use mRNA produced in a different transcription system, e.g., with T7 RNA polymerase. Mix 60 μl reticulocyte lysate mix, 10 μl tRNA mix, 2 μl 50 m*M* magnesium acetate, 18 μl mRNA, 10 μl [^{35}S]methionine, and 2 μl 1 *M* DTT.
3. Incubate this mixture for 60 min at 30°C. Shield it from light to prevent heme-induced photooxidation of the precursor proteins. Remove ribosomes after the translation reaction by centrifugation for 15 min at 150,000 *g* (65,000 rpm with S100 AT3-204 rotor in Sorvall centrifuge). Remove the supernatants, being careful not to disturb the ribosomal pellet.

C. Denaturation of Radiolabeled Precursor Protein

Solutions

1. *Saturated $(NH_4)_2SO_4$ solution*: Weigh 100 g $(NH_4)_2SO_4$ and add H_2O to a final volume of 100 ml. Stir for 30 min at room temperature. The $(NH_4)_2SO_4$ will not dissolve entirely. Remove the supernatant and keep at room temperature.
2. *8M urea*: Dissolve 4.85 g urea in a final volume of 10 ml 25 m*M* Tris–HCl, pH 7.5, containing 25 m*M* DTT.

Steps

1. Proteins synthesized in reticulocyte lysate are either folded or bound to chaperone proteins present in the lysate (Wachter *et al.*, 1994). In order to unfold the protein prior to import, it can be precipitated by high concentrations of ammonium sulfate and subsequently denatured in 8 *M* urea.
2. Add 200 μl of the $(NH_4)_2SO_4$ solution to the 100-μl translation reaction. Mix well and allow precipitation of the protein for 30 min on ice. Collect precipitate by centrifugation in an Eppendorf centrifuge at 20,000 *g* for 10 min.

3. Discard the supernatant and dissolve the pellet in 100 μl of 8 *M* urea solution. Keep the denatured precursor at room temperature for 10–30 min. This precursor solution is used for the import reaction (Section III,D) and preparation of the precursor standard (Section III,G).

D. Import of Denatured Radiolabeled Precursor Proteins

Solutions

For additional solutions required, see Sections III,A and III,B.

1. *2.4M sorbitol*: Dissolve 43.7 g of sorbitol in a final volume of 100 ml H_2O. Autoclave and store at 4°C.
2. *1M KH_2PO_4*: Dissolve 1.36 g KH_2PO_4 in 10 ml H_2O. Filter sterilize and keep at room temperature.
3. *1M HEPES–KOH, pH 7.0*: Dissolve 23.8 g HEPES in 80 ml H_2O and adjust pH to 7.0 with 4*M* KOH. Add H_2O to a final volume of 100 ml. Filter sterilize and store at room temperature.
4. *250 mM EDTA, pH 7.0*: Resuspend 7.3 g of EDTA in 70 ml H_2O. Adjust pH to 7.0 using 5 *M* NaOH. Add H_2O to a final volume of 100 ml. Filter sterilize and keep at room temperature.
5. *2× import buffer*: 1.2*M* sorbitol, 100 m*M* HEPES–KOH, pH 7.0, 100 m*M* KCl, 20 m*M* $MgCl_2$, 5 m*M* EDTA, pH 7.0, 4 m*M* KH_2PO_4, 2 mg/ml BSA, and 1.5 mg/ml methionine. To make 100 ml of 2× import buffer, mix 50 ml 2.4 *M* sorbitol, 400 μl 1*M* KH_2PO_4 solution, 4 ml 2.5 *M* KCl, 10 ml 1*M* HEPES–KOH, pH 7.0, 0.8 ml 2.5 *M* $MgCl_2$, 2 ml 250 m*M* EDTA, pH 7.0, 150 mg methionine, and 200 mg BSA. Adjust pH to 7.0 and add H_2O to 100 ml. Store at −20°C.
6. *1× import buffer minus BSA*: Prepare 2× import buffer, but without BSA. To obtain 1× import buffer minus BSA, mix 2 ml 2× import buffer with 2 ml H_2O.
7. *500 mM NADH*: Dissolve 35.5 mg of NADH in a final volume of 100 μl 20 m*M* HEPES–KOH, pH 7.0. Store at −20°C.
8. *1 mg/ml valinomycin*: Dissolve 2 mg valinomycin in 2 ml ethanol. Store at −20°C.
9. *Purified yeast mitochondria*: 25 mg mitochondrial protein/ml. Store at −70°C in 0.6 *M* sorbitol, 20 m*M* HEPES–KOH, pH 7.4, and 10 mg/ml BSA. Thaw rapidly at 25°C immediately before the experiment. Do not refreeze. A detailed protocol of the purification procedure is given in Glick and Pon, (1995).

Steps

1. Import into the matrix of mitochondria requires a membrane potential across the inner mitochondrial membrane. Therefore, the most thorough control for the specificity of an import reaction is to determine its dependence on a membrane potential. Adding ATP and the respiratory substrate NADH generates this potential. (Note that mammalian mitochondria cannot oxidize added NADH.)
2. Perform two import reactions, one in the absence and one in the presence of valinomycin. Preincubate the import reaction in a 15-ml Greiner tube at 25°C for 1–2 min.

Reaction 1 (+ membrane potential)		Reaction 2 (no membrane potential)	
2× import buffer	500 μl	2× import buffer	500 μl
100 m*M* ATP	20 μl	100 m*M* ATP	20 μl
500 m*M* NADH	4 μl	500 m*M* NADH	4 μl
Yeast mitochondria	20 μl	Yeast mitochondria	20 μl
H_2O	406 μl	H_2O	405 μl
		1 mg/ml valinomycin	1 μl

3. Add 50 μl of the denatured precursor protein solution (see Section III,C) containing denatured malate dehydrogenase to each reaction (reactions 1 and 2). Intact mitochondria should be handled gently. However, it is essential to mix the denatured precursor protein into the import reaction rapidly. Agitate the import reaction gently on a vortex mixer while adding the denatured precursor mixture dropwise. If mixing is performed only after addition, the precursor protein tends to aggregate and becomes import incompetent.
4. Incubate at 25°C for 10 min. Agitate gently every other minute to facilitate gas exchange. Stop the import reaction by transferring the tubes onto ice. Add 1 μl of 1 mg/ml valinomycin to reaction 1.
5. Remove 200 μl each from reactions 1 and 2 and put the samples on ice. Spin down mitochondria in an Eppendorf centrifuge and remove the supernatant (be careful, the pellet will be very small). Resuspend the mitochondrial pellets of reactions 1 and 2 in each 200 μl of 1× import buffer. These samples represent the **total** of the two import reactions (Fig. 9.1, lanes 2 and 5). Add 22 μl 50% TCA to each. Keep on ice and process further after all samples have been acid denatured (for the method of TCA precipitation, see Section III,G).

E. Protease Treatment of Intact Mitochondria

Solutions

1. *10 mg/ml trypsin*: Dissolve 3 mg of trypsin in 300 μl H_2O. Make fresh.
2. *20 mg/ml trypsin inhibitor*: Dissolve 6 mg of trypsin inhibitor in 300 μl H_2O. Make fresh.

Steps

Perform the following steps in parallel with both import reactions.

1. To digest precursor proteins that stick to the surface of the mitochondria, add 8 μl 10 mg/ml trypsin (final concentration 100 μg/ml). Incubate for 30 min on ice.
2. Add 8 μl 20 mg/ml trypsin inhibitor (final concentration 200 μg/ml) and incubate on ice for 5 min.
3. Transfer the sample into a new Eppendorf tube.
4. Spin for 3 min in an Eppendorf microfuge at 10,000 *g*. Remove the supernatant carefully by aspiration.
5. Carefully resuspend the mitochondrial pellet in 800 μl 1× import buffer minus BSA. As it is extremely important to resuspend the pellet completely, it should be done as follows. First add 100 μl of 1× import buffer minus BSA and resuspend mitochondria by pipetting up and down. Then add another 700 μl of 1× import buffer minus BSA to yield 800 μl final volume.
6. Remove 200 μl of each sample and add 22 μl 50% TCA. Keep on ice. These samples represent the material that has crossed the outer membrane completely (**import**, Fig. 9.1, lanes 3 and 6).

F. Inherent Protease Resistance of the Imported Protein

Solutions

1. *1M Tris–HCl stock solution, pH 8.0*: Dissolve 12.1 g Tris in 80 ml H_2O and adjust pH to 8.0 using 5*M* HCl. Add H_2O to a final volume of 100 ml. Autoclave and store at room temperature.
2. *2M $CaCl_2$*: Dissolve 14.7 g $CaCl_2$ in 50 ml H_2O. Autoclave and store at room temperature.
3. *10% Triton X-100 (w/v)*: Dissolve 10 g of Triton X-100 in a final volume of 100 ml H_2O. Store at room temperature in the dark.
4. *10% NaN_3*: Dissolve 1 g NaN_3 in a final volume of 10 ml H_2O. Store at room temperature.

5. *Proteinase K buffer*: 50 m*M* Tris–HCl, pH 8.0, 1 m*M* $CaCl_2$, and 0.02 % NaN_3. To make 10 ml of proteinase K buffer, mix 500 μl 1 *M* Tris–HCl, pH 8.0, 5 μl 2 *M* $CaCl_2$, and 20 μl 10% NaN_3. Add H_2O to a final volume of 10 ml. Store at room temperature.
6. *10 mg/ml proteinase K stock*: Dissolve 5 mg of proteinase K in 500 μl proteinase K buffer. Store at 4°C for up to 1 week without loss of activity.
7. *200 μg/ml proteinase K solution*: Mix 10 μl of 10 mg/ml proteinase K stock with 390 μl H_2O. Add 100 μl 10% Triton X-100. Make fresh.
8. *200 mM PMSF*: Make a fresh solution of PMSF by dissolving 34.85 mg of PMSF in 1 ml of ethanol.

Steps

1. Transfer 200 μl from the remainder of the import reaction into a fresh Eppendorf tube. Add 200 μl of the 200-μg/ml proteinase K solution and mix rapidly. Leave the tube on ice for 15 min.
2. Add 2 μl 200 m*M* PMSF while agitating on a vortex mixer. Keep on ice for 5 min. Add 44 μl 50% TCA. Add 300 μl of acetone to dissolve the Triton X-100 that precipitates in the presence of TCA. These samples measure the fraction of the precursor protein that has completely crossed the outer membrane and has reached the folded state (**folded**, Fig. 9.1, lanes 4 and 7).

G. Final Processing of Samples and Preparation of a Precursor Standard

Steps

1. To inactivate proteases, incubate the TCA-precipitated samples (total, import, folded) at 65°C for 5 min. Place on ice for 5 min and subsequently collect the TCA precipitate by spinning for 10 min at 20,000 *g*.
2. Remove supernatant by aspiration and dissolve the pellets in 30 μl 1× sample buffer. If the sample buffer turns yellow, overlay the sample with NH_3 gas taken from above a 25% NH_3 solution. Agitate to mix the gaseous NH_3 gas into the sample buffer until the color turns blue again.
3. Incubate the samples for 5 min at 95°C.
4. To estimate the efficiency of the import reaction, the amount of precursor protein added to the import reaction has to be determined. The efficiency of import for most precursor proteins is between 5 and 30%. Here we use a 10% standard (Fig. 9.1, lane 1).
5. To obtain a 10% standard, mix 4 μl of purified yeast mitochondria (see Section III,D) with 30 μl 1× sample buffer. Incubate at 95°C for 3 min.
6. Add 1 μl of the precursor protein solution (Section III,C) and incubate for 5 min at 95°C.

H. SDS–Gel Electrophoresis and Processing of the Gel

Solutions

1. *5% TCA*: To make 5 liter, add 250 g of TCA to 5 liter H_2O.
2. *1M Tris base*: Dissolve 121 g of Tris in 1 liter H_2O.
3. *1M sodium salicylate*: Dissolve 160 g of sodium salicylate in 1 liter H_2O.

Steps

1. Run samples on a 10% Tris–tricine gel (Schägger and von Jagow, 1987) stabilized by the addition of 0.26% linear polyacrylamide prior to polymerization.
2. To reduce radioactive background, boil 5% TCA in a beaker under the hood. Add the gel to the boiling TCA and incubate for 5 min.

3. Recover the gel and place it into a tray. Wash briefly with water. Neutralize by incubation in 1*M* Tris-base for 5 min on a shaker.
4. Wash briefly with water. Add 1*M* sodium salicylate and incubate for 20 min on a shaker.
5. Dry gel on a Whatman filter paper and expose to a Kodak X-OMAT X-ray film for the desired time. Exposure time for the experiment shown in Fig. 9.1 was 12 h.

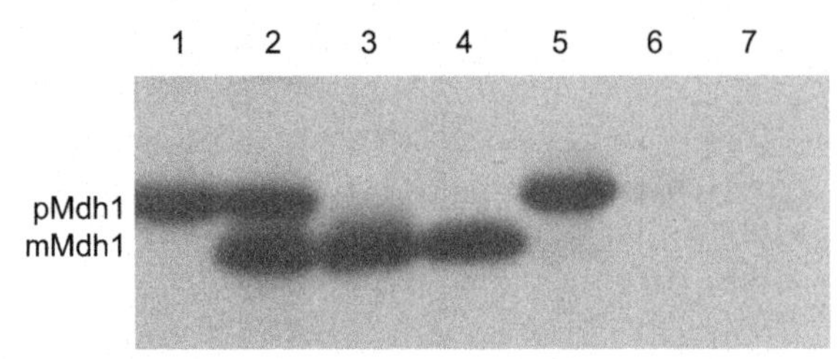

FIGURE 9.1 Import of radiolabeled yeast malate dehydrogenase into isolated yeast mitochondria. Lane 1, 10% of the material added to each import reaction; lanes 2–4, import in the presence of ATP and a membrane potential across the inner mitochondrial membrane; and lanes 5–7, import in the absence of ATP and a membrane potential across the inner mitochondrial membrane. Lanes 2 and 5: total; material isolated together with the mitochondrial pellet. Lanes 3 and 6: import; material protease protected in intact mitochondria. Lanes 4 and 7: folded; material protease resistant even after solubilization of the mitochondria with Triton X-100. pMdh1, precursor form of Mdh1; mMdh1, mature form of Mdh1. For experimental details, see text.

IV. COMMENTS

The method describes a standard experiment to test a precursor protein that has not been used in mitochondrial import before. Most importantly, as demonstrated here for malate dehydrogenase, the protocol will reveal if import is dependent on a membrane potential (compare Fig. 9.1 lanes 3 and 6). This is essential, as sometimes protease-resistant precursor proteins tend to stick to the outside of mitochondria, thereby "mimicking" import.

The efficiency of import can be deduced by a comparison of the amount of imported material with a precursor standard (compare Fig. 9.1 lanes 1 and 3). In addition, the experiment reveals if a precursor protein folds to a protease-resistant conformation after its import into the mitochondrial matrix. Under the conditions chosen here, complete protease resistance was obtained for malate dehydrogenase (Fig. 9.1, lanes 3 and 4).

V. PITFALLS

The quality of the DNA used for transcription is essential for efficiency. Use a clean, RNA- and RNase-free plasmid preparation (e.g., purified with a Qiagen plasmid kit, Qiagen). Linearize plasmid by cutting with a restriction enzyme behind the coding region of the gene of interest. Extract with phenol/chloroform and then with chloroform/isoamylalcohol, precipitate with 100% ethanol, and wash with 70% ethanol. Resuspend the dried pellet in H_2O at a concentration of 1 μg/μl and store at 4°C. Never freeze DNA templates used for transcription.

To avoid RNase contamination, solutions used for transcription and translation have to be prepared with special caution. Always wear gloves even when loading pipette tips into boxes. If initiation at downstream AUG codons is a problem, try diluting the reticulocyte lysate up to fourfold.

It is important to establish that import is linear with time. To establish those conditions it is necessary to perform time course experiments of the import reaction and to try import at different temperatures.

Methods for determining the intramitochondrial localization of an imported precursor protein (Glick, 1991), investigating the energy requirements of mitochondrial import (Glick, 1995), and detecting interaction between imported precursor proteins and matrix chaperones (Rospert and Hallberg, 1995) have been published elsewhere.

References

Dubaquié, Y., Looser, R., Fünfschilling, U., Jenö, P., and Rospert, S. (1998). Identification of *in vivo* substrates of the yeast mitochondrial chaperonins reveals overlapping but non-identical requirement for hsp60 and hsp10. *EMBO J.* **17**, 5868–5876.

Glick, B. S. (1991). Protein import into isolated yeast mitochondria. *Methods Cell Biol.* **34**, 389–399.

Glick, B. S. (1995). Pathways and energetics of mitochondrial protein import in *Saccharomyces cerevisiae. Methods Enzymol.* **260**, 224–231.

Glick, B. S., and Pon, L. A. (1995). Isolation of highly purified mitochondria from *Saccharomyces cerevisiae. Methods Enzymol.* **260**, 213–223.

Jensen, R. E., and Yaffe, M. P. (1988). Import of proteins into yeast mitochondria: The nuclear MAS2 gene encodes a component of the processing protease that is homologous to the MAS1-encoded subunit. *EMBO J.* **7**, 3863–3871.

Melton, D. A., Krieg, P. A., Rebagliati, M. R., Maniatis, T., Zinn, K., and Green, M. R. (1984). Efficient *in vitro* synthesis of biologically active RNA and RNA hybridization probes from plasmids containing a bacteriophage SP6 promoter. *Nucleic Acids Res.* **12**, 7035–7056.

Neupert, W. (1997). Protein import into mitochondria. *Annu. Rev. Biochem.* **66**, 863–917.

Pfanner, N., and Geissler, A. (2001). Versatility of the mitochondrial protein import machinery. *Nature Rev. Mol. Cell. Biol.* **2**, 339–349.

Rospert, S., and Hallberg, R. (1995). Interaction of HSP 60 with proteins imported into the mitochondrial matrix. *Methods Enzymol.* **260**, 287–292.

Schägger, H., and von Jagow, G. (1987). Tricine-sodium dodecyl sulfate-polyacrylamide gel elektrophoresis for the separation of proteins in the range from 1 to 100 kDa. *Anal. Biochemi.* **166**, 368–379.

Wachter, C., Schatz, G., and Glick, B. S. (1994). Protein import into mitochondria: The requirement for external ATP is precursor-specific whereas intramitochondrial ATP is universally needed for translocation into the matrix. *Mol. Biol. Cell.* **5**, 465–474.

CHAPTER

10

Polarographic Assays of Mitochondrial Functions

Ye Xiong, Patti L. Peterson, and Chuan-pu Lee

OUTLINE

I. INTRODUCTION

Mitochondrial metabolic functions are primarily concerned with the conservation of energy liberated by the aerobic oxidation of respiring substrates. Energy is conserved in the form that can subsequently be used to drive various cell functions. In recent years, it has been found that mitochondria also play critical roles in various cellular functions, including electrolyte balance, signal transduction, calcium homeostasis, oxidative stress, immunologic defense, and natural aging and/or apoptosis (cf. Lee, 1994).

Application of the oxygen electrode technique to study mitochondrial respiration and oxidative phosphorylation was first introduced by Chance and Williams in 1955. The kinetic dependence of mitochondrial respiration on the availability of inorganic phosphate and ADP had been earlier reported by Lardy and Wellman (1952). Together, these studies provided the basis of the concept of respiratory control. The polarographic technique for measuring rapid changes in the rate of oxygen utilization by cellular and subcellular systems is now widely used by many laboratories because of its simplicity and ease of performance. Concurrent monitoring of the metabolic changes induced kinetically by various substrates and reagent(s) can provide invaluable information.

This article describes a typical protocol employed in the authors' laboratory for the measurement of respiratory rates and its accompanied oxidative phosphorylation, which are catalyzed by isolated mitochondria. As an example, isolated intact mitochondria were chosen for presentation and discussion. The technique for preparation of isolated intact mitochondria has been described previously (Lee *et al.*, 1993a, b).

II. INSTRUMENTATION AND MATERIALS

The polarographic technique for measuring mitochondrial oxidative changes requires the following four basic components: an oxygen electrode, a closed reaction vessel, a constant voltage source, and a recorder. The Clark-type oxygen electrode is used most commonly (Yellow Springs Instrument Co., Yellow Springs, Ohio, 45387) and consists of a platinum cathode and a Ag/AgCl anode, which is bathed in a half-saturated KCl solution. The tip of the electrode is covered by a polyethylene membrane, which is held firmly over the end of the electrode by a rubber "O" ring. The reaction vessel can be constructed of glass, Plexiglas, or polycarbonate, and the design and the size of the vessel vary, depending on the requirements of the system under investigation. A number of vessels are available commercially. The reaction vessel used routinely in the authors' laboratory was made of polycarbonate and is composed of two parts: a cylindrical open top reaction chamber and a plug that fits in the top of the chamber. The temperature of the reaction chamber is constantly maintained with a water jacket. Constant stirring of the contents of the reaction chamber is accomplished by a disk-shaped, plastic-encased magnet located at the bottom of the reaction chamber. The polycarbonate plug, which is secured with a screw, fits tightly in the top of the reaction chamber. The plug has two vertical openings: the larger is fitted with a screw that holds the Clark electrode securely and the smaller one is used for the delivery of substrate(s) and reagent(s) into the reaction mixture. The physical arrangement of the reaction chamber and the oxygen electrode in the assembled form are shown in Fig. 10.1A. The disassembled form, which shows the details of each individual component, is shown in Fig. 10.1B.

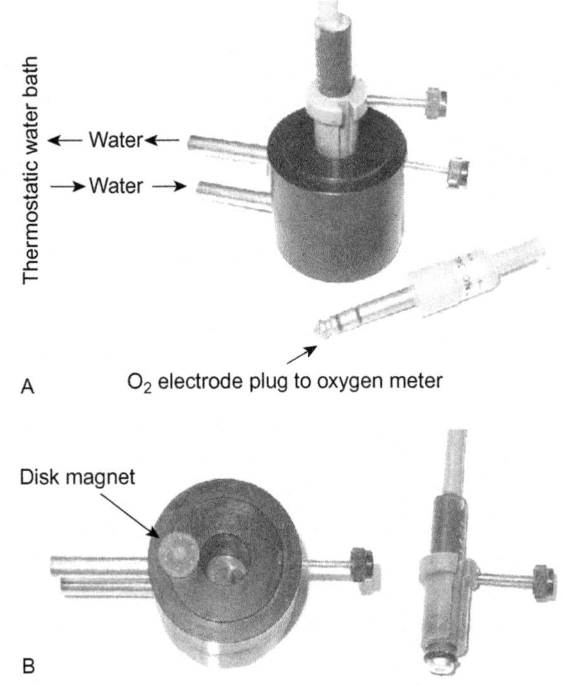

FIGURE 10.1 Reaction vessel for measurement of oxygen utilization with the oxygen electrode. Oxygen electrode and the reaction chamber are displayed in assembled form (A) and in disassembled form (B).

It is imperative that closed reaction vessels are utilized so that air bubbles are not trapped and back diffusion of oxygen is reduced to a minimum. To facilitate this, the bottom of the plug is uneven to ensure that no air bubbles are trapped (cf. Fig. 10.1B). This aids the exit of air bubbles (if any) from the chamber and allows the mixing of added reagents without introducing air bubbles into the reaction mixture. Use of the magnetic stirrer permits continued steady mixing, at a constant temperature, of the reaction mixture. This facilitates the establishment of equilibrium between the oxygen dissolved in solution and the gas diffusing through the polyethylene membrane of the oxygen electrode. The electrode equilibrates with the air-saturated reaction mixture. Reactants and substrates are added through the small narrow opening at the top of the reaction chamber.

At the conclusion of each experiment the reaction chamber is cleaned by first disassembling the plug from the reaction chamber, aspirating the reaction mixture, and washing the plug, oxygen electrode, and the interior of the reaction chamber thoroughly with water. When a water-insoluble reagent is used, initial cleaning is accomplished with ethanol and is then followed with water rinsing. It is essential that no traces of ethanol are left behind as adverse effects may be induced. The reaction vessel is then reassembled and is ready for the next experiment.

When a voltage is imposed across the two electrodes immersed in an oxygen-containing solution, with the platinum electrode negative relative to the reference electrode, oxygen undergoes an electrolytic reduction. When current is plotted as a function of polarizing voltage, a plateau region is observed between 0.5 and 0.8 V. With a polarization voltage of −0.6 V, the current is directly proportional to the oxygen concentration of the solution (Davies and Brink, 1942). The current is generally measured with a suitable amplifier and recorder combination. An oxygen meter (SOM-1, University of Pennsylvania Biomedical Instrumentation Group) and a Varian XY recorder (Model 9176) are used in our laboratory.

Reagents

From Sigma Chemicals Co.: Sucrose (Cat. No. S-978), bovine serum albumin (BSA, Cat. No. A-4378), EGTA (Cat. No. E-4378), EDTA (Cat. No. ED2SS), adenosine 5′-diphosphate (ADP, Cat. No. A-6646), l-malic acid (Cat. No. M-1000), pyruvic acid (Cat. No. P-2256), β-NADH (Cat. No. 340-110), HEPES (Cat. No. H-3375), carbonyl cyanide 4-trifluoromethoxy-phenylhyddrazone (FCCP, Cat. No. C22920), antimycin A (Cat. No. A 8674), oligomycin (Cat. No. O-4876), Tris (Cat. No. T-1503), crystalline *Bacillus subtilis* protease (Nagarse) from Teikoku Chemical Company, Osaka, Japan, or Sigma P-4789, protease type XXVII (7.6 units/mg solid), phospho(enol)pyruvic acid (PEP, Cat. No. P-0564), pyruvate kinase (Cat. No. P-9136), β-hydroxybutyric acid (Cat. No. H 6501), and lactic dehydrogenase (LDH, Cat. No. L-2375).

From other sources: $MgCl_2 \cdot 6H_2O$ (M-33, Fisher Scientific), KH_2PO_4 (P-285, Fisher Scientific), HCl [UN1789, 36.5% (w/w), Fisher Scientific], K_2HPO_4 (3252-2, T. J. Baker), and KCl (No-3040, T. J. Baker), KOH (UN1813, Fisher Scientific), KCN (Lot 6878KBPH, Mallinckrodt, Inc., KT 40361).

III. PROCEDURES

A. Preparation of Solutions

1. *0.25 M sucrose*: Dissolve 171.15 g of sucrose in 1000 ml distilled water, filter through a layer of glass wool, add distilled water to 2 liters in a volumetric flask, mix well, and store at 4°C.
2. *0.5 M Tris–HCl, pH 7.4*: Dissolve 30.28 g Tris into 400 ml distilled water, adjust pH to 7.4 with 4 *N* HCl, add distilled water to 500 ml

in a volumetric flask, mix well, adjust the pH if necessary, and store at 4°C.

3. *0. 5 M $MgCl_2$*: Dissolve 10.17 g of $MgCl_2 \cdot 6H_2O$ in 100 ml distilled water.
4. *0.1 M phosphate buffer, pH 7.4*: To 800 ml distilled water, dissolve 11.83 g of Na_2HPO4 and 2.245 g of KH_2PO4 with the aid of magnetic stirrer. Check the pH and adjust with 0.1 *M* HCl or 0.1 *M* NaOH if necessary. Add distilled water to 1000 ml in a volumetric flask, mix well, and store at 4°C.
5. *0.2 M EDTA, pH 7.4*: Dissolve 3.72 g EDTA in 35 ml of 1 *N* NaOH. Stir until it dissolves completely; adjust pH to 7.4 with 1 *N* HCl. Add distilled water to 50 ml. Divide into 10 × 5.0 ml and store at −20°C.
6. *0.1 M HEPES, pH 7.4*: Dissolve 23.83 g HEPES in 800 ml distilled water. Adjust with 1 *M* NaOH to pH 7.4. Add distilled water to 1000 ml in a volumetric flask. Mix well and store at 4°C.
7. *Sucrose/Tris–Cl (S/T) reaction medium*: 60 ml 0.25 *M* sucrose, 5 ml 0.5 *M* Tris–HCl (pH 7.4), and 35 ml distilled water to a final volume of 100 ml. Make the S/T medium fresh every day.
8. *1.0 M pyruvate, pH 7.4*: Dissolve 5.5 g pyruvic acid in 20 ml of 1 *N* NaOH; adjust with 5 *N* NaOH to pH 7.4. Add distilled water to 50 ml in a volumetric flask. Mix well, transfer into 10 test tubes with each containing 5 ml, and store at −20°C.
9. *0.5 M malate, pH 7.4*: Dissolve 3.35 g of dl-malic acid in 20 ml of 1 *N* NaOH; adjust with 5 *N* NaOH to pH 7.4. Add distilled water to 50 ml in a volumetric flask. Mix well and transfer into 10 test tubes and store at −20°C.
10. *0.1 M ADP, pH 6.8–7.4*: Dissolve 0.48 g ADP in 5 ml distilled water, adjust with 1 *N* NaOH to pH to 6.8–7.4, and add distilled water to 10 ml in a volumetric flask. Mix well, transfer into five test tubes (• 2 ml/tube), and store at −20°C. The concentration of ADP will be determined spectrophotometrically with the pyruvate kinase coupled with lactate dehydrogenase system (see later).
11. *0.1 M phosphoenolpyruvate (PEP), pH. 7.4*: Dissolve 20.8 mg PEP into 0.6 ml 1 *N* KOH, adjust pH to 7.4 with 5 *N* KOH, and add distilled water to 1.0 ml. Mix well, transfer into five vials (0.2 ml/vial), and store at −20°C.
12. *1.0 mg/ml oligomycin*: Dissolve 10 mg oligomycin in 10 ml absolute ethanol. Store at −20°C.
13. *FCCP 1 mM*: Dissolve 5.03 mg FCCP into 20 ml absolute ethanol in a glass tube (either dark glass or covered with aluminum foil to protect against light) and store at −20°C.

B. Enzymatic Assay of ADP Concentration (Jaworek *et al.*, 1974)

This assay is based on a pair of coupled reactions. ADP in the presence of an excess amount of (PEP) and pyruvate kinase will be completely converted into ATP and an equal concentration of pyruvate. The pyruvate will then be converted to lactate in the presence of lactate dehydrogenase and an excess amount of NADH. The NAD^+ formed is equal to the concentration of lactate. The stoichiometry of ADP to NADH bears a 1:1 relationship. The concentration of ADP is therefore equal to that of NADH oxidized. See the following reaction mixture: 2.8 ml S/T medium, 50 μl pyruvate kinase (3 IU), 10 μl lactic dehydrogenase (1 IU), 15 μl NADH (30 m*M*), 20 μl PEP (0.1 *M*), 80 μl $MgCl_2$ (0.25 *M*), and 20 μl KCl (1.0 *M*). The total volume equals 3.0 ml.

Calibrate the spectrophotometer and the settings of the recorder according to the instruction manuals of the instrument and set the wavelength at 340 nm.

1. Place the sample cuvette in the sample chamber with 150 μ*M* NADH (15 μl, 30 m*M*) present; an absorbance in the vicinity of

0.9 is indicated on the instrument and the recorder.

2. Turn on the recorder, let it proceed to a constant reading (a minute or two), add 3 μl of 0.1 *M* ADP, and stir the solution well with a glass or plastic stirrer; the absorbance at 340 nm will decline quickly (when sufficient pyruvate kinase and lactate dehydrogenase are present) followed by a sharp transition to a constant level where there are no more changes in 340 nm absorbance. The decline of 340 nm absorbance reflects the oxidation of NADH. The extent of absorbance change (i.e., $\Delta A = 0.63$) is dependent on the amount of ADP added into the cuvette.
3. The concentration of ADP can therefore be calculated from the absorbance changes at 340 nm. The millimolar absorbance coefficient for NADH is $6.22\,mM^{-1}cm^{-1}$. The ADP concentration = $[0.63/6.22] \times [3.0/0.003] = 101\,mM$.

C. Calibration of Oxygen Concentration

A simple and rapid method of determining the oxygen content of the reaction medium accurately is by using submitochondrial particles (SMP) with a limiting amount of NADH. The high affinity of SMP for NADH permits a stoichiometric titration of oxygen content. NADH concentration can be determined accurately spectrophotometrically (as shown in the previous section). When limiting concentrations of NADH are added, the change in current, which occurs with complete oxidation of the NADH, can be determined directly. A direct calibration can therefore be obtained. For example, add 0.98 ml of air-saturated S/T medium first followed by 10 μl SMP (0.2 mg protein) to the reaction chamber. Allow the reaction mixture to thermo equilibrate; no air bubbles should be trapped in the reaction vessel. Add 5 μl NADH (150 μ*M*) to the chamber. Immediately initiated oxygen uptake with a linear kinetics, followed by a sharp transition to a straight line (i.e., oxygen uptake ceases) when all the added NADH is oxidized. This assay can also be used to estimate of back diffusion of oxygen into the reaction mixture. For instance, if there is a back diffusion of oxygen, an increase in oxygen concentration in the reaction mixture will be noted on the left deflection (a negative slop) of the recorder tracing.

D. Determination of Rate of Oxygen Consumption and ADP/O Ratio of Intact Mitochondria

Figure 10.2A shows a recording tracing from a typical experiment utilizing the oxygen electrode apparatus of the type shown in Fig. 10.1 to determine the respiratory control index (RCI) and P/O ratio of a tightly coupled mitochondrial preparation. Suspend freshly prepared mouse skeletal muscle mitochondria (30–50 μl) in an air-saturated P_i-containing isotonic sucrose medium (0.9 ml); the addition of substrates [5 μl pyruvate (1*M*) + 5 μl malate (0.5*M*)] causes a slow rate of oxygen uptake (state 4). Subsequent addition of 3 μl ADP (100 m*M*) increases the rate (state 3) by 11-fold. Upon the expenditure of the added ADP, as indicated by the sharp transition in the polarographic tracing, the rate of oxygen uptake declines to that observed before the addition of ADP. The duration of the increased rate of oxygen uptake is proportional to the amount of ADP added to the reaction mixture. The concentration of oxygen consumed is proportional to the amount of ADP phosphorylated to ATP. These cycles of stimulation of respiration by the addition of ADP could be repeated several times in a single experiment until all the oxygen in the reaction mixture is consumed (Fig. 10.2A). The ADP/O [P/O] ratio and RCI can be calculated directly from the oxygen electrode tracing. The dependence of substrate oxidation on the presence of ADP indicates that electron transfer of the respiratory chain and

ATP synthesis are coupled to each other. Unless the energy generated during electron transfer is utilized for ATP synthesis or other energy-requiring processes, the oxidation of respiring substrate is restrained.

The recorder deflection "X" (165 mm) represents the total oxygen content of the reaction medium (e.g., 240 μ*M* O_2 at 30°C). Distance "*Y*" is determined by extending the slops of lines A, B, and C and measuring the length of the tracing from the intersects of lines A and B and lines B and C. "*Y*" (34 mm) represents the amount of O_2 utilized by the externally added ADP (300 μ*M*). The ADP/O ratio can therefore be calculated by dividing the amount of ADP by the amount of oxygen utilized. The amount of oxygen utilized can be calculated as follows: Total oxygen content {[240 μ*M*] ÷ "*X*" [165 mm]} × "*Y*" [34 mm] × volume of the reaction mixture [1.0 ml] × 2 = 99 mμatoms oxygen utilized. The amount of ADP was added at the point indicated, e.g., 300 μ*M* ADP or 300 mμmol ADP. The ADP/O ratio is therefore = 300 / 99 = 3.01.

The respiratory control index is defined as the ratio of respiratory rate in the presence of added ADP (state 3) to the rate either before ADP addition or the rate following ADP expenditure (state 4). State 3 and state 4 respiratory rates are 79 and 7 mm/min, respectively. Therefore, the RCI is 79/7 = 11.4. For tightly coupled intact mitochondria as shown in Fig. 10.2A, the respiratory rates before the addition of ADP and then following the expenditure of ADP are virtually identical. However, if mitochondria are loosely coupled and/or contaminated with ATPase, the respiratory rate following the expenditure of ADP is considerable faster than that with substrate alone before the addition of ADP. The quality of the mitochondrial preparation is readily reflected from the polarographic tracing.

Tightly coupled intact mitochondria exhibit a very low (if any) ATPase activity (basal ATPase) that can be stimulated by Mg^{2+} and/or uncouplers. Figure 10.2B illustrates the effects of energy transfer inhibitors and uncoupler on the respiratory rates of intact mitochondria. The best-known energy transfer inhibitors are oligomycin and aurovertin, which primarily inhibit the phosphorylation reaction (e.g., synthesis and hydrolysis of ATP). Electron transfer is inhibited secondarily because of the tight link between oxidation and phosphorylation. Uncoupling agents also prevent the synthesis of ATP, but differ from energy transfer inhibitors in that they do not inhibit phosphorylation directly but rather uncouple phosphorylation from respiration by dissipating the energy generated by substrate oxidation as waste. Thus no energy is available for the synthesis of ATP from ADP and P_i, or any other energy utilizing processes. Uncouplers do not inhibit ATP hydrolysis. In the presence of an uncoupler, respiration proceeds at its maximal rate even in the absence of ADP and P_i. The most commonly used uncouplers are dinitrophenol (DNP) and carbonyl cyanide *p*-trifluoromethoxyphenylhydrazone (FCCP).

As shown in Fig. 10.2B, The ADP-induced state 3 respiratory rate declined to the level comparable to state 4 upon the addition of oligomycin. The oligomycin-induced inhibition was released by the addition of FCCP. The FCCP-induced rate (usually referred to as state 3U) is virtually identical to that at state 3.

IV. COMMENTS, PITFALL, AND RECOMMENDATIONS

1. It is imperative that glass redistilled water be used throughout all the experiments.
2. A closed reaction vessel free of air bubbles and back diffusion of oxygen are essential for the success of polarographic assay of mitochondrial function.
3. It is imperative that the excessive part of the polyethylene membrane, which covers the electrode after secured by a rubber O ring, be removed with a sharp blade, as

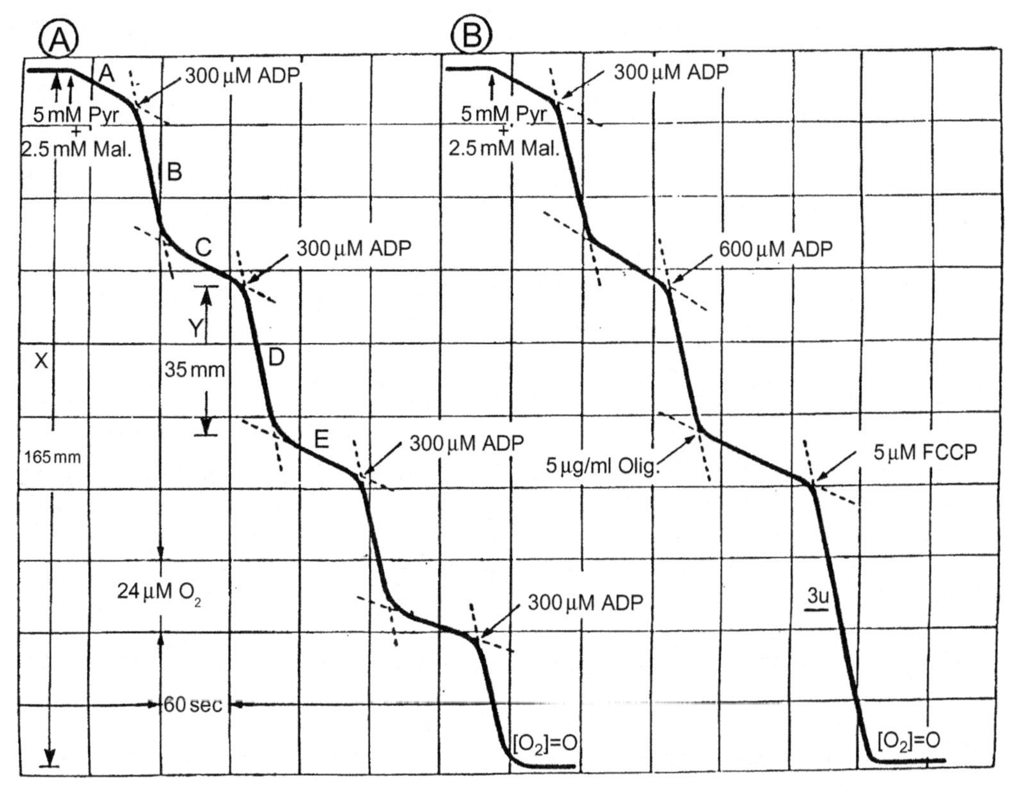

FIGURE 10.2 Polarographic tracing of mouse skeletal muscle mitochondria oxidizing pyruvate plus malate. The reaction mixture consists of 150 m*M* sucrose, 25 m*M* Tri–HCl, and 10 m*M* phosphate, pH 7.5. Additions are as indicated. Volume, 1.0 ml; temperature, 30°C.

the presence of excessive polyethylene membrane may trap materials and make cleaning difficult.

4. All aqueous solutions of medium and reagent should be at neutral pH (i.e., 7.2–7.4). During the assays the temperature of the reaction mixture has to be maintained constant.
5. When water-insoluble reagents are used, at the completion of the experiment, the reaction chamber needs to be cleaned thoroughly first with ethanol followed by distilled water. Contamination of either reagent and/or ethanol of the chamber will result in obscure data, which cannot be interpreted.
6. Tarnish of the Ag/AgCl electrode can be removed by cleaning with a cotton-tipped swab dipped in 4 *N* NH_4OH aqueous solution followed by rinsing thoroughly with distilled water.
7. The polarographic oxygen electrode technique is a convenient method for the determination of P/O ratio and RCI. However, its suitability is dependent on the quality of the mitochondrial preparations. Meaningful results can only be derived from tightly coupled mitochondrial preparations that are free from contamination by other cellular constituents. Consideration of the tissue constituents, i.e., lipid content, and properties will aid in the design of isolation and purification procedures. For instance, as compared to liver and heart, only a small

portion of the total mass of skeletal muscle consists of mitochondria, with the bulk of the tissue being myofibrils. Additionally, skeletal muscle contains a relatively high content of Ca^{2+}, which is capable of damaging mitochondria during the isolation process. Because of the high content of myofibril, the separation of mitochondria from other components is very difficult in nonelectrolyte medium (e.g., the isotonic sucrose used to isolate mitochondria from liver and heart). A similar difficulty results when attempts are made to isolate mitochondria from brain. Because of its high lipid content and high rate of aerobic metabolism, improper isolation may result in data fraught with artifact. In recent years, the polarographic technique has been widely used to evaluate the lesions of mitochondrial functions of skeletal muscle derived from patients suffering from mitochondrial myopathy. A reproducible technique that isolates tightly coupled, intact mitochondria is essential before statements regarding etiology can be made. Impairments in mitochondrial function may actually result from an artifact generated during the isolation procedure rather than a genuine impairment of mitochondrial function caused by the disease.

After many years of experience in our laboratory we have designed a simple method for the isolation and purification of intact skeletal muscle mitochondria (Lee *et al.*, 1993a). This method allows preparation of isolated, intact mitochondria from a small amount (1–3g, wet weight) of muscle biopsy specimens. Additionally, we have designed a simple and fast procedure to isolate intact brain mitochondria from one hemisphere of rat brain (approximately 0.6g wet weight) (Lee *et al.*, 1993b; Xiong *et al.*, 1997a) with excellent yield and stability. This technique has been applied successfully on studies of the impairment of mitochondrial functions induced by head trauma (Xiong *et al.*, 1997a,b, 1998, 1999) and ischemia (Sciamanna *et al.*, 1992; Sciamanna and Lee, 1993).

References

Chance, B., and Williams, G. R. (1955). Respiratory enzymes in oxidative phosphorylation I: Kinetics of oxygen utilization. *J. Biol. Chem.* **217**, 383–393.

Davies, P. W., and Brink, F. J. (1942). Microelectrodes for measuring local oxygen tension in animal tissues. *Rev. Sci. Instr.* **13**, 524–533.

Jaworek, D., Gruber, W., and Bergmeyer, H. U. (1974). Adenosine-5′-diphosphate and adenosine-5′-monophosphate. *In "Methods of Enzymatic Analysis,"* 2nd ed., Vol. 4, pp. 2027–2029.

Lardy, H. A., and Wellman, H. (1952). Oxidative phosphorylations: Role of inorganic phosphate and acceptor systems in control of metabolic rates. *J. Biol. Chem.* **195**, 215–224.

Lee, C. P. (ed.) (1994). Molecular basis of mitochondrial pathology. *Curr. Top. Bioenerg.* **17**, 254.

Lee, C. P., Martens, M. E., and Tsang, S. H. (1993a). Small scale preparation of skeletal muscle mitochondria and its application in the study of human disease. *Methods Toxicol.* **2**, 70–83.

Lee, C. P., Sciamanna, M. A., and Peterson, P. L. (1993b). Intact brain mitochondria from a single animal: Preparation and properties. *Methods Toxicol.* **2**, 41–50.

Sciamanna, M. A., and Lee, C. P. (1993). Ischemia/reperfusion-induced injury of forebrain mitochondria and prevention by ascorbate. *Arch. Biochem. Biophys.* **305**, 215–224.

Sciamanna, M. A., Zinkel, J., Fabi, A. Y., and Lee, C. P. (1992). Ischemia injury to rat forebrain mitochondria and cellular calcium homeostasis. *Biochim. Biophys. Acta* **1134**, 223–232.

Xiong, Y., Gu, Q., Peterson, P. L., Muizelaar J. P., and Lee C. P. (1997a). Mitochondrial dysfunction and calcium perturbation induced by traumatic brain injury. *J. Neurotrauma* **14**, 23–34.

Xiong, Y., Peterson, P. L., and Lee, C. P. (1999). Effect of N-acetylcysteine on mitochondrial function following traumatic brain injury in rats. *J. Neurotrauma* **16**, 1067–1082.

Xiong, Y., Peterson, P. L., Muzelaar, J. P., and Lee, C. P. (1997b). Amelioration of mitochondrial dysfunction by a novel antioxidant U-101033E following traumatic brain injury. *J. Neurotrauma* **14**, 907–917.

Xiong, Y., Peterson, P. L., Verweij, B. H., Vinas, F. C., Muzelaar, J. P., and Lee, C. P. (1998). Mitochondrial dysfunction after experimental traumatic brain injury: Combined efficacy of SNX-111 and U-101033E. *J. Neurotrauma* **15**, 531–544.

CHAPTER

11

Analysis of Nuclear Protein Import and Export in Digitonin-Permeabilized Cells

Ralph H. Kehlenbach, and Bryce M. Paschal

OUTLINE

I. INTRODUCTION

Transport between the nucleus and the cytoplasm is mediated by nuclear pore complexes (NPCs), specialized channels that are embedded in the nuclear envelope membrane. Proteins that undergo nuclear import or nuclear export usually encode a nuclear localization signal (NLS) or a nuclear export signal (NES). These signals are recognized by import or export receptors that, in turn, facilitate targeting of the protein to the NPC and translocation through the central channel of the NPC. Both import and export pathways are regulated by the Ras-related GTPase Ran. Nuclear Ran in its GTP-bound form promotes the release of NLS-proteins from import receptors after the NLS-protein/import receptor has reached the nuclear side of the NPC. In contrast, nuclear Ran in its GTP-bound form promotes the assembly of NES-containing proteins with export receptors by forming an NES-protein/export receptor/RanGTP. Export complexes are then disassembled on the cytoplasmic side of the NPC through the action of a Ran GTPase-activating protein that stimulates GTP hydrolysis by Ran. More specific information regarding import and export complex assembly, models for translocation, and additional aspects of nuclear transport regulation are described elsewhere (Steggerda and Paschal, 2002; Weis, 2003 and references therein).

Digitonin-permeabilized cells have become one of the most widely used experimental systems for studying nuclear transport (Adam *et al.*, 1990). They have been used to analyze nuclear import and export signals (Pollard *et al.*, 1996), to purify nuclear transport factors from cell extracts (Görlich, 1994; Paschal and Gerace, 1995; Kehlenbach *et al.*, 1998), and to measure nuclear transport kinetics (Görlich and Ribbeck, 2003). In all of these applications, the principle is that treating mammalian cells with a defined concentration of digitonin results in selective perforation of the plasma membrane, leaving the nuclear membrane intact. Thus, soluble transport factors are released from the cytoplasmic compartment and low molecular weight transport factors such as Ran (24 kDa) and NTF2 (28 kDa) are released from the nuclear compartment by diffusion through the NPC. Nuclear import and export can subsequently be reconstituted in digitonin-permeabilized cells by the addition of transport factors, an energy-regenerating system, and inclusion of a fluorescent NLS or NES reporter protein to monitor transport (Adam *et al.*, 1990). Transport factors can be supplied as unfractionated cytosol from HeLa cells or as commercially available rabbit reticulocyte lysate. The level of import is then measured by following the accumulation of an NLS reporter in the nucleus or, in the case of export, by the loss of an NES reporter from the nucleus.

This article describes assays for both nuclear import and export (Fig. 11.1). The nuclear import assay can be performed with various cell lines and analyzed by fluorescence microscopy. The nuclear export assay is described for HeLa cells expressing a GFP-tagged reporter protein, the nuclear factor of activated T cells (GFP-NFAT). NFAT is a transcription factor that contains defined nuclear localization and nuclear export signals and that shuttles between the cytoplasm and the nucleus in a phosphorylation-dependent manner (for review, see Crabtree and Olsen, 2002). We take advantage of the tight regulation of nucleocytoplasmic transport of GFP-NFAT to induce nuclear import in intact cells, followed by export under controlled conditions from nuclei of permeabilized cells (Kehlenbach *et al.*, 1998). The efficiency of nuclear export can be analyzed either qualitatively by fluorescence microscopy (see Fig. 11.3) or quantitatively by flow cytometry (see Fig. 11.4). In the latter case, nuclear import of a fluorescently labeled reporter protein can be analyzed in parallel.

II. MATERIALS AND INSTRUMENTATION

Rabbit recticulocyte lysate (L4151) is from Promega. HeLa S3 cells (CCL-2.2) and NIH 3T3 cells are from the American Type Culture Collection. Joklik's modified minimum essential medium for suspension cell culture (JMEM; Cat. No. M0518) is from Sigma. Newborn calf serum (Cat. No. 16010-159), penicillin–streptomycin (Cat. No. 15140-122), and trypsin EDTA (Cat. No. 25200-056) are from GIBCO/Invitrogen. ATP (Cat. No. A-2383), creatine phosphate (Cat. No. P-7936), creatine phosphokinase (Cat. No. C-7886), sodium bicarbonate (Cat. No. S-4019), HEPES (Cat. No. H-7523), potassium acetate (Cat. No. P-5708), magnesium acetate (Cat. No. M-2545), EGTA (Cat. No. E-4378), dithiothreitol (DTT, Cat. No. D-5545), NaCl (Cat. No. S-9763), potassium phosphate, monobasic (Cat. No. P-0662), phenylmethylsulfonyl fluoride (PMSF, Cat. No. P-7626), dimethyl sulfoxide (DMSO, Cat.

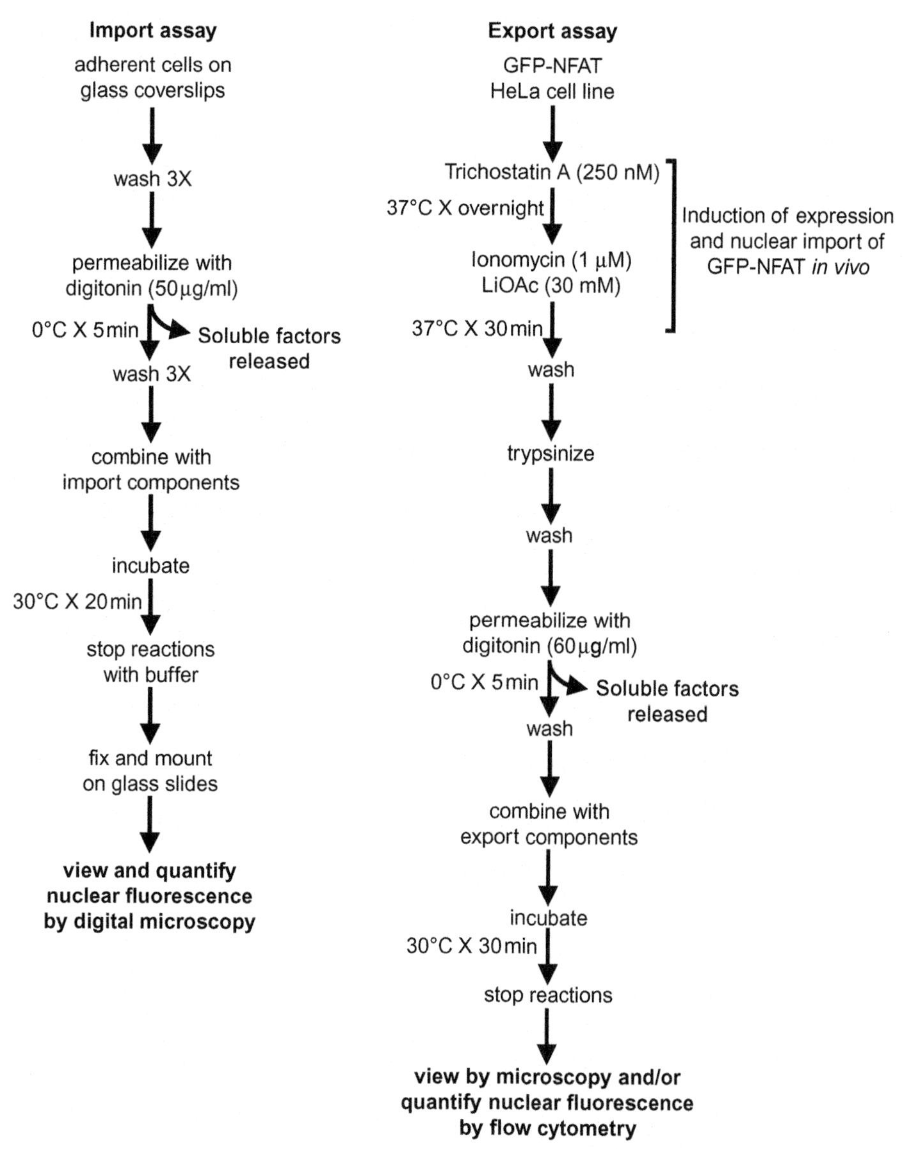

FIGURE 11.1 Overview of assays for measuring nuclear import and export in digitonin-permeabilized cells. See text for details.

No. D-8779), sodium carbonate (Cat. No. S-7795), trypan blue (Cat. No. T-8154), ionomycin (Cat. No. I-0634), LiOAc (Cat. No. L-4158), Hoechst 33258 (Cat. No. B-2883), 4′,6-diamidino-2-phenyindole, dilactate (DAPI; Cat. No. 9564), and trichostatin A (Cat. No. T-8552) are from Sigma. High-purity bovine serum albumin (BSA, Cat. No. 238031), aprotinin (Cat. No. 236624), leupeptin (Cat. No. 1017101), and pepstatin (Cat. No. 253286) are from Roche. The FITC isomer I (Cat. No. F-1906) and sulfo-SMCC (Cat. No. 22322) are from Molecular Probes and Pierce, respectively.

PD-10 columns (Cat. No. 17-0851-01), Cy2 (Cat. No. PA-22000), and Cy5 (Cat. No. PA-25001) are from Amersham Pharmacia Biosciences. Centricon filters (30-kDa cutoff, Cat. No. 4208) are from Amicon. High-purity digitonin (Cat. No. 300410) is from Calbiochem. Vectashield mounting medium (Cat. No. H-1000) is from Vector Laboratories. Six-well dishes (Cat. No. 3516) are from Corning. Glass coverslips (#1 thickness; Cat. No. 12-548A) and the polystyrene tubes used for the transport assays (Cat. No. 2058) are from Fisher. The microscope used for measuring nuclear import in adherent cells is a Nikon E800 equipped with a charge-coupled device camera. The system is linked to a MacIntosh computer running OpenLab software for image acquisition. An Olympus IX70 inverted fluorescence microscope is used for analysis of nuclear export. Images from import and export assays are processed using Adobe Photoshop. The flow cytometric analysis system is the FACScan unit from Beckton-Dickinson.

The following equipment required for growing and harvesting HeLa cells is from Bellco Glass. The 250-ml spinner flask (Cat. No. 1965-00250) and stir plate (Cat. No. 7760-0600) are used for the continuous culture of HeLa cells. Additional spinner flasks are required to scale up the preparation to 15 liters. These include 100 ml (Cat. No. 1965-01000), 3000 ml (Cat. No. 1965-03000), and 15 liter (Cat. No. 7764-00110), cap assembly (Cat. No. 7764-10100), and Teflon paddle assembly (Cat. No. 1964-30015).

Equipment for centrifugation includes the JS5.2 swinging bucket rotor and J6B centrifuge, the JA-20 fixed angle rotor and J2 centrifuge, and the type 60 Ti fixed angle rotor and L7 centrifuge, all from Beckman. Cytosol dialysis is carried out using the collodion vacuum dialysis apparatus (Cat. No. 253310), and 10,000-Da cutoff membranes (Cat. No. 27110) are from Schleicher and Schuell. The homogenizer used for cell disruption is a 0.02-mm-clearance stainless-steel unit (Cat. No. 885310-0015) from Kontes.

III. PROCEDURES

A. Preparation of Cytosol

The rabbit reticulocyte lysate contains all the soluble factors necessary to reconstitute import and export in digitonin-permeabilized cells (Adam *et al.*, 1990). The only preparation involved is dialysis against 1× transport buffer containing 2 m*M* DTT and 1 μg/ml each of aprotinin, leupeptin, and pepstatin (two buffer changes). Reticulocyte lysate is used at 50% (by volume) of transport reactions. This results in a relatively high total protein concentration (~25–40 mg/ml) in transport assays because the reticulocyte lysate contains a high concentration of hemoglobin. A more economical approach, especially if larger quantities of cytosol are needed for protein purification, is to prepare cytosol from suspension culture HeLa cells. In this case, only 1–2 mg/ml (final assay concentration) of HeLa cell cytosol is required to obtain a maximum level of protein import or protein export in digitonin-permeabilized cells.

Solutions

1. *10× transport buffer*: 200 m*M* HEPES, pH 7.4, 1.1*M* potassium acetate, 20 m*M* magnesium acetate, and 5 m*M* EGTA. To make 1 liter, dissolve 47.6 g HEPES, 107.9 g potassium acetate, 4.3 g magnesium acetate, and 1.9 g EGTA in 800 ml distilled water. Adjust the pH to 7.4 with 10 *N* NaOH and bring the final volume to 1 liter. Sterile filter and store at 4°C.
2. *1× transport buffer*: 20 m*M* HEPES, pH 7.4, 1.1*M* potassium acetate, 2 m*M* magnesium acetate, and 0.5 m*M* EGTA. To make 1 liter, add 100 ml of 10× transport buffer to 900 ml distilled water.
3. *1M HEPES stock, pH 7.4*: To make 500 ml, dissolve 119.1 g HEPES (free acid) in 400 ml distilled water. Adjust the pH to 7.4 with 10 *N* NaOH and bring the

final volume to 500 ml. Sterile filter and store at 4°C.

4. *1M potassium acetate stock*: To make 500 ml, dissolve 49 g potassium acetate in 400 ml distilled water. Bring the final volume to 500 ml, sterile filter, and store at 4°C.
5. *1M magnesium acetate stock:* To make 500 ml, dissolve 107.2 g magnesium acetate (tetrahydrate) in 400 ml distilled water. Bring the final volume to 500 ml, sterile filter, and store at 4°C.
6. *0.2M EGTA stock*: To make 500 ml, dissolve 38 g EGTA (free acid) in 400 ml distilled water. Adjust the pH to ~7.0 with 10 *N* NaOH and bring the final volume to 500 ml. Store at 4°C.
7. *Cell lysis buffer*: 5 m*M* HEPES, pH 7.4, 10 m*M* potassium acetate, 2 mM magnesium acetate, and 1 m*M* EGTA. To make 500 ml, combine 2.5 ml 1*M* HEPES, pH 7.4, 5 ml 1*M* potassium acetate, 1 ml 1*M* magnesium acetate, 2.5 ml 0.2*M* EGTA, and 489 ml distilled water. Store at 4°C.
8. *Phosphate-buffered saline (PBS)*: To make 1 liter, dissolve 8 g sodium chloride, 0.2 g potassium chloride, 1.44 g sodium phosphate (dibasic), and 0.24 g potassium phosphate (monobasic) in 900 ml distilled water. Adjust the pH to 7.4 and bring the final volume to 1 liter. Store at 4°C.

Steps

1. Grow HeLa cells at a density of 2–7 × 10^5 cells per milliliter in a spinner flask (30–50 rpm) in a 37°C incubator (CO_2 is not required). The medium is JMEM containing 2.0 g sodium bicarbonate and 2.38 g HEPES per liter. Adjust the pH to 7.3, sterile filter, and store at 4°C. Before use, supplement the medium with 10% newborn calf serum and 1% penicillin–streptomycin. The cells should have a doubling time of approximately 18 h, making it necessary to dilute the culture with fresh, prewarmed medium every 1–2 days.
2. HeLa cells from a 250-ml culture provide the starting point for scaling up the preparation to 15 liters. This is carried out by sequential dilution of the culture into larger spinner flasks. The culture should not be diluted to a density below 2 × 10^5 cells/ml. The spinner flasks used for scaling up preparation are 250 ml (1 each), 1 liter (1 each), 3 liters (2 each), and 15 liters (1 each). This process generally takes 5 days.
3. Perform the cell harvest and subsequent steps at 0–4°C. Collect the cells by centrifugation (300*g* for 15 min) in 780-ml conical glass bottles in a Beckman J6B refrigerated centrifuge equipped with a JS5.2 swinging bucket rotor. The cell harvest takes about 1 h.
4. Wash the cells by sequential resuspension and centrifugation. Two washes are carried out in ice-cold PBS (1 liter each) and one wash is carried out in 1× transport buffer containing 2 m*M* DTT. The yield from a 15-liter culture should be approximately 40 ml of packed cells.
5. Resuspend the cell pellet using 1.5 volume of lysis buffer, supplemented with 3 μg/ml each aprotinin, leupeptin, pepstatin, 0.5 m*M* PMSF, and 5 m*M* DTT. Allow the cells to swell on ice for 10 min.
6. Disrupt the cells by two to three passes in a stainless-steel homogenizer. Monitor the progress of homogenization by trypan blue staining and phase-contrast microscopy. The goal is to obtain ~95% cell disruption. Excessive homogenization should be avoided because it results in nuclear fragmentation and the release of nuclear contents into the soluble fraction of the preparation.
7. Dilute the homogenate with 0.1 volume of 10× transport buffer and centrifuge in a fixed angle rotor such as the Beckman JA-20 (40,000 *g* for 30 min).
8. Filter the resulting low-speed supernatant fraction through four layers of cheesecloth.

Subject the filtered low-speed supernatant to ultracentrifugation using a fixed angle rotor such as the Beckman type 60 Ti (150,000*g* for 60 min).

9. Dispense the resulting high-speed supernatant fraction (~50 ml, protein concentration ~5 mg/ml) into 1- and 4-ml aliquots, flash freeze in liquid N_2, and store at −80°C indefinitely.
10. HeLa cell cytosol is generally subjected to a rapid dialysis step before use in transport reactions. Thaw a 4-ml aliquot of cytosol at 0–4°C and dialyze for 3 h in 1× transport buffer containing 2 m*M* DTT and 1 μg/ml each of aprotinin, leupeptin, and pepstatin (two buffer changes). We use a vacuum apparatus and a collodion membrane to achieve a twofold concentration of the sample (10 mg/ml). Dispense the dialyzed, concentrated cytosol into 100-μl aliquots, flash freeze in liquid N_2, and store at −80°C.

B. Preparation of FITC-BSA-NLS Import Substrate

Synthetic peptides containing an NLS can be used to direct the nuclear import of a variety of fluorescent reporter proteins. FITC- or Cy2-BSA-NLS and Cy5-BSA-NLS are all suitable for measuring import in the fluorescence microscope or in the flow cytometer. Because excitation and emission spectra of Cy5 are distinct from GFP, Cy5-BSA-NLS is ideal for measuring import and GFP-NFAT export in the same cells. Importantly, the average size of the protein conjugate (>70 kDa) is too large to allow diffusion through the NPC and it displays low nonspecific binding to the permeabilized cell. Preparation of fluorescent import ligands is carried out in three steps: fluorescent labeling of BSA, modification of the fluorescent BSA with the heterobifunctional cross-linker sulfo-SMCC, and attachment of NLS peptides. Sulfo-SMCC provides a covalent linkage between primary amines on BSA and cysteine present on the N terminus of the NLS peptide.

Solutions

1. *PBS*: To make 1 liter, dissolve 8 g sodium chloride, 0.2g potassium chloride, 1.44 g sodium phosphate (dibasic), and 0.24 g potassium phosphate (monobasic) in 900 ml distilled water. Adjust the pH to 7.4 and bring the final volume to 1 liter. Store at 4°C.
2. *0.1 M* sodium carbonate: To prepare 250 ml, dissolve 3.1 g sodium carbonate in 200 ml distilled water, adjust the pH to 9.0, and bring the final volume to 250 ml. Store at 4°C.
3. *1.5 M* hydroxylamine: To prepare 100 ml, dissolve 10.4 g in a total volume of 100 ml distilled water. Store at room temperature.
4. *10 mg/ml FITC*: Add 1 ml DMSO to 10 mg FITC in an amber vial and vortex to dissolve.
5. *20 mM sulfo-SMCC*: Prepare a 20 m*M* stock of sulfo-SMCC in the following manner. Preweigh a microfuge tube on a fine balance and use a small spatula to add approximately 1–2 mg of sulfo-SMCC to the microfuge tube. Reweigh the tube containing the sulfo-SMCC and add DMSO for a final concentration of 8.7 mg/ml.

Steps

1. Dissolve 10 mg high-purity BSA in 1 ml sodium carbonate buffer, pH 9.0.
2. Stir the BSA solution in a glass test tube with a microstir bar and add 0.1 ml of 10 mg/ml FITC. Cover with foil and stir for 60 min at room temperature.
3. Stop the reaction by adding 0.1 ml 1.5*M* hydroxylamine.
4. Separate FITC-labeled BSA from unincorporated FITC by desalting on a PD-10 column equilibrated in PBS, collecting

0.5-ml fractions. The bright yellow FITC-BSA will elute in the void volume of this column.

5. Pool the four or five most concentrated fractions, dispense into 1-mg aliquots, and freeze in foil-wrapped microfuge tubes at −20°C.
6. Combine 1 mg of FITC-BSA with 50 μl of freshly prepared 20 m*M* sulfo-SMCC and mix end over end for 45 min at room temperature.
7. Separate the sulfo-SMCC-activated FITC–BSA from unincorporated sulfo-SMCC by desalting on a PD-10 column equilibrated in PBS. After loading the sample, fill the buffer reservoir of the column with PBS and collect 0.5-ml fractions. The bright yellow FITC–BSA will elute in the void volume as before.
8. Pool the three most concentrated fractions of sulfo-SMCC-activated FITC–BSA and combine with 0.3 mg of NLS peptide (CGGGPKKKRKVED). Mix end over end in a foil-wrapped microfuge tube overnight at 4°C.
9. Remove unincorporated NLS peptide by subjecting the sample to four cycles of centrifugation and resuspension in 1× transport buffer using a 2-ml 30-kDa cutoff Centricon filter. Follow the manufacturer's recommendations for centrifugation conditions.
10. Adjust the FITC–BSA–NLS conjugate to a final concentration of 2 mg/ml, dispense into 50-μl aliquots, flash freeze in liquid N_2, and store at −80°C.

C. Preparation of Cy2- or Cy5-BSA-NLS Import Substrate

The preparation of Cy2- and Cy5-labeled import substrate is very similar to the method described earlier for the FITC-labeled substrate.

Steps

1. Dissolve 2.5 mg BSA in 1 ml 0.1*M* sodium carbonate. Use one vial-activated Cy2 or Cy5 for coupling. Incubate for 40 min at room temperature.
2. Separate the CyDye-BSA conjugate from the free dye by chromatography on a PD-10 column equilibrated with PBS.
3. To activate CyDye-BSA, add sulfo-SMCC to a final concentration of 2 mM. Incubate for 30 min at room temperature. Remove free cross-linker using a PD-10 column as described earlier.
4. Dissolve 1 mg of NLS-peptide (CGGGPKKKRKVED) with activated CyDye-BSA and incubate the solution overnight at 4°C. Remove free peptide and adjust protein concentration as described previously.
5. Freeze aliquots in liquid nitrogen and store at −80°C. After thawing, the import substrate can be kept at 4°C in the dark for a few weeks.

D. Nuclear Protein Import Assay

Solutions

1. *10× and 1× transport buffer*: See Section III,A.
2. *Complete transport buffer*: 1× transport buffer containing 1 μg/ml each aprotinin, leupeptin, pepstatin, and 2 m*M* DTT.
3. *10% digitonin*: To make 2 ml, add 0.2 g high-purity digitonin to 1.7 ml DMSO and dissolve by vigorous vortexing, Dispense into 20-μl aliquots and freeze at −20°C.
4. *100 mM MgATP*: To make 5 ml, add 0.5 ml 1*M* magnesium acetate and 0.1 ml 1*M* HEPES, pH 7.4, to 4 ml distilled water. Add 275.1 mg ATP, dissolve by vortexing, and bring the final volume to 5 ml. Dispense into 20-μl aliquots and freeze at −80°C.
5. *250 mM creatine phosphate*: To make 5 ml, add 0.32 g creatine phosphate to 4 ml distilled water, dissolve by vortexing, and bring the final volume to 5 ml. Dispense into 20-μl aliquots and freeze at −20°C.

6. *2000 U/ml creatine phosphokinase*: To make 5 ml, dissolve 10,000 U creatine phosphokinase in 20 m*M* HEPES, pH 7.4, containing 50% glycerol. Dispense into 1-ml aliquots and store at −20°C.

Steps

1. Plate NIH 3T3 cells onto glass coverslips in six-well dishes at a density of ~5 × 10^4 cells/well and grow overnight.
2. Place the six-well dishes on ice. Aspirate media and gently replace with 2 ml ice-cold, complete transport buffer. Aspirate the transport buffer and replace with fresh transport buffer twice, taking care not to disturb the cells. Complete transport buffer in this and subsequent steps refers to ice-cold, 1× transport buffer supplemented with DTT and protease inhibitors.
3. To permeabilize the cells, aspirate the transport buffer from each well and immediately add 0.05% digitonin diluted into complete transport buffer. Incubate for 5 min on ice.
4. Stop the permeabilization reaction by aspirating the digitonin solution and replacing with complete transport buffer. Wash the cells twice by alternate steps of aspiration and buffer addition.
5. Assemble the import reactions in 0.6-ml microfuge tubes on ice. Each reaction contains (final concentration given) unlabeled BSA (5 mg/ml), FITC-BSA-NLS (25 μg/ml), MgATP (1 m*M*), MgGTP (1 m*M*), creatine phosphate (5 m*M*), creatine phosphpkinase (20 U/ml), rabbit reticulocyte lysate (25 μl), and complete transport buffer in a total volume of 50 μl.
6. Create an incubation chamber by lining a flat-bottomed, air-tight box with parafilm and include a moistened paper towel in the chamber as a source of humidity. Place the chamber on ice.
7. Using fine forceps, remove each coverslip, wick excess buffer using filter paper, and place cells side up on the parafilm. Pipette the import reaction onto the coverslip surface without introducing bubbles.
8. Float the incubation chamber on a 30°C water bath for 20 min.
9. Using fine forceps, remove each coverslip, wick most of the import reaction using filter paper, and immediately place back into the wells of the six-well dish.
10. Wash the coverslips twice by alternate steps of aspiration and complete transport buffer addition.
11. Fix the coverslips by aspirating the complete transport buffer, adding formaldehyde (3.7%) diluted into PBS, and incubating for 15 min at room temperature.
12. Remove each coverslip, submerge in distilled water briefly, wick excess water, and mount on glass slides using Vectashield. Seal the edges with clear nail polish.
13. View the cells by fluorescence microscopy, and quantify the nuclear fluorescence in 50–100 cells per condition using image analysis software such as OpenLab.

E. Nuclear Protein Export Assay

Solutions and Reagents

1. The HeLa cell line stably expressing GFP-NFAT, which is used for the export assay, has been described in detail (Kehlenbach *et al.*, 1998) and is available upon request.
2. *1 mM trichostatin A*: Dissolve 1 mg trichostatin A in 3.3 ml ethanol. Store in aliquots at −20°C. Trichostatin A is an inhibitor of histone deacetylases and promotes the expression of GFP-NFAT.
3. *1 mM ionomycin*: Dissolve 1 mg ionomycin in 1.34 ml DMSO. Store in aliquots at −20°C.

Nuclear accumulation of the reporter protein is induced by the calcium ionophore ionomycin.

4. *Double-stranded oligonucleotides*: Dissolve oligonucleotides (5′AGAGGAAAATTTGTTTCATA and 5′TATGAAACAAATTTTCCTCT), each at 200 μ*M*, in 40 m*M* Tris, pH 7.4, 20 m*M* $MgCl_2$, and 50 m*M* NaCl. Anneal by heating to 65°C for 5 min and slow cooling to room temperature. Freeze in aliquots and store at −20°C. The oligonucleotide sequence corresponds to a DNA-binding site of NFAT. It stimulates export of GFP-NFAT about twofold, probably by releasing the protein from chromatin.
5. *1× transport buffer with LiOAc*: 20 m*M* HEPES, pH 7.4, 80 m*M* potassium acetate, 2 m*M* magnesium acetate, and 0.5 m*M* EGTA. To make 1 liter, dissolve 4.76 g HEPES, 7.85 g potassium acetate, 3.06 g lithium acetate dihydrate, 0.43 g magnesium acetate tetrahydrate, and 0.19 g EGTA in 800 ml distilled water. Adjust the pH to 7.4 with 1 *N* KOH and bring the final volume to 1 liter. Before use, add 1 μg/ml each of aprotinin, leupeptin, pepstatin, and 2 m*M* DTT.
6. *Complete transport buffer*: See Section III,D.

Steps

1. To stimulate expression of GFP-NFAT, add 250 n*M* trichostatin A to stably transfected HeLa cells and incubate overnight. One 15-cm dish containing ~10^6 cells is sufficient for 30 reactions.
2. Induce nuclear import of GFP-NFAT by adding 1 μ*M* ionomycin and 30 m*M* LiOAc directly to the culture media and return the cells to the 37°C incubator for 30 min. Lithium inhibits one of the kinases involved in nuclear phosphorylation of NFAT, a step that is required for efficient export *in vivo*.
3. Rinse cells with PBS and remove from dish by adding trypsin EDTA containing 1 μ*M* ionomycin and 30 m*M* LiOAc. Transfer the cells to 50 ml of cold transport buffer with 5% newborn calf serum. Centrifuge for 5 min at 300*g* at 4°C and wash once in 50 ml transport buffer.
4. Resuspend cells in complete transport buffer at 10^7/ml. Add digitonin to 100 μg/ml (1 μl of a 10% stock per 10^7 cells). Leave on ice for 3 min and check permeabilization with trypan blue. Dilute the cells to 50 ml with transport buffer to release soluble transport factors and collect by centrifugation as described previously.
5. Preincubation (optional): Resuspend cells in transport buffer containing 30 m*M* LiOAc at 10^7/ml and add MgATP (1 m*M*), creatine phosphate (5 m*M*) and creatine phosphokinase (20 U/ml). Incubate for 15 min in a 30°C water bath. Wash cells with transport buffer. This step results in the depletion of additional transport factors such as CRM1, rendering them rate limiting in the subsequent reaction. The reporter protein largely remains in the nucleus under these conditions.
6. Resuspend cells in transport buffer at 3 × 10^7/ml. Assemble 40-μl transport reactions in FACS tubes: 300,000 permeabilized cells (10 μl), ATP-regenerating system (1 m*M* MgATP, 5 m*M* creatine phosphate, 20 U/ml creatine phosphokinase), Cy5-BSA-NLS (25–50 μg/ml), 1 μ*M* annealed oligonucleotide, and HeLa cytosol (2 mg/ml).
7. Incubate tubes in a 30°C water bath for 30 min. Control reactions contain the same components but are kept on ice, as nuclear import and export are temperature dependent.
8. Stop the reaction by adding 4 ml of cold transport buffer. Centrifuge for 5 min at 400 *g* and 4°C. Remove most of supernatant by aspiration and resuspend cells in residual buffer. Proceed to step 9 or fix cells for microscopic analysis by the

addition of 2 ml of formaldehyde (3.7% in PBS). Incubate cells for 15 min at room temperature, add 1 μl Hoechst 33258 (10 mg/ml in H_2O), and incubate for 5 more minutes. Collect cells by centrifugation (300*g* for 5 min), wash twice with 1 ml PBS, and resuspend the final cell pellet in 15 μl PBS. Apply the cell suspension to a glass slide, cover with a coverslip, and seal with nail polish.

9. Measure the fluorescence of 10,000 cells by flow cytometry. In Becton-Dickinson instruments, GFP-NFAT (or BSA-NLS, labeled with FITC or Cy2, if only import is analyzed) is detected in FL1 and Cy5-BSA-NLS in FL4.
10. Normalize the mean fluorescence values with respect to a reaction kept on ice.

IV. COMMENTS

Several of the basic controls for assaying nuclear import in digitonin-permeabilized cells are shown (Fig. 11.2). Nuclear import in digitonin-permeabilized cells is stimulated by the addition of HeLa cytosol or reticulocyte lysate, which provides a source of factors, including import receptors and Ran. A low level of cytosol-independent import is usually observed because digitonin permeabilization does not result in the quantitative release of transport factors. Nuclear import is inhibited at low temperature or by incubation with wheat germ agglutinin (0.2–0.5 mg/ml), a lectin that binds to NPC proteins and blocks translocation through the pore. The level of nuclear import is quantified by measuring the nuclear fluorescence in 50–100 cells per condition using a commercially available program such as OpenLab or using a public domain program such as ImageJ (http://rsb.info.nih.gov/ij/).

The same controls that apply to nuclear import are also used to validate nuclear export reactions (Fig. 11.3). Export of GFP-NFAT is stimulated by the addition of cytosol or reticulocyte lysate and is inhibited at low temperature or by the addition of wheat germ agglutinin.

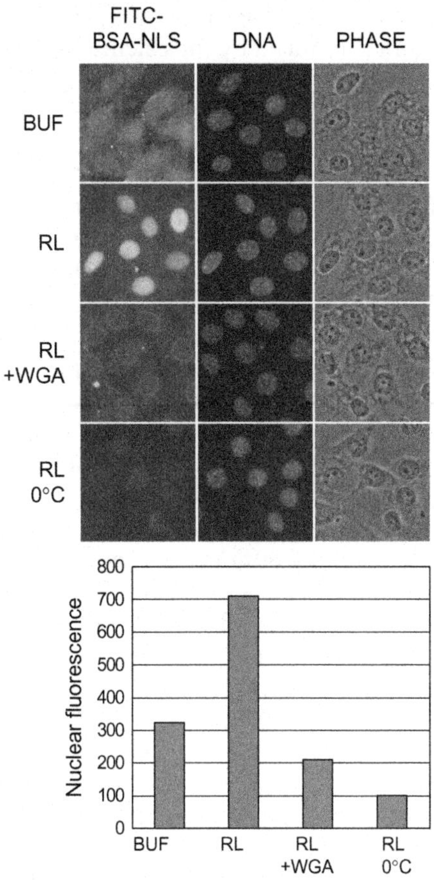

FIGURE 11.2 Nuclear import of FITC-BSA-NLS analyzed by fluorescence microscopy. (Top) NIH 3T3 cells were digitonin permeabilized and incubated with the indicated components (BUF, buffer; RL, reticulocyte lysate; WGA, wheat germ agglutinin). DAPI staining of DNA and phase-contrast (PHASE) images are also shown. Note that FITC-BSA-NLS binds nonspecifically to cells in the absence of cytosol. The rim fluorescence in the presence of WGA reflects the arrest of FITC-BSA-NLS in import complexes at the NPC. (Bottom) The level of nuclear import observed under each condition was quantified from 12-bit images. Values reflect the average fluorescence intensity per pixel from at least 100 nuclei per condition. Data courtesy of Leonard Shank (University of Virginia).

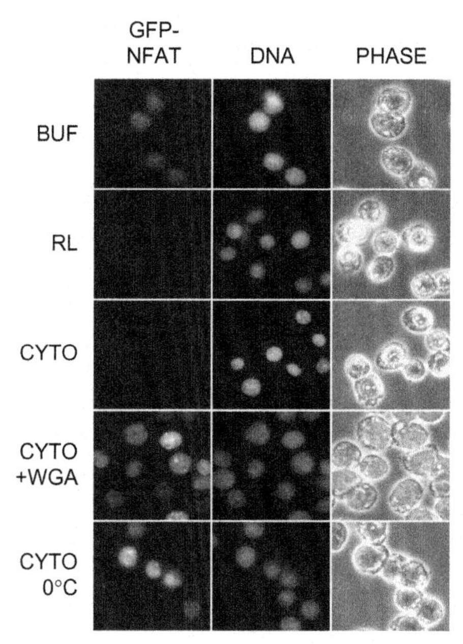

FIGURE 11.3 Nuclear export of GFP-NFAT was analyzed by fluorescence microscopy. GFP-NFAT cells were treated with trichostatin A and ionomycin to induce expression and nuclear import of GFP-NFAT. After digitonin permeabilization, the cells were incubated with the indicated components (BUF, buffer; RL, reticulocyte lysate; CYTO, HeLa cytosol; WGA, wheat germ agglutinin). Hoechst staining of DNA and phase-contrast (PHASE) images are also shown.

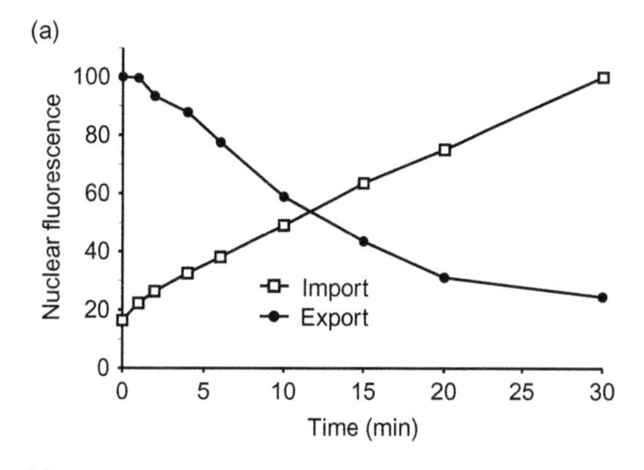

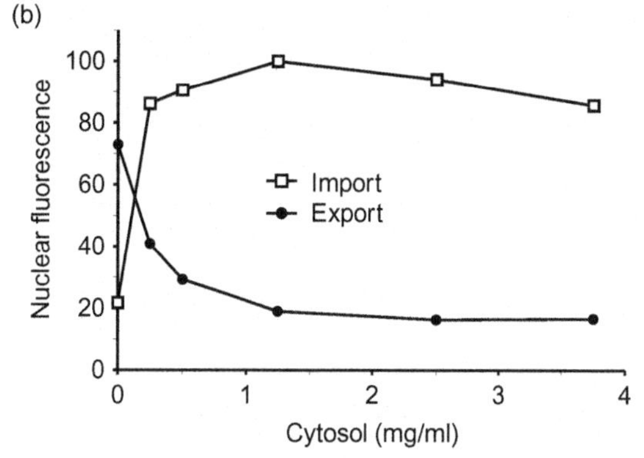

FIGURE 11.4 Nuclear export of GFP-NFAT and nuclear import of Cy5-BSA-NLS were analyzed in parallel by flow cytometry. (a) Time course of nuclear transport. All reactions contained 2 mg/ml of cytosol and 25 μg/ml of recombinant Ran. (b) Cytosol dependence of nuclear transport. Reactions contained 50 μg/ml of recombinant Ran and the indicated amounts of cytosol. All reactions were performed using preincubated cells (see step 5 in Section III,E) to enhance the cytosol dependence of transport. Portions of this figure were reprinted by permission from the *J. Cell Biol.*, Rockefeller University Press.

Note that the preexport level of GFP-NFAT fluorescence, measured in cells that have been kept on ice, will vary depending on the cellular expression level. Therefore, a reliable quantification of nuclear export requires analysis of a large number of cells in order to obtain statistically meaningful results. To this end, we use flow cytometry to measure the residual fluorescence in 10,000 cells (Fig. 11.4). This approach allows rapid analysis of a large number of samples, e.g., for determination of the transport kinetics (Fig. 11.4a) or the cytosol dependence of transport (Fig. 11.4b). Nuclear import of a fluorescently labeled import substrate can be analyzed simultaneously, allowing a direct comparison between different transport pathways. Nuclear import and nuclear export can be reconstituted using recombinant factors instead of cytosol, an approach that allows the contributions of individual transport factors to be analyzed (Görlich *et al.*, 1996; Black *et al.*, 2001; Kehlenbach and Gerace, 2002).

V. PITFALLS

Any compromise in the integrity of the nuclear envelope renders permeabilized cell assays uninterpretable. This could occur if cells are overpermeabilized with digitonin. Under such a condition, NLS-containing reporters can appear to undergo nuclear import when they are, in fact, simply binding to DNA after leakage through a permeabilized nuclear envelope. Likewise, NES-containing reporters can appear to undergo export due to simple leakage from the nucleus. The easiest way to establish that nuclear import or export is mediated by the NPC is to test for inhibition by WGA. Alternatively, the intactness of the nuclear envelope can be demonstrated by showing that a fluorescently labeled dextran (⩾70 kDa) is excluded from the nucleus. Thus, it is helpful to test a range of digitonin concentrations (25–100 μg/ml) and stain with trypan blue to optimize permeabilization of the plasma membrane.

The assays described in this article feature NIH 3T3 and HeLa cells; however, these methods should be applicable to virtually any mammalian cell line. Digitonin permeabilization on adherent cells works best when the cells are 40–70% confluent and poorly if the cells are approaching confluence. Cells that are not well adhered may detach during the permeabilization and wash steps, a problem that usually can be overcome by coating coverslips with poly-d-lysine or by plating the cells 2 days before permeabilization. Also, because cells near the edge of the coverslip may be subject to evaporation artifacts even in a humid chamber, it is best to restrict analysis to the central region of the coverslip.

References

Adam, S. A., Sterne-Marr, R. E., and Gerace, L. (1990). Nuclear import in permeabilized mammalian cells requires soluble factors. *J. Cell Biol.* **111**, 807–816.

Black, B. E., Holaska, J. M., Lévesque, L., Ossareh-Nazari, B., Gwizdek, C., Dargemont, C., and Paschal, B. M. (2001). NXT1 is necessary for the terminal step of Crm1-mediated nuclear export. *J. Cell Biol.* **152**, 141–155.

Crabtree, G. R., and Olson, E. N. (2002). NFAT signaling: Choreographing the social lives of cells. *Cell* **109**, 67–79.

Görlich, D., Panté, N., Kutay, U., Aebi, U., and Bischoff, F. R. (1996). Identification of different roles for RanGDP and RanGTP in nuclear protein import. *EMBO J.* **15**, 5584–5594.

Görlich, D., Prehn, S., Laskey, R. A., and Hartmann, E. (1994). Isolation of a protein that is essential for the first step of nuclear protein import. *Cell* **79**, 767–778.

Kehlenbach, R. H., Dickmanns, A., and Gerace, L. (1998). Nucleocytoplasmic shuttling factors including Ran and CRM1 mediate nuclear export of NFAT *in vitro*. *J. Cell Biol.* **141**, 863–874.

Kehlenbach, R. H., and Gerace, L. (2002). Analysis of nuclear protein import and export *in vitro* using fluorescent cargoes. *Methods Mol. Biol.* **189**, 231–245.

Paschal, B. M., and Gerace, L. (1995). Identification of NTF2, a cytosolic factor for nuclear import that interacts with nuclear pore complex protein p62. *J. Cell Biol.* **129**, 925–937.

Pollard, V. W., Michael, W. M., Nakielny, S., Siomi, M. C., Wang, F., and Dreyfuss, G. (1996). A novel receptor-mediated nuclear import pathway. *Cell* **86**, 985–994.

Ribbeck, K., and Görlich, D. (2001). Kinetic analysis of translocation through nuclear pore complexes. *EMBO J.* **20**, 1320–1330.

Steggerda, S. M., and Paschal, B. M. (2002). Regulation of nuclear import and export by the GTPase Ran. *Int. Rev. Cytol.* **217**, 41–91.

Weis, K. (2003). Regulating access to the genome: Nucleocytoplasmic transport throughout the cell cycle. *Cell* **112**, 441–451.

CHAPTER

12

Heterokaryons: An Assay for Nucleocytoplasmic Shuttling

Margarida Gama-Carvalho, and Maria Carmo-Fonseca

OUTLINE

I. INTRODUCTION

Interspecies heterokaryon assays are used to analyse the shuttling properties of proteins that localise predominantly in the nucleus at steady state. Conventional localisation methods do not allow the detection of small amounts of protein transiently present in a cellular compartment. Thus, demonstration of a predominantly nuclear (or cytoplasmic) localisation for a given protein frequently obscures the existence of a constant shuttling activity between the two major cell compartments. Indeed, the number of proteins shown to possess nucleocytoplasmic shuttling activity has increased enormously in the past few years, highlighting the importance of shuttling in the regulation of many cellular processes (reviewed by Gama-Carvalho and Carmo-Fonseca, 2001).

Identification of shuttling cytoplasmic proteins can be achieved through analysis of the localisation pattern of deletion mutants affecting export signals or by analysing the effect of inhibition of the major protein export pathway by the drug leptomycin B. In either case, the observation of a shift in the steady-state localisation of the protein from the cytoplasm to the nucleus suggests that the protein under study shuttles continuously between both compartments and that the mutation or drug treatment has interfered with its export pathway.

Identification of nuclear shuttling proteins can be achieved by a number of assays. Using yeast cells, identification of nuclear proteins that shuttle to the cytoplasm can be performed with temperature-sensitive nuclear import mutant strains. In this case, cytoplasmic accumulation of the protein of interest at the restrictive temperature indicates shuttling activity (Lee *et al.*, 1996). In higher eukaryotes, the original demonstration that nuclear proteins shuttle between the nucleus and the cytoplasm was based on nuclear transfer experiments whereby the nucleus from a [^{35}S] methionine-labeled cell was transferred to an unlabeled cell and the appearance of radioactivity in the unlabeled nucleus was monitored (Goldstein, 1958). However, the identification of specific shuttling proteins had to await the development of antibody microinjection or interspecies heterokaryon assays (Borer *et al.*, 1989).

In the microinjection assay, antibodies against the protein of interest are introduced in the cytoplasm of the cell and their appearance in the nucleus is monitored in the presence of protein synthesis inhibitors. As immunoglobulins do not cross the nuclear envelope unless they are bound to a protein that is targeted to the nucleus via a receptor-mediated import pathway, the detection of the antibody in the nucleus in the absence of protein synthesis indicates that the target protein has the ability to shuttle between the nucleus and the cytoplasm. Although virtually any cultured cell type may be used, microinjection assays are often performed in *Xenopus laevis* oocytes, a widely used model system to study transport across the nuclear envelope. A major advantage of this model, in addition to the ease of microinjection, is the possibility to dissect and analyse the contents of the cytoplasm and germinal vesicle (i.e., the nucleus) of a single oocyte, allowing for quantitative biochemical assays. However, it is noteworthy that shuttling assays using *X. laevis* oocytes have provided contradictory results to microinjection or heterokaryon assays performed with human HeLa cells (e.g., Bellini and Gall, 1999; Almeida *et al.*, 1998; Calado *et al.*, 2000). One possibility is that the same protein may have different shuttling properties in *X. laevis* oocytes and human somatic cells. Alternatively, oocytes may contain a surplus of nuclear proteins stored in the cytoplasm that are not normally present in somatic cells.

Interspecies heterokaryons, which constitute the most widely used shuttling assay, involve the formation of hybrid cells containing at least two nuclei from different species. In an interspecies heterokaryon, a cell line containing the protein of interest—donor cell—is fused with a cell line from a distinct species that does not contain the same protein—receptor cell. The appearance in the "receptor" nucleus of the protein of interest that was originally found exclusively in the "donor" nucleus is monitored by fluorescence microscopy. The assay is performed in the presence of translation inhibitors to ensure that the imported protein has not been newly synthesised in the cytoplasm but rather has been exported from the donor cell nucleus (Fig. 12.1).

Two different approaches may be taken to label the protein of interest present in donor nuclei. If a species-specific antibody is available, the endogenous protein can be identified by immunostaining of the heterokaryons. For example, when a monoclonal antibody directed against a human protein does not recognise its homologue in *Drosophila melanogaster*, heterokaryons can be made by fusion of human HeLa and *D. melanogaster* SL2 cell lines. These heterokaryons have the advantage that human and fly nuclei are easily distinguished by their different sizes. However, the small size of SL2 nuclei may limit the use of a *D. melanogaster*-specific antibody in a reverse assay, where the larger mammalian nuclei serve as receptors. In this case, the concentration of the target protein

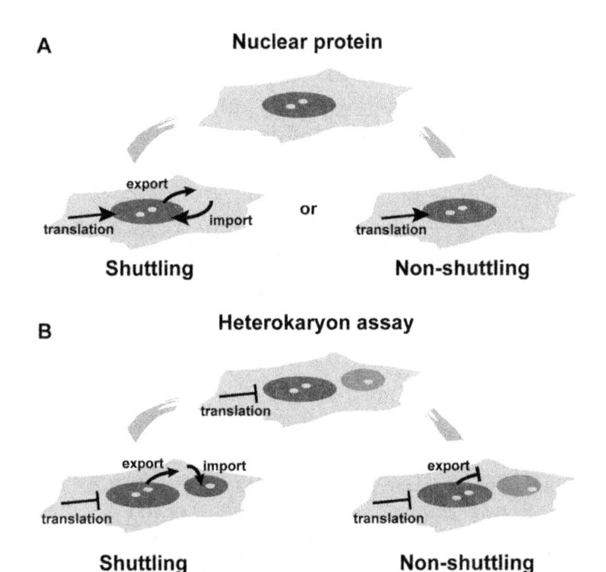

FIGURE 12.1 **The heterokaryon assay.** (**A**) Proteins may accumulate in the nucleus as the result of a unidirectional import pathway (nonshuttling nuclear proteins). Alternatively, proteins may be targeted simultaneously by import and export pathways, resulting in a dynamic cycling (shuttling) between the nucleus and the cytoplasm. (**B**) Heterokaryon assays distinguish between shuttling and nonshuttling nuclear proteins. If a protein originally present in the donor (large) nucleus shuttles, then it will appear in the receptor (small) nucleus. Nonshuttling proteins are not exported from the donor nucleus and are never detected in the receptor nucleus. Assays have to be performed in the presence of protein synthesis inhibitors in order to prevent the import of newly synthesised protein into the receptor nucleus.

in the receptor nucleus may be below the detection limit of the fluorescence microscope. Depending on the specificity of the antibody available, other donor/receptor nuclei combinations are possible, namely human × mouse heterokaryons. An alternative approach to identify the protein of interest involves transfection of a vector encoding a tagged form of the protein into the donor cell line, followed by fusion with a nontransfected receptor cell line. Both green fluorescent protein or amino acid epitope tags have proven to work successfully in these assays. In this case, heterokaryon assays are usually performed with human HeLa and murine NIH 3T3 cell lines. Murine nuclei are easily distinguished by their typical heterochromatin staining pattern.

When a shuttling protein is identified, heterokaryon assays may be further employed to dissect the signals and pathways involved in the transport. More recently, heterokaryon assays have been used to characterise the intranuclear pathway of newly imported proteins (Leung and Lamond, 2002). In this case, a donor cell line transfected with a nucleolar shuttling protein was used as a source of labeled protein, whose entry into the receptor nuclei could be followed in a time course assay from the moment of cell fusion. As receptor nuclei do not contain this protein, the intranuclear pathway followed after import until it accumulated in the nucleolus could be determined.

To perform heterokaryons assays, knowledge of basic tissue culture, transient transfection, and immunofluorescence microscopy techniques is required and will not be addressed here.

II. MATERIALS AND INSTRUMENTATION

Human HeLa cells

Murine NIH 3T3 and/or *D. melanogaster* SL2 cells

Minimum essential medium (MEM) with Glutamax-I (Cat. No. 41090-028) and MEM nonessential amino acids (Cat. No. 11140-035) (Gibco BRL)

Dulbeco's modified Eagle Medium (D-MEM) (Cat. No. 41966-029) or *D. melanogaster* Schneiders medium (Cat. No. 21720-024) and 200 m*M* l-glutamine (Cat. No. 25030-032) (Gibco BRL)

Fetal bovine serum (FBS, Cat. No. 10270-1064, Gibco BRL)

Basic tissue culture facility with a 37°C, 5% CO_2 incubator and a 29°C, 5% CO_2 incubator

(required only for HeLa × SL2 heterokaryons)
35 × 10-mm (P35) tissue culture dishes and general tissue culture material required for cell line maintenance
10 × 10-mm glass coverslips
Curved tweezer
Transfection reagents: Lipofectin (Cat. No. 18292-037, Invitrogen) or Fugene (Cat. No. 1814443, Roche Applied Science)
Protein synthesis inhibitors: anisomycin (Cat. No. A9789), cycloheximide (Cat. No. L9535) or emetine dihydrochloride hydrate (Cat. No. E 2375) (all from Sigma-Aldrich)
Polyethylene glycol (PEG) 1500 (Cat. No. 783 641, Roche Applied Science)
Primary antibody specific for the human–target protein or protein–fusion construct in eukaryotic expression vector with GFP or amino acid epitope tags (e.g., Ha, FLAG, His) and anti-tag primary antibody
Anti-hnRNP C mAb 4F4 (Choi and Dreyfuss, 1984) and anti-hnRNP A1 9H10mAb (Pinol-Roma *et al.*, 1988)
DAPI, dilactate (Cat. No. D 9564, Sigma-Aldrich)
Fluorochrome-conjugated secondary antibodies (Jackson Laboratories)
Fluorescence microscope

III. PROCEDURES

A. HeLa × SL2 Heterokaryons

Cell Culture

1. Maintain HeLa cells routinely in MEM supplemented with nonessential amino acids and 10% FBS and grow in a 37°C, 5% CO_2 incubator. Cells should be split the day before the heterokaryon assay is performed.
2. Grow SL2 *D. melanogaster* cells in Schneider's medium supplemented with 12% FBS and 2mM glutamine at 25°C without CO_2 (a clean laboratory drawer will provide a convenient "incubator"). This is a suspension cell line that can be induced to adhere to the coverslips in the absence of serum.
3. Heterokaryon medium: HeLa × SL2 heterokaryons should be maintained in MEM supplemented with nonessential amino acids and 12% FBS in a 29°C, 5% CO_2 incubator.

Solutions

1. *Protein synthesis inhibitors*: Prepare a stock solution at 10mg/ml in 50% ethanol and store at −20°C. Final use concentrations should be established for the cell lines used; most widely used concentrations range between 20 and 100μg/ml for cicloheximide or anisomycin and 0.5 and 20μg/ml for emetine.
2. *Phosphate-buffered saline PBS*: 137mM NaCl, 2.68mM KCl, 8.06mM Na_2HPO_4, and 1.47mM KH_2PO_4. Prepare by weighing 8g NaCl, 0.2g KCl, 1.14g Na_2HPO_4, and 0.2g KH_2PO_4 for 1 liter of solution. Sterilise by autoclaving.

Steps

1. All the steps of the procedure should be performed in a laminar flow hood using sterile cell culture material and solutions. The procedure is adapted for cells grown on 10 × 10-mm coverslips placed in P35 dishes.
2. HeLa cells should be grown to subconfluent density on 10 × 10-mm coverslips. A 35 × 10-mm tissue culture petri dish can accommodate four of these coverslips.
3. Resuspend exponentially growing *D. melanogaster* SL2 cells in serum-free HeLa culture medium to a concentration of 3–4 × 10^7 cells/ml.
4. Remove the medium from the HeLa cell culture and overlay coverslips with 500μl of

the SL2 suspension. Incubate for 20 min in a 29°C, 5% CO_2 incubator to induce adherence of SL2 cells.

5. Replace medium with 1.5 ml heterokaryon medium and add the protein synthesis inhibitor. Place cells in the 29°C, 5% CO_2 incubator for at least 3 h to inhibit protein synthesis. A control experiment replacing the protein synthesis inhibitor with a similar volume of 50% ethanol should be performed.
6. For the fusion procedure, place a 50-μl drop of PEG 1500 prewarmed to 29°C in a sterile petri dish.
7. Remove the HeLa/SL2 coculture from the incubator and wash twice with PBS prewarmed to 29°C.
8. With a sterile forceps, remove a coverslip and invert it over the PEG drop for 2 min (cell side down). Then, place the coverslip (cell side up) in a new P35 dish and rinse twice with PBS prewarmed to 29°C. Replace the PBS with 1.5 ml of prewarmed heterokaryon medium with a protein synthesis inhibitor and place the heterokaryons in the incubator for the duration of the shuttling assay (usually 6 h or more).
9. Fix and immunostain cells using standard procedures, following a time course with 1-h intervals. Positive and negative control experiments should be performed using monoclonal antibodies directed against human hnRNPA 1 (a protein that shuttles between the nucleus and the cytoplasm; Pinol-Roma *et al.*, 1988) and human hnRNPC (a protein that does not shuttle; Choi and Dreyfuss, 1984) (Fig. 12.2).

B. HeLa × 3T3 Heterokaryons

Cell Culture

1. Maintain HeLa cells routinely in MEM supplemented with nonessential amino acids and 10% FBS and grow in a 37°C, 5% CO_2 incubator. Cells should be split to petri dishes 2 days before the heterokaryon assay is performed and transfected on the following day.

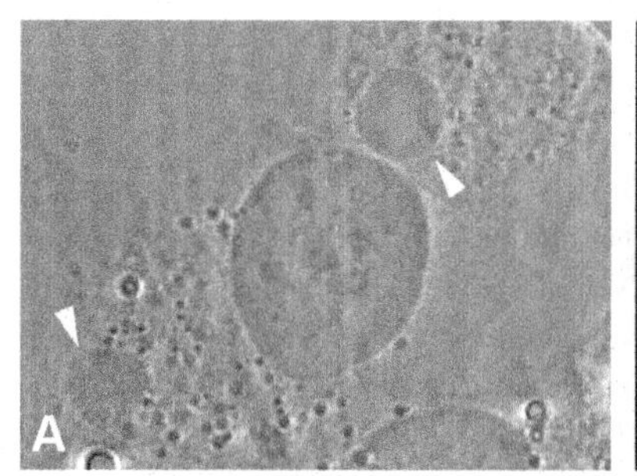

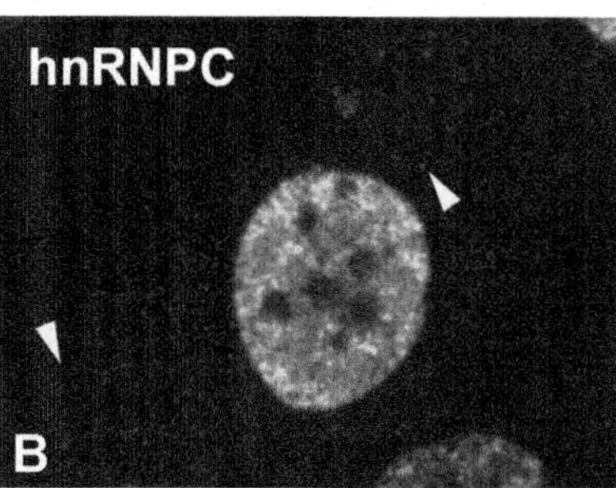

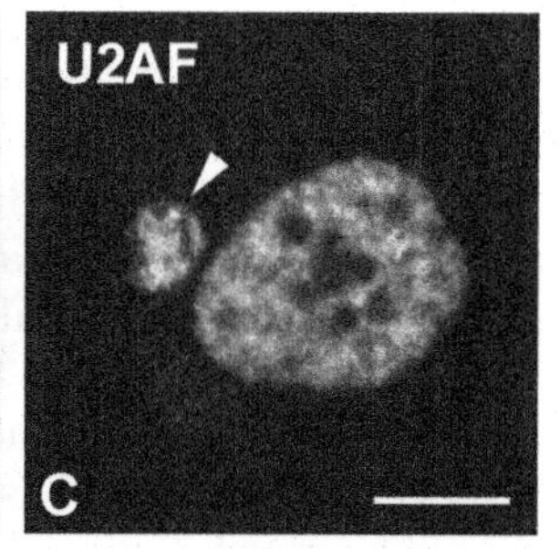

FIGURE 12.2 **HeLa × SL2 heterokaryons.** Monoclonal antibodies directed against a human protein often do not cross-react with its *Drosophila melanogaster* homologue. In this case, shuttling of the endogenous human protein can be analysed in a HeLa × SL2 heterokaryon. (**A**) Phase-contrast image of a HeLa × SL2 heterokaryon. Arrowheads label small SL2 nuclei from cells that have fused with the HeLa cytoplasm. (**B**) The heterokaryon shown in A has been immunostained with a monoclonal antibody specific for the human nonshuttling protein hnRNPC (mAb 4F4; Choi and Dreyfuss, 1989), showing it to be restricted to the HeLa nucleus in the presence of protein synthesis inhibitors (20 μg/ml emetine). Arrowheads point to the position of SL2 nuclei. (**C**) Immunostaining of a HeLa × SL2 heterokaryon with a monoclonal antibody specific for human $U2AF^{65}$ (mAb MC3; Gama-Carvalho *et al.*, 1997) shows that the human protein is present in the SL2 nucleus (arrowhead) in the presence of protein synthesis inhibitors, demonstrating the shuttling activity of this protein (Gama-Carvalho *et al.*, 2001).

2. Grow mouse 3T3 cells in D-MEM supplemented with 10% FBS at 37°C, 5% CO_2.
3. Heterokaryon medium: HeLa × 3T3 heterokaryons should be maintained as 3T3 cells.

Solutions

1. *PBS*: 137 m*M* NaCl, 2.68 m*M* KCl, 8.06 m*M* Na_2HPO_4, and 1.47 m*M* KH_2PO_4. Prepare by weighing 8 g NaCl, 0.2 g KCl, 1.14 g Na_2HPO_4, and 0.2 g KH_2PO_4 for 1 liter of solution. Sterilise by autoclaving.
2. *DAPI*: Prepare a stock solution at 1 mg/ml and store at −20°C protected from light; staining solution: dilute stock to 1 μg/ml in PBS and store at +4°C protected from light for several weeks.
3. *Protein synthesis inhibitors*: Prepare a stock solution at 10 mg/ml in 50% ethanol and store at −20°C. Final use concentrations should be established for the cell lines used; most widely used concentrations range between 20 and 100 μg/ml for cycloheximide or anisomycin and 0.5 and 20 μg/ml for emetine.

Steps

1. All the steps of the procedure should be performed in a laminar flow hood using sterile cell culture material and solutions. The procedure is adapted for cells grown on 10 × 10 mm coverslips placed in P35 dishes.
2. The evening before the assay, HeLa cells should be transfected with the desired construct and left to grow to subconfluent density on 10 × 10-mm coverslips. A 35 × 10-mm tissue culture petri dish can accommodate four of these coverslips. Common transfection reagents such as Lipofectin or Fugene work well with this procedure and can be used according to the instructions of the manufacturer. The number of hours between transfection and the heterokaryon assay should be determined for each construct to avoid overexpression of the exogenous protein.
3. Tripsinize and count 3T3 cells. For each P35 HeLa culture, prepare 1.5 ml of a 3T3 suspension with 1×10^6 cells/ml in DMEM with 10% FBS and protein synthesis inhibitor.
4. Remove the medium from the HeLa cell culture and overlay coverslips with the 3T3 suspension. Incubate for 3 h in a 37°C, 5% CO_2 incubator to allow 3T3 cells to adhere and inhibit protein synthesis.
5. For the fusion procedure, place a 50-μl drop of prewarmed PEG 2000 in a sterile petri dish per coverslip.
6. Remove the HeLa/3T3 coculture from the incubator and wash twice with prewarmed PBS.
7. With a sterile forceps, remove a coverslip and invert it over the PEG drop for 2 min. Then place the coverslip (cell side up) in a new P35 dish and rinse twice with prewarmed PBS. Replace the PBS with 1.5 ml of prewarmed heterokaryon medium with protein synthesis inhibitor and place the heterokaryons in the incubator for the duration of the shuttling assay (usually from 1 to 5 h).
8. Fix and stain cells using standard procedures, following a time course with 1-h intervals. In addition to the procedure (if any) necessary for detection of the exogenous protein, cells should be stained with DAPI to allow identification of heyterokaryon nuclei. Positive and negative control experiments should be performed using monoclonal antibodies directed against human hnRNPA 1 (a protein that shuttles between the nucleus and the cytoplasm; Pinol-Roma *et al.*, 1988) and human hnRNPC (a protein that does not shuttle; Choi and Dreyfuss, 1984) (Fig. 12.3).

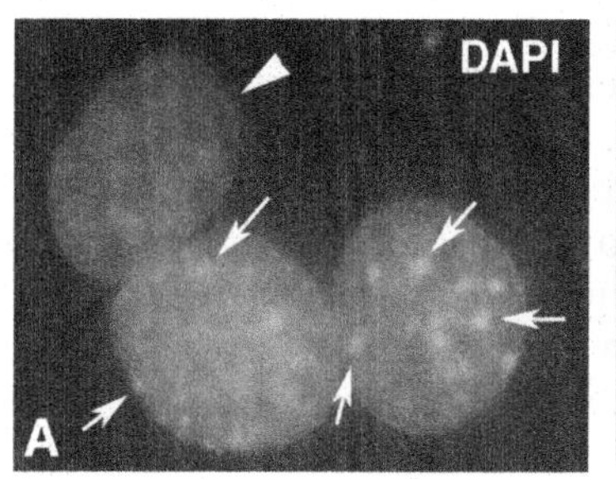

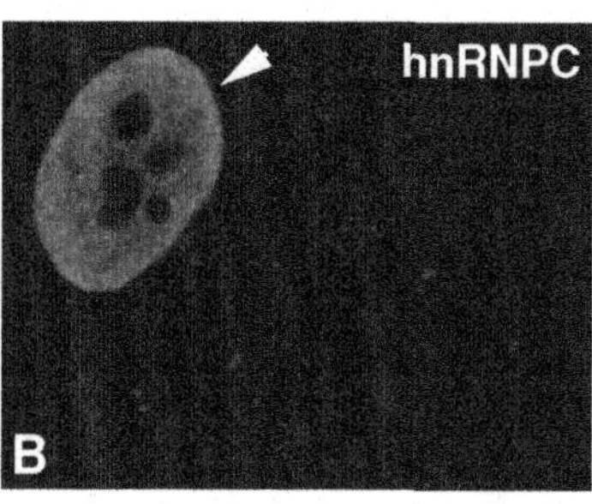

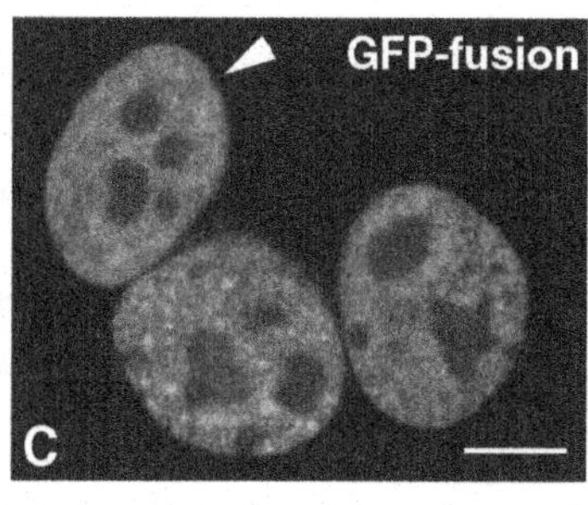

FIGURE 12.3 **HeLa × 3T3 heterokaryons.** HeLa cells were transiently transfected with a vector encoding GFP-U2AF^{65} (Gama-Carvalho *et al.*, 2001) and fused with murine NIH 3T3 cells in the presence of 20μg/ml emetine. A single heterokaryon labeled with DAPI (A), anti-hnRNPC mAb (B), and GFP (C) is shown. (**A**) DAPI staining of the HeLa × 3T3 heterokaryon. Murine cells are easily distinguished by the presence of brightly stained blocks or pericentric heterochromatin (arrows). (**B**) The heterokaryon shown in A was stained with an antibody specific for the nonshuttling human hnRNP C protein (mAb 4F4; Choi and Dreyfuss, 1989). hnRNP C is restricted to the HeLa cell nucleus (arrowhead), confirming that there is an efficient inhibition of protein synthesis. (**C**) Detection of GFP in the HeLa × 3T3 heterokaryon reveals that the fusion protein originally present in HeLa cells shuttles between the nucleus and the cytoplasm. Bar: 10μm.

IV. COMMENTS

Shuttling rates vary significantly from protein to protein, with some being extremely slow (e.g., nucleolin takes up to 16h to be detected in the receptor nuclei) and others very fast (e.g., transport receptors are detected in receptor nuclei within minutes after the fusion). The ideal time period for a shuttling assay should thus be determined for each case by performing a time course analysis. As a general rule, detection of the protein in receptor nuclei in HeLa × 3T3 heterokaryon assays involving transient expression of exogenous protein is significantly faster than HeLa × SL2 assays with the endogenous protein. For example, for the same protein, equilibrium between donor and receptor nuclei may be achieved in 90min in the first case and take up to 7h in the second.

Proper interpretation of results from a heterokaryon assay requires the use of adequate controls. Whenever possible, shuttling activity should be demonstrated for both endogenous and exogenously expressed proteins, as amino acid tags and protein overexpression may modify the results. When assaying for shuttling of the endogenous protein, the specificity of the antibody has to be demonstrated carefully. The efficiency of protein synthesis inhibition must be controlled properly as different cell strains may show a wide variation in the sensitivity to protein synthesis inhibitors. Demonstration of efficient translation inhibition can be performed by staining the heterokaryons with a monoclonal antibody specific for a human nuclear nonshuttling protein, such as the hnRNP C protein (mAb 4F4, Choi and Dreyfuss, 1984). If cytoplasmic protein synthesis is going on, this protein will be detected in the nucleus of the receptor cell. Whenever possible, heterokaryons should be double stained for the protein of interest and hnRNPC. Alternatively, when the protein of interest is also detected with a mAb, a parallel assay should be performed in which the heterokaryons are labeled with the anti-hnRNP C antibody.

If a shuttling protein is identified, it is convenient to determine whether it is diffusing through the nuclear pores passively or being exported from the nucleus through a receptor-mediated pathway. For this purpose, donor cells should be placed at 4°C for the same period of time as used for the heterokaryon assay in culture medium supplemented with the protein synthesis inhibitor and 20 m*M* HEPES. Immunostaining for the protein of interest should then be performed. At low temperature, receptor-mediated nuclear import and export processes do not occur. Thus, if the protein is exported via a receptor-mediated pathway, it should be retained in the nucleus. In contrast, if the protein can leak through the nuclear pores passively, it will accumulate in the cytoplasm. If translation is not inhibited, newly synthesised proteins will also accumulate in the cytoplasm, invalidating the assay. When performing shuttling analysis based on transient transfection, it is important to note that accumulation of exogenous protein in the cytoplasm and passive leakage from the nucleus are common events in cells expressing the protein at high levels. Thus, analysis of exogenous proteins should always be centred on cells with low expression levels. A control experiment to determine the degree of exogenous protein leakage to the cytoplasm should also be performed (e.g., a low temperature shift experiment).

The use of interspecies heterokaryons in association with transient transfection opens the possibility to analyse the protein domains involved in the export pathway by the use of mutant forms of the protein. However, this analysis can only be performed with mutants that retain full nuclear localisation. Moreover, leakage of small deletion mutants to the cytoplasm should be monitored carefully.

V. PITFALLS

The confluence of HeLa cells is critical for the procedure. If the density of the culture is too low, few heterokaryons will form. If cells have reached confluence, there will be no space left for the SL2 or 3T3 cells to adhere and often their nuclei will be above HeLa nuclei, making analysis of the results impossible. HeLa coverslips should be checked one by one to choose those with an appropriate subconfluent density.

Often during the fusion procedure there is a significant loss of cells. This may result from the mechanical stress generated when the coverslips that were turned over the PEG droplet are picked. To avoid this, simply pipette PBS beneath the coverslip to make it float and then pick it up with a forceps. In addition, it is possible that the protein synthesis inhibitor is inducing apoptosis of a high percentage of cells in culture. Indeed, some cell strains are highly sensitive to these inhibitors at concentrations that have no negative effects for other cells. Thus, the choice of inhibitor and concentration should be assayed carefully for the particular cell strain used in the assay. Cycloheximide is the most widely used inhibitor, although some cell lines may be more sensitive to emetine. In some cases, efficient protein synthesis inhibition may be hard to achieve, with a significant proportion of cells in the culture showing significant levels of newly translated protein. To prevent this from influencing the interpretation of the assay, perform double labeling of heterokaryons with the anti-hnRNP C 4F4 mAb whenever possible and consider only heterokaryons in which this protein is restricted to the human nuclei.

Finding HeLa × 3T3 heterokaryons in the preparation may present some difficulty, as the observation in phase contrast does not always allow a clear visualisation of the cytoplasm. The best approach is to look for two nuclei that are very close together, which is done most easily when visualising DAPI staining. The identification of murine nuclei by staining with DAPI also requires some training. Staining with the 4F4 mAb can serve as an aid, as it will only label human cells when translation is inhibited.

As discussed earlier, when analysing the shuttling ability of transiently transfected proteins, it is crucial to consider only low expressing cells. A good fluorescence microscope is essential for detecting low levels of expression. Often, multinucleated heterokaryons form. These should not be considered in the shuttling assay, as it is hard to identify the donor nucleus and access the original expression level of the protein (as it has spread out to many nuclei).

References

Almeida, F., Saffrich, R., Ansorge, W., and Carmo-Fonseca, M. (1998). Microinjection of anti-coilin antibodies affects the structure of coiled bodies. *J. Cell Biol.* **142**, 899–912.

Bellini, M., and Gall, J. G. (1999). Coilin shuttles between the nucleus and cytoplasm in *Xenopus laevis* oocytes. *Mol. Biol. Cell* **10**, 3425–3434.

Borer, R. A., Lehner, C. F., Eppenberger, H. M., and Nigg, E. A. (1989). Major nucleolar proteins shuttle between nucleus and cytoplasm. *Cell* **56**, 379–390.

Calado, A., Kutay, U., Kuhn, U., Wahle, E., and Carmo-Fonseca, M. (2000). Deciphering the cellular pathway for transport of poly(A)-binding protein II. *RNA* **6**, 245–256.

Choi, Y. D., and Dreyfuss, G. (1984). Monoclonal antibody characterization of the C proteins of heterogeneous nuclear ribonucleoprotein complexes in vertebrate cells. *J. Cell Biol.* **99**, 1997–2204.

Gama-Carvalho, M., and Carmo-Fonseca, M. (2001). The rules and roles of nucleocytoplasmic shuttling proteins. *FEBS Lett.* **498**, 157–163.

Gama-Carvalho, M., Carvalho, M. P., Kehlenbach, A., Valcarcel, J., and Carmo-Fonseca, M. (2001). Nucleocytoplasmic shuttling of heterodimeric splicing factor U2AF. *J. Biol. Chem.* **276**, 13104–13112.

Gama-Carvalho, M., Krauss, R. D., Chiang, L., Valcarcel, J., Green, M. R., and Carmo-Fonseca, M. (1997). Targeting of U2AF65 to sites of active splicing in the nucleus. *J. Cell Biol.* **137**, 975–987.

Goldstein, L. (1958). Localization of nucleus-specific protein as shown by transplantation experiments in *Amoeba proteus*. *Exp. Cell Res.* **15**, 635–637.

Lee, M. S., Henry, M., and Silver, P. A. (1996). A protein that shuttles between the nucleus and the cytoplasm is an important mediator of RNA export. *Genes Dev.* **10**, 1233–1246.

Leung, A. K., and Lamond, A. I. (2002). *In vivo* analysis of NHPX reveals a novel nucleolar localization pathway involving a transient accumulation in splicing speckles. *J. Cell Biol.* **157**, 615–629.

Pinol-Roma, S., Choi, Y. D., Matunis, M. J., and Dreyfuss, G. (1988). Immunopurification of heterogeneous nuclear ribonucleoprotein particles reveals an assortment of RNA-binding proteins. *Genes Dev.* **2**, 215–227.

SECTION IV

MEASURING INTRACELLULAR SIGNALING EVENTS USING FLUORESCENCE MICROSCOPY

CHAPTER

13

Ca^{2+} as a Second Messenger: New Reporters for Calcium (Cameleons and Camgaroos)

Klaus P. Hoeflich, Kevin Truong, and Mitsuhiko Ikura

OUTLINE

I. INTRODUCTION

The intracellular Ca^{2+} ion concentration has been found to be associated with a wide variety of cellular processes (Carafoli, 2003). These include diverse events such as secretion, fertilization, cleavage, nuclear envelope breakdown, and apoptosis. Several diseases, including types of muscular dystrophy, diabetes, and leukemia, involve proteins that directly respond to or control Ca^{2+}. Indeed, it may be more difficult to find cellular processes that do not involve Ca^{2+} than ones that do. From decades of research we have learned that the process of Ca^{2+} signaling consists, in general terms, of molecules for Ca^{2+} signal production, spatial and temporal shaping, sensors, and targets that elicit changes in biological function. Hence, this has prompted the development of sensitive signaling techniques to measure and image submicromolar levels of $[Ca^{2+}]$ and decode the dynamic Ca^{2+} messages throughout the propagation of the signal.

Ca^{2+} transients have traditionally been measured using synthetic fluorescent chelators (such as Fura-2 and Quin2) or recombinant aequorin (Grynkiewicz *et al.*, 1985: Montero *et al.*, 1995). Synthetic molecules provide a bright fluorescent signal, but these dyes are not easy to load and gradually leak out of cells at physiological temperatures. Cellular targeting is also not specific and some chemical indicators have been shown not to accumulate well in certain organelles. Aequorin is targeted easily but it requires incorporation of the cofactor coelenterazine, is irreversibly consumed by Ca^{2+}, and is very difficult to image due to low bioluminescence. By comparison, green fluorescent protein (GFP) and calmodulin (CaM)-based "cameleon" probes have been developed and retain several of the benefits of the aforementioned indicators, yet also provide significant improvements for *in vivo* imaging (Miyawaki, 2003; Zhang *et al.*, 2002; Truong and Ikura, 2001).

The use of cameleon indicators is gradually becoming more common within the Ca^{2+} signaling community and the literature is rich with examples of various applications. For instance, fusion of cameleons to specific signal sequences has successfully sorted them to nuclei, endoplasmic reticulum, caveolae, and secretory granule membranes (Isshiki *et al.*, 2002; Demaurex and Frieden, 2003; Emmanouilidou *et al.*, 1999). In addition to their use in detecting rapid stimulus-induced $[Ca^{2+}]$ transients, genetic studies in which cameleons were stably expressed in *Arabidopsis* stomatal guard cells (Allen *et al.*, 1999), nematode pharyngeal muscle (Kerr *et al.*, 2000), or larval thermoresponsive neurons of *Drosophila* (Liu *et al.*, 2003) show that the sensors are also applicable to long-term monitoring of Ca^{2+} concentration. This has been demonstrated further for murine cells where the circadian rhythm of cytosolic but not nuclear Ca^{2+} in hypothalmic suprachiasmatic neurons was demonstrated (Ikeda *et al.*, 2003).

It is hoped that this article provides some explicit and practical information relevant to the laboratory use of cameleon fluorescence resonance energy transfer (FRET) indicators. Our aim is that it will benefit and be of interest to colleagues both unfamiliar or experienced in using fluorescent Ca^{2+} indicators.

II. MATERIALS AND INSTRUMENTATION

A. Expression and Purification of Cameleons

Enhanced cyan fluorescent protein (ECFP) and enhanced yellow fluorescent protein (EYFP) expression constructus (Clontech Cat. No. 6075-1 and 6004-1, respectively); calmodulin cDNA (M. Ikura); pRSETB prokaryotic expression vector (Invitrogen Cat. No. V351-20); Luria broth (LB) media; isopropyl-β-d-thiogalactopyranoside

(IPTG, Fermentas Cat. No. R0391); complete protease inhibitor cocktail tablets (Roche Cat. No. 1697498); Ni-NTA agarose (Qiagen Cat. No. 1018240); *Escherichia coli* BL21 (DE3) strain (Stratagene Cat. No. 200133); sonicator; EGTA buffer: 100 m*M* KCl, 50 m*M* HEPES (pH 7.4), and 10 m*M* EGTA; $CaC1_2$ buffer: 100 m*M* KCl, 50 m*M* HEPES (pH 7.4), 10 m*M* EGTA, and 10 m*M* $CaCl_2$.

B. *In Vitro* Fluorescence Quantitation

Shimadzu spectrofluorometer RF5301; 10-mm path-length quartz cuvette.

C. *In Vitro* Imaging of Cameleons

pcDNA3 eukaryotic transient expression vector (Invitrogen Cat. No. V790-20); uncoated; γ-irradiated, 35-mm tissue culture dishes with glass bottom No. 0 (MatTek Cat. No. P35G-0-10-C); Dulbecco's modified Eagle medium (DMEM) supplemented with 10% dialyzed fetal bovine serum (FBS, Invitrogen Cat. No. 26400044), Hanks' balanced salts solution (HBSS) with Ca^{2+} (Invitrogen Cat. No. 14170120); 37°C CO_2 incubator; HeLa cells or appropriate eukaryotic strain; Lipofectamine (Invitrogen Cat. No. 18324012) and PLUS (Invitrogen Cat. No. 11514015) reagents; histamine, ionomycin, ethylene glycol-bis(2-aminoethylether)-*N,N,N′,N′*-tetraaceticacid (EGTA) and 1,2-bis(2-aminophenoxy)ethane-*N,N,N′,N′*-tetraacetic acid tetrakis(acetoxymethyl ester) (BAPTA-AM) (Sigma Cat. No. H7125, I0634, E0396, and A1076, respectively); Olympus IX70 inverted epifluorescence microscope; Olympus Xenon lamp; MicroMax 1300YHS CCD camera and Sutter Lambda 10-2 filter changers controlled by Metafluor 4.5r2 software (Universal Imaging); ECFP-EYFP FRET filter set (Omega Optical); 440AF21 excitation filter (ECFP excitation), 455DRLP dichroic mirror, 480AF30 emission filter (ECFP emission), and 535AF26 emission filter (EYFP emission); neutral density (ND) filter set (Omega Optical); UApo 40xOil Iris/340 objective (Olympus); U-MNIBA bandpass mirror cube unit (Olympus).

III. PROCEDURES

A. Engineering Cameleon Constructs

The history of cameleon engineering is reflected in its nomenclature (Table 13.1). Optimization of Ca^{2+} affinities, pH dependency, maturation time, and other parameters is by no means complete. However, we present construction of a general cameleon designed in our laboratory, YC6.1, to serve as a reference point for future work (Truong *et al.*, 2001). Molecular biology techniques for manipulating recombinant DNA are not given as they can be obtained from common reference books.

In these constructs, ECFP and EYFP function as a donor–acceptor pair for nonradiative, intramolecular FRET. During FRET, excitation of the donor (cyan) leads to emission from the acceptor (yellow), provided that the molecules are close enough (within 80Å) and in a parallel orientation. In this way, on binding Ca^{2+} the CaM wraps around its adjacent CKKp target peptide and ECFP and EYFP are brought closer to each other and FRET increases (Fig. 13.1).

1. A successful construction of GFP-fused protein indicators is facilitated greatly by careful inspection of available three-dimensional structure information of the protein or protein domain used for sensing functions. Typically, such structural analysis can be done using SwissPDBviewer (Windows) and MODELLER (Unix), which allow the measurement of atomic distances and the molecular modeling of fusion proteins, respectively. This was also the case for designing YC6.1. We found that by virtue of the hairpin-like complex structure

TABLE 13.1 Properties of Commonly Used Cameleon Indicators

Name	pK_a	[Ca^{2+}] range (μ*M*)	Comments	Reference
Cameleon-1	6.9	0.1–10	BFP/GFP hybrid with CaM-M13 tandem	Miyawaki *et al.* (1997)
YC2	6.9	0.1–10	Use of ECFP/EYFP for donor/acceptor	Miyawaki *et al.* (1997)
YC2.1	6.1	0.1–10	V68L/Q69K for increased pH resistance	Miyawaki *et al.* (1999)
YC2.12	6.0	0.1–10	"Venus" quick-maturation derivative	Nagai *et al.* (2002)
YC3	6.9	0.5–1000	E104Q low-affinity indicator	Miyawaki *et al.* (1997)
YC3.3	5.7	0.5–1000	Q69M "citrine" for added pH resistance	Griesbeck *et al.* (2001)
YC4	6.9	10–1000	E31Q low-affinity indicator	Miyawaki *et al.* (1997)
YC6.1	6.1	0.1–1	Internal CKKp CaM recognition peptide	Truong *et al.* (2001)
SapRC2	5.5	0.2–0.4	Sapphire/DsRed cameleon	Mizuno *et al.* (2001)

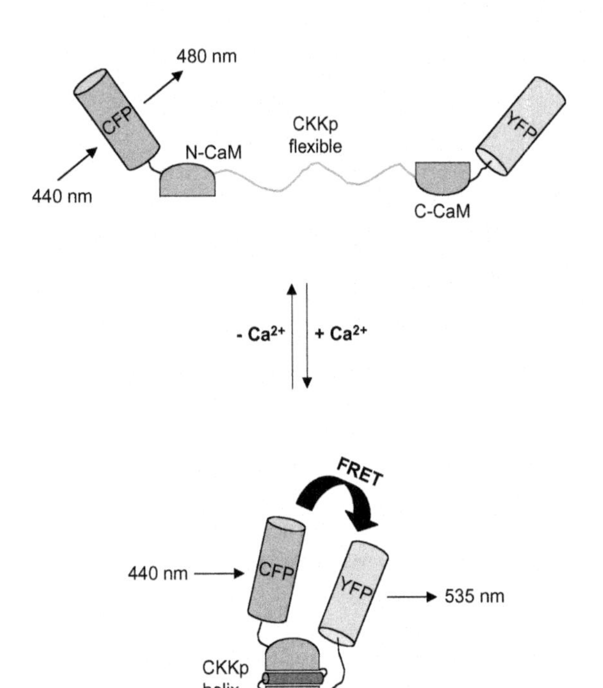

FIGURE 13.1 Schematic depiction of YC6.1 FRET in response to [Ca^{2+}]. See text for details.

of CKKp, the peptide can be inserted into the domain linker region (residues 78–81) of CaM.

2. Insert a CaM-binding peptide derived from CaM-dependent protein kinase kinase (CKKp; residues 438–463) between the terminal EF hand Ca^{2+}-binding domains (N-CaM and C-CaM) within the CaM linker domain (between CaM residues 79 and 80) and connect via two Gly-Gly linkers.
3. Add ECFP- and EYFP-encoding open reading frames to termini of the (N-CaM)-GG-CKKp-GG-(C-CaM) module by recombinant DNA methods. Expression vectors for many GFP family members are available through Clontech. Cameleon constructs can benefit from truncation of the last 11 C-terminal amino acids of ECFP (the minimal region to form GFP) to reduce the relative tumbling of the fluorophores. Additionally, linkers introduced between CaM domains and the target peptide can be optimized for complex formation.

4. Sequence construct to ensure polymerase chain reaction errors are not present. Note that oligonucleotides designed to the 5′ or 3′ end of ECFP open reading frames will also recognize the counterpart sequences in EYFP due to high sequence identity.
5. Subclone this cameleon domain into either pRSETB plasmid, for prokaryotic protein expression sufficient for biochemical and biophysical characterization, or pcDNA3.1 for mammalian expression and *in vivo* Ca^{2+}-imaging experiments. If optimal expression is not crucial, we also found that the pTriEx3 (Novagen) vector is convenient for expression in both prokaryotic and eukaryotic systems.
6. In addition to this cytoplasmic version of cameleon YC6.1, nucleus- and endoplasmic reticulum-targeted versions (YC6.1nu and YC6.2er, respectively) can be constructed by the addition of appropriate signal sequences to termini.

B. Overexpression and Purification of Cameleons

1. From a single colony of newly transformed *E. coli* strain BL21(DE3), grow liquid cultures at 37°C in LB medium containing 100 μg/ml ampicillin.
2. At OD_{600}, induce cultures with 0.5 m*M* IPTG at 15°C overnight.
3. Harvest cells by centrifugation at 6500 rpm for 20 min at 4°C.
4. Resuspend cell pellets in 1/20 culture volume of lysis buffer [50 m*M* HEPES (pH 7.4), 10% glycerol, 100 m*M* KCl, 1 m*M* $CaCl_2$, and 1 m*M* phenylmethyl sulfonyl fluoride (PMSF)], sonicate, and centrifuge at 15,000 rpm for 30 min to remove debris.
5. Incubate the supernatant with nickel chelate agarose for 1 h at 4°C and wash with of 50 m*M* HEPES (pH 7.4), 100 m*M* KCl, and 5 m*M* imidazole.
6. Elute YC6.1 with 300 m*M* imidazole in the aforementioned buffer. Proteolytic cleavage to remove the $(His)_6$ tag is not necessary as it does not interfere with YC6.1 fluorescent properties.
7. *Optional*: The eluant can be then purified further on a Superdex 75 HR 10/30 FPLC column using 20 m*M* HEPES, pH 7.5, 150 m*M* KCl, 5% glycerol, 5 m*M* dithiothreitol, and 1 m*M* PMSF.
8. Use the fraction that has the highest FRET ratio for fluorescence experiments.
9. Dialyze the sample against 2 liter of 50 m*M* HEPES (pH 7.4), 100 m*M* KCl at 4°C.
10. *Optional*: Glycerol can be added to the sample at a final concentration of 20%, and aliquot, flash freeze in liquid N_2, and store the YC6.1 at −70°C.

If necessary, it is possible to perform the characterization using a mammalian cell lysate. For harvesting, cells should be transfected in several 100-mm-diameter culture dishes, washed thoroughly to remove traces of phenol red and serum, and lysed in a hypotonic lysis buffer [50 m*M* HEPES (pH 7.4) 100 m*M* KCl, 5 m*M* $MgCl_2$, and 0.5% Triton X-100]. Following removal of cellular debris by centrifugation, dialyze the supernatant in 2 liter of buffer [50 m*M* HEPES (pH 7.4) and 100 m*M* KCl]. Finally, the sample can be used for characterization as described.

C. *In Vitro* Cameleon Fluorescence Spectroscopy

1. Record fluorescence spectra on a Shimadzu spectrofluorometer RF5301 using a 10-mm path-length quartz cuvette at room temperature.
2. Dilute cameleon chimeric proteins in 50 m*M* HEPES (pH 7.4), 100 m*M* KCl, and 20 μ*M* EGTA to a final concentration of 60 nM. Although many dilution factors are

acceptable, intermolecular ECFP-EYFP FRET will occur at higher concentrations, thereby inflating the emission signal falsely.

3. Excitate at 433 nm and monitor the fluorescence emission between 450 and 570 nm with excitation and emission slit widths of 5 nm (Table 13.2).
4. Record the fluorescence emission spectra of buffer (background) and cameleon protein solutions.
5. Subtract the background spectrum for buffer alone from the cameleon sample to find spectra of the cameleon in the absence of Ca^{2+}.
6. Determine the fluorescence emission ratio (*R*) by dividing the integration of fluorescence intensities of the FRET acceptor (for EYFP, between 520 and 536 nm) by that of the FRET donor (for ECP, between 470 and 485 nm).
7. Determine R_{min} from this spectrum. A key parameter for a Ca^{2+} indicator is its dynamic range in response to $[Ca^{2+}]$. The dynamic range of a cameleon is defined as the division of the maximum ratio, R_{max}, by the minimum ratio, R_{min}.
8. In the presence of 1 m*M* $CaCl_2$, repeat to find spectra of the cameleon in the presence of saturating amounts of Ca^{2+}. Determine R_{max} from this spectrum.

The Ca^{2+}-binding curve is used to assess the effective range of $[Ca^{2+}]$ measurement. Ca^{2+}/EDTA and Ca^{2+}/EGTA buffers are used as standards because even trace Ca^{2+} contaminants can significantly distort $[Ca^{2+}]_{free}$ values at low $[Ca^{2+}]$ (Bers *et al.*, 1994; Miyawaki *et al.*, 1997).

1. Prepare EGTA buffer [100 m*M* KCl, 50 m*M* HEPES (pH 7.4), 10 m*M* EGTA] and $CaC1_2$ buffer [100 m*M* KCl, 50 m*M* HEPES (pH 7.4), 10 m*M* EGTA, 10 m*M* $CaCl_2$]. pH should be held constant.
2. Use a 1-ml cuvette and dilute the sample in the EGTA buffer. Record the fluorescence emission spectrum from 450 to 570 nm at 433-nm excitation. Determine the emission ratio.
3. To obtain the Ca^{2+}-binding curve, add successive fractions of the $CaC1_2$ solution to the sample and determine the emission ratio. Given that the experiment is performed in 20°C with these EGTA and $CaCl_2$ buffers, the free calcium can be calculated by solving the quadratic equation: $[Ca^{2+}]_{free}{}^2 + (10{,}000{,}060.5 - [Ca^{2+}]_{total}) * [Ca^{2+}]_{free} - 60.5 * [Ca^{2+}]_{total} = 0$.
4. To produce the Ca^{2+}-binding curve, plot the $[Ca^{2+}]_{free}$ versus emission ratio change (percentage of maximum). Initial cameleons show biphasic Ca^{2+} dependency, whereas YC6.1 has a monophasic response.

TABLE 13.2 Fluorescent Properties of Fluorescent Protein Pairs Used for FRET Studies

Donor	Acceptor	Donor excitation wavelength (nm)	Donor emission wavelength (nm)	Acceptor emission wavelength (nm)
EBFP	EGFP	370	440	510
ECFP	EYFP	440	480	535
EYFP	mRFP1	510	535	607
Sapphire	DsRed	400	510	580

5. Extract the apparent dissociation constant (K'_d) and Hill coefficient (n) from the fitted curves.
6. The emission ratio can then be transformed to [Ca^{2+}]

D. Live Cell Cameleon Fluorescence Imaging

This section describes a Ca^{2+}-imaging experiment using HeLa cells; however, with minor modifications the method can be applied to other cellular and physiological contexts.

1. Plate HeLa cells on 35-mm-diameter glass-bottom dishes with DMEM–10% FBS media.
2. Incubate the cells at 37°C (5% CO_2) until cells are 50–80% confluent.
3. Transfect cells with the mammalian expression plasmid containing your cameleon using Lipofectamine and PLUS reagents (Invitrogen) according to the manufacturer's instructions.
4. Remove the transfection mixture after 5–18h and replace with fresh 1.5ml of DMEM–10% FBS media.
5. Incubate the cells at 37°C (5% CO_2) for 24h. The cells are ready to perform the Ca^{2+}-imaging experiment.
6. All data acquisition should be performed in a dark room to reduce background light.
7. Wash with 1ml of HBSS (+$CaCl_2$) and add 1ml of fresh HBSS (+$CaCl_2$).
8. Put the cells on the stage of the microscope. Cells are viable and healthy for at least 60 minutes at ambient conditions. A CO_2 box and temperature controller are required for long-term/extended time course experiments.
9. The MetaFluor software controls the shutters, filter exchangers, and camera during data acquisition.
10. Screen for EYFP fluorescence (which is brighter and distinguished more easily than ECFP fluorescence) using the eyepiece to find transfected cells expressing the cameleon construct. Turn the filter turret to the U-WNIBA band-pass mirror cube and use ND filter in theA range of 0.1–10% to reduce photobleaching depending on the intensity of fluorescence emission (Table 13.3).
11. Using the 40× oil objective, center the microscope viewing area on a cell that

TABLE 13.3 Recommended Optical Components[a]

	FRET system			
	BFP/GFP	**ECFP/EYFP**	**ECFP/EYFP/RFP**	**Sap/DsRed**
Excitation filter	365WB50	440AF21	440DF21 (ECFP) 510DF23 (EYFP)	400DF15
Dichroic mirror	400DCLP	455DRLP	450-520-590TBDR	455DRLP
Emission filter, donor	450DF65	480AF30	480AF30 (ECFP) 535AF26 (EYFP)	480DF30
Emission filter, acceptor	535AF45	535AF26	535AF26 (EYFP) 600ALP (RFP)	535DF25

[a] The components are from Omega Optical but can be substituted with several equivalent products from other companies.

has a healthy morphology and displays a strong cytosolic fluorescence.

12. Acquire single images on the computer screen using MetaFluor while adjusting the focus until you have the sharpest screen image. Focus through the eyepiece and the CCD usually vary slightly.
13. Turn the filter turret to the cube with the 455DRLP dichroic mirror. This allows visible light shorter and longer than ~455 nm to be reflected as excitation to the sample and collected as emission from the sample, respectively. Consult spectra from Omega Optical for relatively wavelength transmission efficiencies.
14. To monitor the peak emissions of ECFP and EYFP as a result of ECFP peak excitation over time, set the data acquisition conditions as follows: time interval to every 10 s and exposure time to 200 ms for ECFP and EYFP. Longer exposures improve resolution at the expense of bleaching the fluorescent signal. MetaFluor will display the emission ratio over time.
15. Draw a region of interest on the field of view of the CCD. Usually, this region will outline the whole cell, but one can specify only a portion of a cell if desired (e.g., the nucleus). It is important that the stage or the cell does not move during the observation period as the region initially drawn may drift from the region of interest. Additionally, the intensity in the region (the signal) should be at least five times the intensity of the background.
16. The emission intensities of both ECFP and EYFP will decrease over the course of the experiment due to some unavoidable photobleaching; however, the effect on the emission ratio should be negligible. To reduce photobleaching, decrease exposure time and excitation light intensity. Also, binning can sum the signal from multiple pixels on the CCD camera so that less light is required while keeping a good signal-to-noise ratio.
17. MetaFluor will record the fluorescence emission intensities of the ECFP and EYFP, together with their emission ratios in the regions over time.
18. When the emission ratio reaches a steady state, add 50 μl of 2 m*M* histamine to the culture dish for a final concentration of 100 μ*M*. Be careful not to move the culture dish in this process. The histamine binds to cell receptors on the plasma membrane that set off a signaling cascade, resulting in the release of Ca^{2+} from the endoplasmic reticulum through the inositol-1,4,5-triphosphate receptor. This should cause a conformational change in the cameleon that can be observed by a rise in emission intensity of EYFP and a decline in ECFP intensity. Therefore, the emission ratio should increase. The EYFP/ECFP emission ratio should return to steady-state levels when the effect of the histamine wanes. Figure 13.2 shows a representative example of MetaFluor software as it is collecting data.
19. In order to correlate the emission ratio to $[Ca^{2+}]_{cytosolic}$, it is necessary to determine R_{min} and R_{max} so that emission ratios can be mapped to the Ca^{2+}-binding curve. Add 50 μl of 20 μ*M* ionomycin for a final concentration of 1 μ*M*. Ionomycin open pores on the plasma membrane to allow permeability to Ca^{2+} ions. Because the medium is saturated with $CaCl_2$, the ratio will rise to R_{max}. To determine R_{min}, add 50 μl of 100 m*M* EGTA and 600 μ*M* BAPTA-AM for a final concentration of 5 m*M* and 30 μ*M*, respectively. The emission ratio should drop to R_{min}.

IV. OTHER APPROACHES

Several other Ca^{2+} probes are also being developed, including the so-called "camgaroos" and "pericams." Camgaroos take an alternative

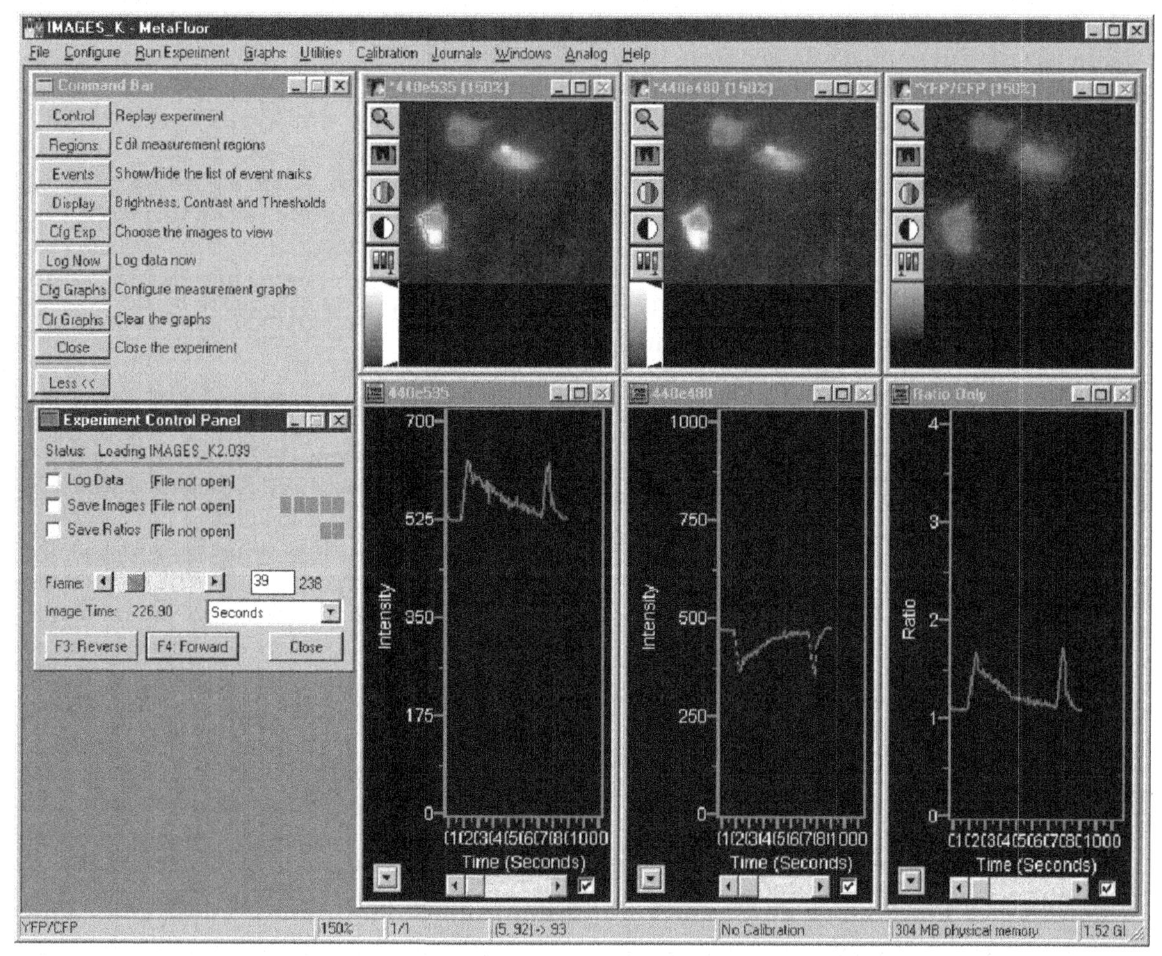

FIGURE 13.2 Example of Ca^{2+} imaging experiment using MetaFluor software. The region of observation is highlighted in the 440e535 panel. The 440e535 panel and graph plot the change in EYFP fluorescence as a result of ECFP excitation; the 440e480 panel and graph plot the change in ECFP fluorescence as a result of ECFP excitation; the EYFP/ECFP panel and graph are the ratio of 440e535 and 440e480 panels and graphs. In this experiment, the graphs display a sharp rise in $[Ca^{2+}]_c$ from the initial stimulation with histamine followed by a slow decline in $[Ca^{2+}]_c$ to baseline levels. The second stimulation with histamine causes a significantly more rapid return to baseline levels.

approach to designing fluorescent Ca^{2+} sensors based on CaM and GFP family members (Baird *et al.*, 1999). While ECFP and EYFP in cameleons are appended to the amino and carboxyl termini of CaM and Ca^{2+} binding is detected by FRET, camgaroo indicator proteins take advantage of the robust structure and profound fluorescence sensitivity of GFP to altered pK_a values and chromophore orientation. Circular permutations and insertion of whole CaM in place of Tyr-145 within EEYFP thereby render this indicator responsive to Ca^{2+} binding. As a result, both excitation and emission spectra of camgaroo simply increase in amplitude by up to seven-fold upon saturation with Ca^{2+}, without any significant shift in peak wavelength. This Ca^{2+}-dependent fluorescence enhancement is substantially larger than other published genetically encoded fluorescent indicators. However, camgaroos are limited by pH sensitivity inherent in the current mechanism of modulating fluorescence via changes in pK_a of the chromophore and have to date only been preliminarily subjected to systematic mutational improvement.

Pericams, in which EYFP is circularly fused to CaM and the M13 myosin light chain kinase peptide, improve approximately 10-fold upon the low affinity of camgaroo for Ca^{2+} (K_d = 7 μM) and are thereby better capable of sensing low physiological changes in intracellular [Ca^{2+}] (Nagai *et al.*, 2001). Taken together, these strategies offer alternatives complementary to cameleons for creating genetically encoded, physiological Ca^{2+} indicators.

V. OTHER CONSIDERATIONS AND PITFALLS

A. Ca^{2+} Ion Sensitivity

Although cameleon probes vary in their Ca^{2+} affinities, most are suitable for monitoring [Ca^{2+}] between 0.5 and 100 μM. This poses a problem for examining the relatively high [Ca^{2+}] found in the endoplasmic reticulum of resting cells (approximately 500 μM). However, CaM mutagenesis studies have shown that substitution of a conserved glutamic acid residue at the 12th position of each Ca^{2+}-binding loop abolishes its Ca^{2+}-binding ability (Zhu *et al.*, 1998). The effect of combinations of these mutations on cameleon Ca^{2+} range is currently being examined further in our laboratory.

B. Maturation

GFP variants have been developed in which chromophore oxidative maturation (and thereby become fluorescent) occurs more quickly and efficiently at 37°C. It would be advantageous to utilize efficiently folding versions, such as the recently developed "Venus" form of EYFP (F46L/F64L/M153T/V163A/S175G). These EYFP mutations confer an eight-fold increase of fluorescence intensity when expressed in mammalian cells (Nagai *et al.*, 2002; Rekas *et al.*, 2002). This will enable assay of cells 24 h postrecovery transfection, if so desired.

C. pH Sensitivity

The hydrogen bond network within the β barrel of the chromophore is sensitive to external pH. Hence, in order to analyze Ca^{2+} levels in acidic organelles (such as secretory vesicles), the FRET donor/acceptor pair must be engineered so that it is pH resistant. Two mutations within EEYFP (V68L and Q69K) have been shown to decrease its pK_a to 6.1 (Miyawaki *et al.*, 1999). The pH sensitivity was improved further via a Q69M or "citrine" mutation (pK_a 5.7; Griesbeck *et al.*, 2001). Another approach would be to change the donor/acceptor pair to the pH-insensitive sapphire-red cameleon probe (SapRC2). Although this construct has a tendency to aggregate and form homotetramers, the recent engineering of a monomeric mRFP1 offers an interesting alternative (Campbell *et al.*, 2002).

D. Oligomerization

GFP family proteins have been observed to form obligate dimers and may thereby generate false-positive FRET signals. However, this aggregation problem need not preclude their use in biological systems, even when present in higher local concentrations. Nonoligomerizing mutants of EYFP have been suggested from its crystal structure (Wachter *et al.*, 1998; Rekas *et al.*, 2002), but these have yet to be validated experimentally. Alternatively, using a monomeric version of the evolutionary distinct RFP (mRFP1) in tandem with a EYFP donor would also serve to eliminate this issue (Campbell *et al.*, 2002).

E. Influences on Biological Systems

It is very important to consider potential competition of the FRET indicator for native CaM or CaM-dependent enzymes. Previous comparisons of the effect of recombinant CaM and cameleon chimeras on prototypical CaM-dependent enzymes have revealed that the primary effect of cameleons is on buffering $[Ca^{2+}]$ and not interfering with CaM-mediated signaling (Miyawaki *et al.*, 1999). This could be due to the CaM component of the cameleon being inhibited by the adjoining CKKp efficiently occupying its substrate-binding site. Also, as the YC6.1 CKKp is embedded within the cameleon polypeptide, it is not likely to interact with endogenous CaM proteins.

F. Interpretation of Live Cell FRET Data

There are two commonly used simple and practical approaches. The first, measurement of donor emission quenching and acceptor emission enhancement by using three filter sets and then mathematical processing to determine emission/FRET ratios, is described in Section III. The alternative approach is by detection of donor dequenching following acceptor bleaching. We find bleaching with a minimum of 200 ms, compared to 1 ms for control excitation, is sufficient.

G. Additional Ways to Do Ratio Imaging

FRET requires rapid intensity measurements at different wavelengths. Switching time (in the millisecond range) may be an important parameter for some applications. New imaging systems have become available in the marketplace, notably TILLvisION (T.I.L.L. Photonics) and AquaCosmos (Hamamatsu), which allow for excellent time resolution for fluorescence–intensity ratio imaging.

H. FRET Using Red Fluorescence Protein

For YC6.1 applications, the ECFP-EYFP FRET filter set is sufficient. However, if you are using RFP for FRET, the following dichoric mirrors and filters from Omega Optical (or equivalents) will be needed: 450-520-590TBDR for the dichoric mirror; 440DF21 for ECFP excitation; 510DF23 for EYFP excitation; 575DF26 for RFP excitation; 480AF30 for ECFP emission; 535AF26 for EYFP emission; and 600ALP for RFP emission. This filter set allows you to excite or acquire emission from ECFP, EYFP, and RFP individually, albeit with a tradeoff in efficiency.

I. Imaging Systems

Using a confocal microscope is the best way to increase spatial resolution for FRET experiments. Confocal YC6.1 measurements can be performed with single-photon excitation using the 458-nm line of an argon laser, but much more efficient excitation of ECFP is attained with the 442-nm line of a HeCd laser. Two-photon excitation microscopy, in addition to providing optical sections of a specimen as with confocal microscopy, offers certain advantages. Its applicability to cameleons has been demonstrated using video-rate scanning instrumentation (Fan *et al.*, 1999). This latter imaging approach may not be readily available to most laboratories, however.

J. Preparation of CaM^{2+}/EGTA Buffers

The accuracy of the Ca^{2+}-binding curve depends on accurate preparation of the Ca^{2+}/EGTA and Ca^{2+}/HEEDTA systems below

$10^{-5}M$ free Ca^{2+} and unbuffered Ca^{2+} above (Bers *et al.*, 1994). The purity of EGTA, temperature, and pH are all practical issues.

Acknowledgments

We are grateful to Atsushi Miyawaki for his help in setting up a FRET microscope system in our laboratory, as well as for much advice on the use of GFP variants. This work was supported by grants from the Cancer Research Society Inc. and the Institute for Cancer Research of the Canadian Institutes of Health Research (CIHR). K.P.H. is a recipient of a NCIC Research Fellowship, K.T. holds a CIHR scholarship, and M.I. is a CIHR senior investigator.

References

Allen, G. J., Kwak, J. M., Chu, S. P., Llopis, J., and Tsien, R. Y. (1999). Cameleon calcium indicator reports cytoplasmic calcium dynamics in Arabidopsis guard cells. *Plant J.* **19**, 735–747.

Baird, G. S., Zacharias, D. A., and Tsien, R. Y. (1999). Circular permutation and receptor insertion within green fluorescent proteins. *Proc. Natl. Acad. Sci. USA* **96**, 11241–11246.

Bers, D. M., Patton, C. W., and Nuccitelli, R. (1994). A practical guide to the preparation of Ca^{2+} buffers. *Methods Cell Biol.* **40**, 3–29.

Campbell, R. E., Tour, O., Palmer, A. E., Steinbach, P. A., Baird, G. S., Zacharias, D. A., and Tsien, R. Y. (2002). A monomeric red fluorescent protein: *Proc. Natl. Acad. Sci. USA* **99**, 7877–7882.

Carafoli, E. (2003). The calcium-signalling saga: Tap water and protein crystals. *Nature Rev. Mol. Cell Biol.* **4**, 326–332.

Demaurex, N., and Frieden, M. (2003). Measurements of the free luminal ER Ca(2+) concentration with targeted "cameleon" fluorescent proteins. *Cell Calcium* **34**, 109–119.

Emmanouilidou, E., Teschemacher, A. G., Pouli, A. E., Nicholls, L. I., Seward, E. P., and Rutter, G. A. (1999). Imaging Ca^{2+} concentration changes at the secretory vesicle surface with a recombinant targeted cameleon. *Curr. Biol.* **9**, 915–918.

Fan, G. Y., Fujisaki, H., Miyawaki, A., Tsay, R. K., Tsien, R. Y., and Ellisman, M. H. (1999). Video-rate scanning two-photon excitation fluorescence microscopy and ratio imaging with cameleons. *Biophys. J.* **76**, 2412–2420.

Griesbeck, O., Baird, G. S., Campbell, R. E., Zacharias, D. A., and Tsien, R. Y. (2001). Reducing the environmental sensitivity of yellow fluorescent protein: Mechanism and applications. *J. Biol. Chem.* **276**, 29188–29194.

Grynkiewicz, G., Poenie, M., and Tsien, R. Y. (1985). A new generation of Ca^{2+} indicators with greatly improved fluorescence properties. *J. Biol. Chem.* **260**, 3440–3450.

Ikeda, M., Sugiyama, T., Wallace, C. S., Gompf, H. S., Yoshioka, T., Miyawaki, A., and Allen, C. N. (2003). Circadian dynamics of cytosolic and nuclear Ca^{+} in single suprachiasmatic nucleus neurons. *Neuron* **38**, 253–263.

Isshiki, M., Ying, Y. S., Fujita, T., and Anderson, R. G. (2002). A molecular sensor detects signal transduction from caveolae in living cells. *J. Biol. Chem.* **277**, 43389–43398.

Kerr, R., Lev-Ram, V., Baird, G., Vincent, P., Tsien, R. Y., and Schafer, W. R. (2000). Optical imaging of calcium transients in neurons and pharyngeal muscle of *C. elegans*. *Neuron* **26**, 583–594.

Liu, L., Yermolaieva, O., Johnson, W. A., Abboud, F. M., and Welsh, M. J. (2003). Identification and function of thermosensory neurons in Drosophila larvae. *Nature Neurosci.* **6**, 267–273.

Miyawaki, A. (2003). Visualization of the spatial and temporal dynamics of intracellular signaling. *Dev. Cell* **4**, 295–305.

Miyawaki, A., Griesbeck, O., Heim, R., and Tsien, R. Y. (1999). Dynamic and quantitative Ca^{2+} measurements using improved cameleons. *Proc. Natl. Acad. Sci. USA* **96**, 2135–2140.

Miyawaki, A., Llopis, J., Heim, R., McCaffery, J. M., Adams, J. A., Ikura, M., and Tsien, R. Y. (1997). Fluorescent indicators for Ca^{2+} based on green fluorescent proteins and calmodulin. *Nature* **388**, 882–887.

Mizuno, H., Sawano, A., Eli, P., Hama, H., and Miyawaki, A. (2001). Red fluorescent protein from Discosoma as a fusion tag and a partner for fluorescence resonance energy transfer. *Biochemistry* **40**, 2502–2510.

Montero, M., Brini, M., Marsault, R., Alvarez, J., Sitia, R., Pozzan, T., and Rizzuto, R. (1995). Monitoring dynamic changes in free Ca^{2+} concentration in the endoplasmic reticulum of intact cells. *EMBO J.* **14**, 5467–5475.

Nagai, T., Ibata, K., Park, E. S., Kubota, M., Mikoshiba, K., and Miyawaki, A. (2002). A variant of yellow fluorescent protein with fast and efficient maturation for cell-biological applications. *Nature Biotechnol.* **20**, 87–90.

Nagai, T., Sawano, A., Park, E. S., and Miyawaki, A. (2001). Circularly permuted green fluorescent proteins engineered to sense Ca^{2+}. *Proc. Natl. Acad. Sci. USA* **98**, 3197–31202.

Rekas, A., Alattia, J. R., Nagai, T., Miyawaki, A., and Ikura, M. (2002). Crystal structure of venus, a yellow fluorescent protein with improved maturation and reduced environmental sensitivity. *J. Biol. Chem.* **277**, 50573–50578.

Truong, K., and Ikura, M. (2001). The use of FRET imaging microscopy to detect protein-protein interactions and protein conformational changes in vivo. *Curr. Opin. Struct. Biol.* **11**, 573–578.

Truong, K., Sawano, A., Mizuno, H., Hama, H., Tong, K. I., Mal, T. K., Miyawaki, A., and Ikura, M. (2001). FRET-based *in vivo* Ca^{2+} imaging by a new calmodulin-GFP fusion molecule. *Nature Struct. Biol.* **8**, 1069–1073.

Wachter, R. M., Elsliger, M. A., Kallio, K., Hanson, G. T., and Remington, S. J. (1998). Structural basis of spectral shifts in the yellow-emission variants of green fluorescent protein. *Structure* **6**, 1267–1277.

Zhang, J., Campbell, R. E., Ting, A. Y., and Tsien, R. Y. (2002). Creating new fluorescent probes for cell biology. *Nature Rev. Mol. Cell. Biol.* **3**, 906–918.

Zhu, T., Beckingham, K., and Ikebe, M. (1998). High affinity Ca^{2+} binding sites of calmodulin are critical for the regulation of myosin Ibeta motor function. *J. Biol. Chem.* **273**, 20481–20486.

CHAPTER

14

Ratiometric Pericam

Atsushi Miyawaki

OUTLINE

I. INTRODUCTION

Our understanding of the structure–photochemistry relationships of green fluorescent protein (GFP) (Tsien, 1998) has enabled the development of genetic calcium probes based on a circularly permuted GFP (cpGFP) in which the amino and carboxyl portions have been interchanged and reconnected by a short spacer between the original termini (Baird *et al.*, 1999). The resulting new amino and carboxyl termini of the cpGFP have been fused to calmodulin and its target peptide M13, generating a chimeric protein named pericam (Nagai *et al.*, 2001). This new protein was fluorescent, and its spectral properties changed reversibly with Ca^{2+} concentration, probably due to the interaction between calmodulin and M13, which alters the environment surrounding the chromophore. Three types of pericam have been obtained by mutating several amino acids adjacent to the chromophore. Of these, "flash pericam" becomes fluorescent with increasing Ca^{2+}, whereas "inverse pericam" dims. However, "ratiometric pericam" has an excitation wavelength that changes in a Ca^{2+}-dependent manner, thereby enabling dual-excitation ratiometric Ca^{2+} imaging. Ratiometric dyes permit quantitative Ca^{2+} measurements by minimizing the effects of several artifacts that are unrelated to changes in the concentration of free Ca^{2+} ($[Ca^{2+}]$), such as uneven loading or partitioning of dye within the cell or varying cell thickness. This article presents an outline of an imaging experiment using HeLa cells expressing ratiometric pericam to measure receptor-stimulated changes in intracellular $[Ca^{2+}]$ ($[Ca^{2+}]_i$) (Protocol 1).

In contrast to cameleons (Miyawaki *et al.*, 1997), which are fluorescence resonance energy transfer-based Ca^{2+} indicators, pericams can be easily targeted into the mitochondria matrix using an upstream targeting sequence encoding subunit IV of cytochrome *c* oxidase. Ratiometric pericam has been used successfully to monitor changes in $[Ca^{2+}]$ in mitochondria ($[Ca^{2+}]_m$) (Nagai *et al.*, 2001). In most dual-excitation imaging experiments, the excitation wavelength is alternated using a rotating wheel containing two band-pass filters. However, it is also possible to use a high-speed grating monochromator to increase the rate at which the ratio measurement is conducted to approximately 10 Hz. The latter instrumentation enables the measurement in spontaneously contracting cardiac myocytes of beat-to-beat changes in $[Ca^{2+}]_m$, contributing to the demonstration that $[Ca^{2+}]_m$ oscillates synchronously with cytosolic $[Ca^{2+}]$ during beating (Robert *et al.*, 2001). This article discusses factors to be considered when using such a monochromator with ratiometric pericam.

Although the aforementioned measurements are typically performed using conventional microscopy, the monitoring of changes in $[Ca^{2+}]$ is often severely limited by the poor spatiotemporal resolution of such wide-field techniques. To obtain a more reliable representation of changes in subcellular $[Ca^{2+}]$, it is necessary to increase the *z*-axis resolution and the speed of production and collection of the ratios of the excitation peaks. This article provides a detailed protocol describing a modified laser-scanning confocal microscopic (LSCM) system for ratiometric pericam (Protocol 2). In our experiment (Shimozono *et al.*, 2002), fast exchange between two laser beams was achieved using acousto-optic tunable filters (AOTFs). Samples were scanned on each line sequentially by a violet laser diode (408 nm) and a diode-pumped solid-state laser (488 nm). In this way, the ratios of the excitation peaks can be obtained at a frequency of up to 200 Hz.

II. MATERIALS

HeLa cells (ATCC # CCL-2.2)
Culture medium: Add fetal bovine serum (Gibco BRL, Life Technologies) to Dulbecco's modified Eagle's medium (DMEM) (Sigma Aldrich) to a final concentration of 10% (v/v)
Hank's balanced salt solution (HBSS): Dissolve $CaCl_2$ (0.14 g), KCl (0.40 g), KH_2PO_4 (0.06 g), $MgCl_2 \cdot 6H_2O$ (0.10 g), $MgSO_2 \cdot 7H_2O$ (0.10 g), NaCl (8.00 g), $NaHCO_3$ (0.35 g), Na_2HPO_4 (0.048 g), and d-glucose (1.00 g) in 800 ml of distilled H_2O, adjust the pH to 7.4, and then adjust the volume to 1 liter.
Histamine solution: Dissolve 1.841 mg of histamine dihydrochloride (Sigma Aldrich, H7125) in 1 ml of HBSS to make a 10 m*M* stock solution.
Ratiometric pericam cDNA
Ratiometric pericam mitochondrial cDNA
35-mm glass-bottom cell culture plates (Matsunami-glass, Osaka, Japan)

Interference Filters

400DF15, XF1006, Omega
485DF15, XF1042, Omega
535AF45, XF3084, Omega
505DRLP, XF2010, Omega

505DRLP-XR, XF2031, Omega
500ALP, XF3092, Omega
DM420SP, Olympus
FV5-DM442, Olympus
BA505-525, Omega

III. INSTRUMENTATION

A. Conventional Microscopy for Time-Lapse $[Ca^{2+}]_i$ Imaging

Dual-excitation imaging with ratiometric pericam uses two excitation filters (485DF15 and 400DF15), which are alternated by a filter changer (Lambda 10-2, Sutter Instruments, Novato, CA), a 505DRLP-XR dichroic mirror, and a 535AF45 emission filter. The excitation and emission spectra of ratiometric pericam in the presence and absence of Ca^{2+} with passbands of the emission filter and two excitation filters are shown in Fig. 14.1A. Also, the transmittance of the dichroic mirror (505DRLP-XR) is superimposed. The dichroic mirror has an eXtended Reflection region below 505 nm; it was designed originally for the measurement of Ca^{2+} (fura-2) and pH (BCECF). In this situation, the broad reflection of a dichroic mirror is imperative. A more common long-pass dichroic mirror (505DRLP) cannot be used; superimposition of its transmission spectrum (Fig. 14.1B) makes one notice a complex transmission band at short wavelengths, which prevents reflection of the 400-nm light. The author strongly recommends that researchers make graphs of spectra plotted as percentage transmittance for the interference filters (excitation filters, emission filters, and dichroic mirrors) that are actually used, together with excitation and emission spectra of the relevant fluorescent dyes. The transmittance curves for the filters (normal incidence) and dichroic mirrors (45° incidence) can be measured easily using a conventional spectrophotometer. This measurement also provides a chance for researchers to check the quality of their interference filters, which may deteriorate with time.

B. A Modified LSCM System for Ratiometric Pericam

Although the following procedure assumes familiarity with LSCM, and with the Olympus LSCM in particular, the procedure can be adapted easily to other LSCM types. The microscope and lasers we use are described

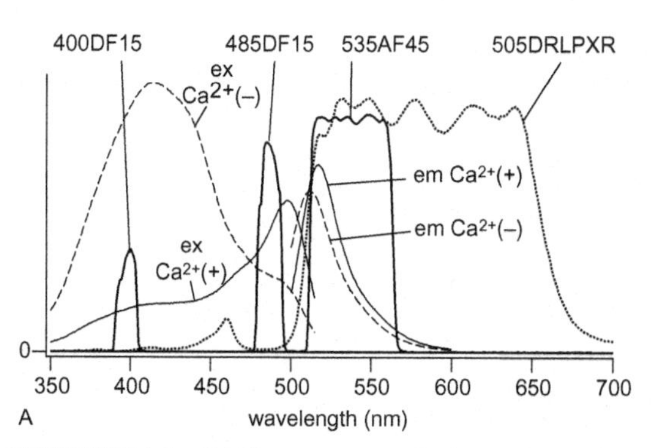

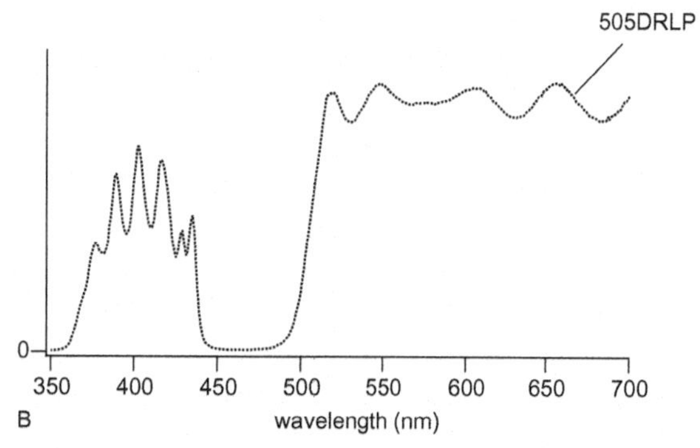

FIGURE 14.1 (A) Fluorescence excitation and emission spectra of ratiometric pericam in the presence (solid line) and absence (broken line) of Ca^{2+}. Transmittance spectra for filters (400DF15, 485DF15, and 535AF45) and the dichroic mirror (505DRLPXR) are shown with solid and dotted lines, respectively. (B) The transmittance spectrum for a 505DRLP dichroic mirror.

in Fig. 14.2. The beam from a diode-pumped solid-state laser (Sapphire 488-20, COHERENT) directly enters an AOTF (AOTF1; AOTF.8C, AA Opto-Electronic). The light is relayed through an optical fiber (FV5-FUR, Olympus) to a confocal microscope scan head and passes through a dichroic mirror (DM1; DM420SP, Olympus). A laser diode (NLHV3000E, NICHIA) is mounted on a heat sink with temperature control, which consists of an LD mount and heat sink [F125-4A (Suruga Seiki)], LD current driver [LDX-3525 (ILX Lightwave)], and LD temperature controller [LDT5525 (ILX Lightwave)]. The output of the laser diode is collimated (made parallel) using multiple lenses, and its power is stabilized by a feedback-regulated device. The beam is directed into another AOTF (AOTF2; AOTF.4C UV, AA Opto-Electronic). The diffracted light is sent through another fiber-optic line (FV5-FUR-UV, Olympus) to the scan head and is then reflected on DM1. The two AOTFs are controlled electronically, allowing the reciprocal choice of one of the two laser lines. The chosen laser beam is reflected on a second dichroic mirror (DM2; FV5-DM442, Olympus). The scan head is coupled to an inverted microscope (IX70, Olympus) through its right-side port. The objective lens used is a PlanApo 60× N.A. 1.00

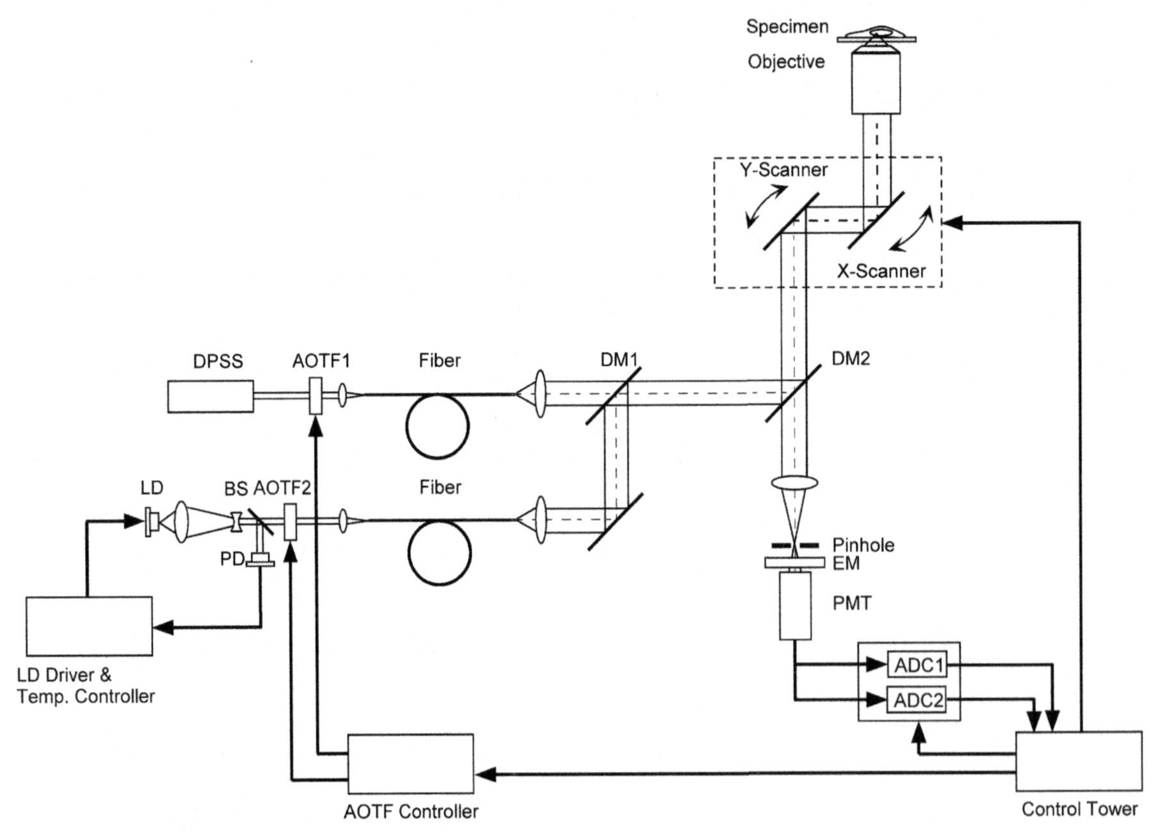

FIGURE 14.2 Schematic diagram of the laser-scanning confocal microscopy system for fast dual-excitation ratiometric imaging. DPSS, diode pumped solid-state laser; LD, laser diode; PD, photodiode; BS, beam splitter; DM, dichroic mirror; EM, emission filter; PMT, photomultiplier tube; ADC, analog-to-digital converter.

WLSM (Olympus). The fluorescent light is descanned, passed through DM2, a pinhole, and an emission filter (BA505-525 (Omega)), and detected by a photomultiplier. Fluorescence signals with excitation wavelengths of 408 and 488 nm are distributed into two channels during analog-to-digital conversion. The pair of galvanometer mirrors, the digitized detector output, and the AOTF controller are orchestrated. The excitation and emission spectra of ratiometric pericam in the presence and absence of Ca^{2+}, together with the passband of the emission filter (BA505-525) and the two laser lines, are shown in Fig. 14.3.

IV. PROCEDURES

Protocol 1

The procedure for time-lapse $[Ca^{2+}]_i$ imaging in HeLa cells is as follows.

1. Plate HeLa cells or cells of interest onto a 35-mm glass-bottom dish with culture medium.

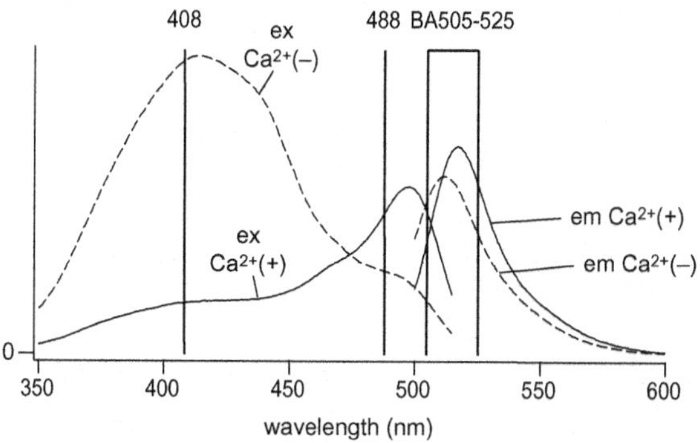

FIGURE 14.3 Fluorescence excitation and emission spectra of ratiometric pericam in the presence (solid line) and absence (broken line) of Ca^{2+}. The wavelengths of the two laser lines, 408 nm (laser diode) and 488 nm (diode-pumped solid state), are indicated by vertical lines. The wavelengths that pass through the emission filter BA505–525 are shown by a box. Modified with permission from T. Nagai, A. Sawano, E. S. Park, and A. Miyawaki, Proc. Natl. Acad. Sci. U.S.A. 98, 3197 (2001). Copyright (2001) National Academy of Sciences, U.S.A.

2. Transfect the cells with 1 μg per dish of cDNA encoding ratiometric pericam mitochondria using Lipofectin according to the manufacturer's instructions.
3. Between 2 and 10 days after cDNA transfection, image HeLa cells on an inverted microscope (IX7) with a cooled CCD camera (MicroMax or Cool Snap HQ, Roper Scientific, Tucson, AZ). Expose cells to reagents in HBSS containing 1.26 m*M* $CaCl_2$. Image acquisition and processing are controlled by a personal computer connected to a camera and a filter wheel (Lambda 10-2, Sutter Instruments, San Rafael, CA) using the program MetaFluor (Universal Imaging, West Chester, PA). The excitation filter wheel in front of the xenon lamp (Lambda 10-2, Sutter Instruments, San Rafael, CA) is also under computer control. Excitation light from a 75-W xenon lamp is passed through a 400DF15 (400 ± 7.5) or 485DF15 (485 ± 7.5) excitation filter. The light is reflected onto the sample using a 505-nm long-pass (505DRLP-XR) dichroic mirror with an extended reflection. The emitted light is collected with a 40X (numerical aperture: 1.35) objective and passed through a 535 ± 22.5-nm band-pass filter (535AF45). Interference filters are from Omega Optical or Chroma Technologies (Brattleboro, VT).
4. Define several factors for image acquisition, including (i) excitation power, which depends on the type of light source and neutral density filter, (ii) numerical aperture of the objective, (iii) time of exposure to the light, (iv) image acquisition interval, and (v) binning. The last three factors should be considered in terms of whether temporal or spatial resolution is pursued.
5. Choose moderately bright cells. Select regions of interest so that pixel intensities are averaged spatially.
6. At the end of an experiment, convert fluorescence signals into values of $[Ca^{2+}]$.

R_{max} and R_{min} can be obtained as follows. To saturate the intracellular indicator with Ca^{2+}, increase the extracellular $[Ca^{2+}]$ to 10–20 m*M* in the presence of 1–5 μ*M* ionomycin. Wait until the fluorescence intensity reaches a plateau. Then, to deplete the Ca^{2+} indicator, wash the cells with Ca^{2+}-free medium (1 μ*M* ionomycin, 1 m*M* EGTA, and 5 m*M* $MgCl_2$ in nominally Ca^{2+}-free HBSS). The *in situ* calibration for $[Ca^{2+}]$ uses the equation $[Ca^{2+}] = K'_d[(R - R_{min})/(R_{max} - R)]^{(1/n)}$, where K'_d is the apparent dissociation constant corresponding to the Ca^{2+} concentration at which R is midway between R_{max} and R_{min} and n is the Hill coefficient. The Ca^{2+} titration curve of ratiometric pericam can be fitted using a single K'_d of 1.7 μ*M* and a single Hill coefficient of 1.1 (Fig. 14.4). A typical time course of $[Ca^{2+}]_i$ reported by ratiometric pericam is shown in Fig. 14.5.

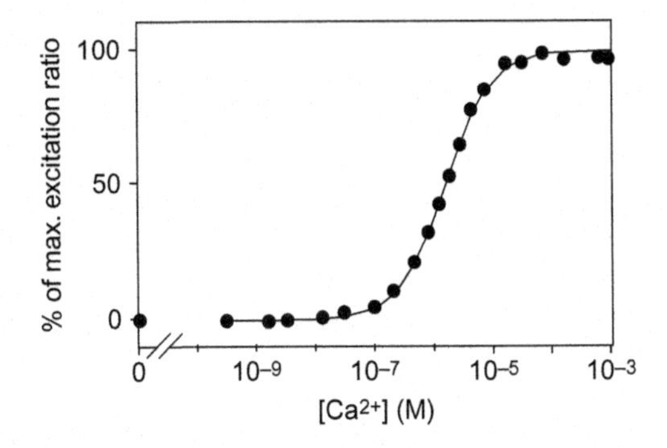

FIGURE 14.4 The Ca^{2+} titration curve of ratiometric pericam. Modified with permission from T. Nagai, A. Sawano, E. S. Park, and A. Miyawaki, Proc. Natl. Acad. Sci. U.S.A. 98, 3197 (2001). Copyright (2001) National Academy of Sciences, U.S.A.

Protocol 2

The following is a procedure for confocal imaging of $[Ca^{2+}]_m$ in HeLa cells using the modified LSCM system.

1. Plate HeLa cells or cells of interest onto a 35-mm glass-bottom dish with culture medium.
2. Transfect the cells with 1 μg per dish of cDNA encoding ratiometric pericam mitochondria using Lipofectin according to the manufacturer's instructions.
3. Incubate the cells in a humidified atmosphere at 37°C under 5% CO_2 for 1 to 3 days.
4. Replace the culture medium with HBSS.
5. Observe the cells on an inverted microscope (IX70, Olympus) for green fluorescence of the indicator (ratiometric pericam-mt) using a 490-nm excitation light generated using 490DF10.
6. Choose moderately bright cells in which the fluorescence is well distributed in the mitochondria.
7. Switch on the lasers, AOTF controller, scanning head, and the computer. The lasers need 30 min to warm up and stabilize.
8. Start the scanning process with only excitation from the 408-nm laser diode. Set the scan mode to "Normal (unidirectional)" and scan speed to "Fast." While looking at the acquired fluorescence images, adjust the laser intensity to the minimum level that allows easy identification of individual cells, adjust the sensitivity of the photomultiplier tube (PMT) for an optimal signal-to-noise ratio, and adjust the size of the pinhole for acceptable depth of the image. There is a trade-off between the scan speed and the pixel size: faster scanning speeds decrease resolution because fewer pixels of information are collected.
9. Adjust the intensity of the 488-nm laser by setting the microscope to "Normal (unidirectional)" scan mode and "Fast" scan speed. While looking at the acquired fluorescence images, adjust the laser intensity to a minimum level that allows easy identification of individual cells and adjust the sensitivity of the PMT for an optimal signal-to-noise ratio and the size of

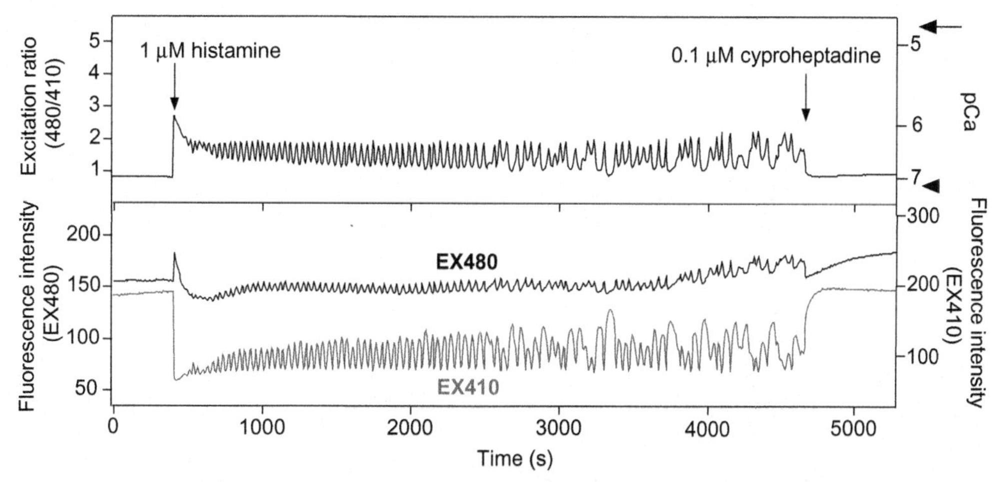

FIGURE 14.5 Typical $[Ca^{2+}]_i$ transients and oscillations induced by receptor stimulation in HeLa cells expressing ratiometric pericam. The sampling interval was 3–5 s. (Top) Excitation ratios of 480 and 410 nm. The right-hand ordinate indicates $[Ca^{2+}]_i$ in μM with R_{max} and R_{min} indicated by an arrow and arrowhead, respectively. (Bottom) Excitations of 480 nm (black line, left-hand scale) and 410 nm (gray line, right-hand scale). Modified with permission from T. Nagai, A. Sawano, E. S. Park, and A. Miyawaki, Proc. Natl. Acad. Sci. U.S.A. 98, 3197 (2001). Copyright (2001) National Academy of Sciences, U.S.A.

the pinhole for acceptable depth of the image. Establishing the microscope settings described in steps 8 and 9 is a process that must be iterated to produce adequate ratio images.

10. Start the control software for line-sequential dual-excitation ratio analysis. The "Normal" scan mode and "Fast" scan speed, with adequate pixel size, enable rapid exchange between the two lasers on a line every 5 to 20 ms.
11. Determine the desired scanning field (image size) and reduce the height of the image so that 5 to 10 frames can be obtained per second.
12. Start to acquire images every 100 to 200 ms for 1 to 5 min.
13. During image acquisition, add histamine solution to a final concentration of 10 μM for receptor stimulation.
14. Determine the image ratio by dividing the images acquired with excitation at 488 nm by those acquired at 408 nm.

V. COMMENTS

A. pH and Photochromism as Practical Considerations

Ratiometric pericam is sensitive to changes in pH in both the presence and the absence of Ca^{2+} (Fig. 14.6). pH-related artifacts were not an issue in experiments that used HeLa cells because agonist-induced $[Ca^{2+}]_c$ mobilization did not induce any intracellular pH changes detectable by the pH indicator, BCECF (data not shown). However, ratiometric pericam expressed in dissociated hippocampal neurons was perturbed by acidification following depolarization or glutamate stimulation (data not shown).

Pericams are derived from yellow fluorescent protein (YFP). If YFP is excited too strongly, its fluorescence will be reduced. This apparent bleaching is actually photochromism because the fluorescence recovers to some extent spontaneously and can be restored further by UV illumination (Dickson *et al.*, 1997). Intense excitation of ratiometric pericam also causes photochromism, which results in a decrease in the 490/410-nm excitation ratio independent of Ca^{2+} change. The extent of photochromism is dependent on excitation power, numerical aperture of the objective, and exposure time. Therefore, it is necessary to optimize these factors for each cell sample in order to minimize photochromism while preserving a high signal-to-noise ratio. A good solution is to bin pixels at the cost of spatial resolution. The increased signal-to-noise ratio permits a decrease in intensity of the excitation light with a neutral-density filter and the observation of $[Ca^{2+}]_i$ oscillations without significant photochromism of the indicators.

B. Use of a High-Speed Grating Monochromator

Instead of a filter wheel containing 400DF15 and 485DF15 excitation filters, a fast wavelength exchanger based on a grating monochromator (U7773-XX and U7794-16, Hamamatsu Photonics) can be used with a fast-acquisition CCD camera (HiSCA, C6790-81, Hamamatsu Photonics). The spectrum of the excitation light when the wavelength was set at 410 or 490 nm is shown in Fig. 14.7. Because light acquired with the setting at 490 nm spills over into the observing optical path, it is preferable to eliminate the unwanted light by putting a short-pass filter like 500SP (Fig. 14.7, broken line) in front of the microscope.

C. Calcium Transients in Motile Mitochondria

When changes in $[Ca^{2+}]_m$ are monitored by alternating the excitation wavelength automatically with wide-field conventional microscopy, the rate of acquisition of the excitation ratio is about 10 Hz, which is identical to the frame rate. Despite this rapid rate of data collection, the $[Ca^{2+}]_m$ measurements are often affected adversely by the active motion of mitochondria, especially at warmer temperatures. Using our LSCM technique, we attempted to increase the speed of the alternation of excitation wavelengths so that it was faster than the movement of mitochondria. With this method, the frame rate was 5 Hz and the rate of ratio acquisition was 200 Hz. Although the frame rate did not

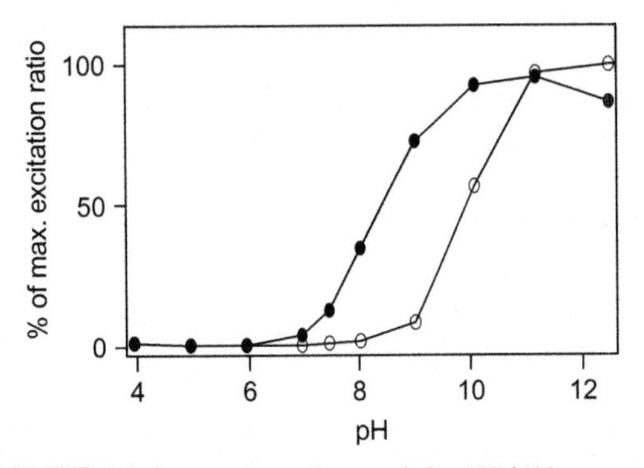

FIGURE 14.6 pH dependency of the 495/410-nm excitation ratio in the presence (●) and absence (○) of Ca^{2+}. Modified with permission from T. Nagai, A. Sawano, E. S. Park, and A. Miyawaki, Proc. Natl. Acad. Sci. U.S.A. 98, 3197 (2001). Copyright (2001) National Academy of Sciences, U.S.A.

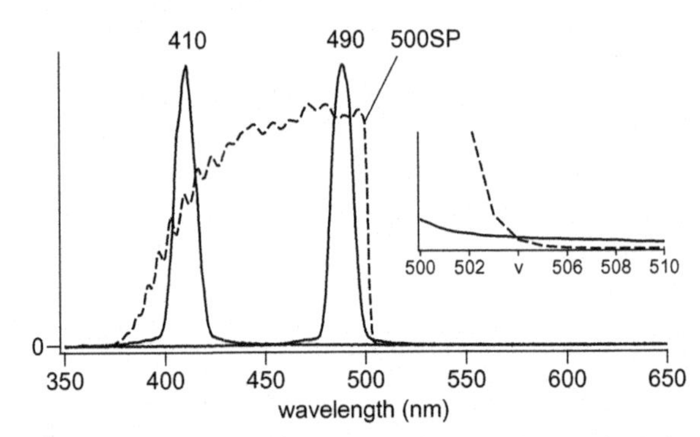

FIGURE 14.7 Spectrum of the 410- or 490-nm light selected by the monochromator (U7773-XX and U7794-16, Hamamatsu Photonics) (solid line) and transmittance of a 500SP short-pass filter (Omega, 3rd Milenium) (broken line). (*Inset*) The crossing of the two curves for 490-nm light and 500SP on an expanded scale.

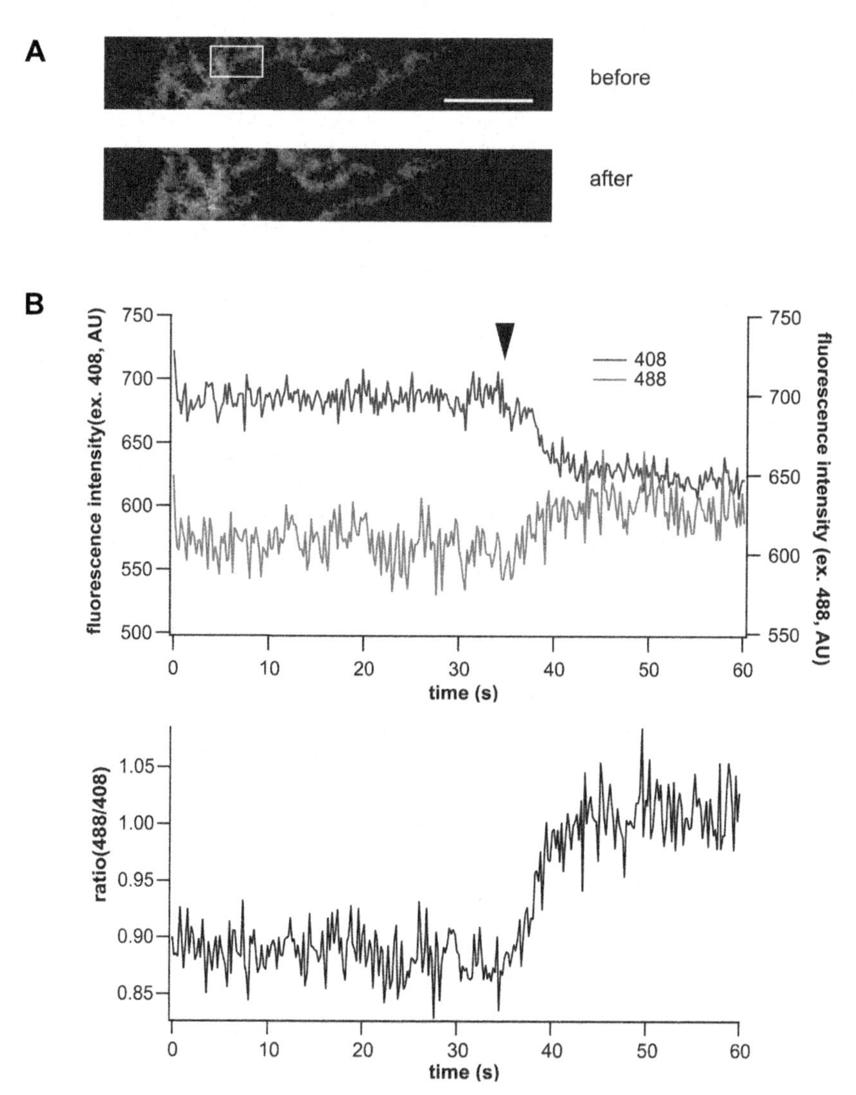

FIGURE 14.8 Confocal and dual-excitation imaging of $[Ca^{2+}]_m$ using ratiometric pericam mitochondia. (A) Ratio images before and after application of 10 μ*M* histamine. Scale bar: 5 μm. (B) Time course of averaged fluorescence signals from the white box in A with excitation at 488 and 408 nm (top) and their ratio (bottom). The arrowhead indicates the time when histamine was applied.

allow us to fully follow the rapid movement of mitochondria, the fast rate of ratio acquisition minimized the time lag between the two measurements used to calculate the ratiometric signal. We feel that this method of imaging $[Ca^{2+}]_m$ effectively corrects for mitochondrial movement both laterally and into and out of the optical section. After the application of histamine, spots of $[Ca^{2+}]_m$ increase within a single mitochondrion were identifiable (Fig. 14.8A) and the global increase in $[Ca^{2+}]_m$ was found to occur relatively slowly (Fig. 14.8B).

References

Baird, G. S., Zacharias, D. A., and Tsien, R. Y. (1999). Circular permutation and receptor insertion within green fluorescent proteins. *Proc. Natl. Acad. Sci. USA* **96**, 11241–11246.

Dickson, R. M., Cubitt, A. B., Tsien, R. Y., and Moerner, W. E. (1997). On/off blinking and switching behaviour of single molecules of green fluorescent protein. *Nature* **388**, 355–358.

Miyawaki, A., Llopis, J., Heim, R., McCaffery, J. M., Adams, J. A., Ikura, M., and Tsien, R. Y. (1997). Fluorescent indicators for Ca^{2+} based on green fluorescent proteins and calmodulin. *Nature* **388**, 882–887.

Nagai, T., Sawano, A., Park, E. S., and Miyawaki, A. (2001). Circularly permuted green fluorescent proteins engineered to sense Ca^{2+}. *Proc. Natl. Acad. Sci. USA* **98**, 3197–3202.

Robert, V., Gurlini, P., Tosello, V., Nagai, T., Miyawaki, A., Di Lisa, F., and Pozzan, T. (2001). Beat-to-beat oscillations of mitochondrial $[Ca^{2+}]$ in cardiac cells. *EMBO J.* **20**, 4998–5007.

Shimozono, S., Fukano, T., Nagai, T., Kirino, Y., Mizuno, H., and Miyawaki, A. (2002). Confocal imaging of subcellular Ca^{2+} concentrations using a dual-excitation ratiometric indicator based on green fluorescent protein. *Sci. STKE* http://www.stke.org/cgi/content/full/OC_sigtrans;2002/125/pl4.

Tsien, R. Y. (1998). The green fluorescent protein. *Annu. Rev. Biochem.* **67**, 509–544.

SECTION V

IN SITU CELL ELECTROPORATION TO STUDY OF SIGNALING PATHWAYS

CHAPTER

15

Dissecting Pathways; *In Situ* Electroporation for the Study of Signal Transduction and Gap Junctional Communication

Leda Raptis, Adina Vultur, Heather L. Brownell, and Kevin L. Firth

OUTLINE

I. INTRODUCTION

Electroporation has been used for the introduction of DNA and proteins, as well as various nonpermeant drugs and metabolites into cultured mammalian cells (reviewed in Neumann *et al.*, 2000; Chang *et al.*, 1992). Most electroporation techniques for adherent cells involve the delivery of the electrical pulse while the cells are in suspension. However, the detachment of these cells from their substratum by trypsin or EDTA can cause significant metabolic alterations (Matsumura *et al.*, 1982). In addition, the efficient incorporation of proteins, peptides, or drugs without cellular damage is an especially crucial requirement, since contrary to DNA, no convenient large-scale method exists for the selection of viable from damaged cells or cells where no introduction took place after electroporation, for most proteins or drugs of interest. For these reasons, a number of approaches have been taken to bypass this problem (Kwee *et al.*, 1990; Yang *et al.*, 1995).

This article describes a technique where cells are grown on a glass surface coated with electrically conductive, optically transparent indium-tin oxide (ITO) at the time of pulse delivery. This coating promotes excellent cell adhesion and growth, allows direct visualization of the electroporated cells, and offers the possibility of ready examination due to their extended morphology. The procedures described are applicable to a wide variety of nonpermeant molecules, such as peptides (Giorgetti-Peraldi *et al.*, 1997; Boccaccio *et al.*, 1998; Bardelli *et al.*, 1998), oligonucleotides (Boccaccio *et al.*, 1998; Gambarotta *et al.*, 1996), radioactive nucleotides (Boussiotis *et al.*, 1997), proteins (Nakashima *et al.*, 1999), DNA (Raptis and Firth, 1990), or drugs (Marais *et al.*, 1997). These compounds can be introduced alone or in combination, at the same or different times, in growth-arrested cells or cells at different stages of their division cycle. After introduction of the material, cells can be either extracted and biochemically analysed or their morphology and biochemical properties examined *in situ*. In a modified version, this assembly can be used for the study of intercellular, junctional communication. The instant introduction of the molecules into essentially 100% of the cells makes this technique especially suitable for kinetic studies of effector activation. Unlike other techniques of cell permeabilization, under the appropriate conditions, *in situ* electroporation does not affect cell morphology, the length of the G1 phase of serum-stimulated cells (Raptis and Firth, 1990), the activity of the extracellular signal regulated kinase (Erk1/2 or Erk), or two kinases commonly activated by a number of stress-related stimuli, JNK/SAPK and $p38^{hog}$ (Robinson and Cobb, 1997), presumably because the pores reseal rapidly so that the cell interior is restored to its original state (Brownell *et al.*, 1998).

II. MATERIALS AND INSTRUMENTATION

The purity of the material to be electroporated is of paramount importance. Substances such as detergents, preservatives, or antibiotics could kill the cells into which they are electroporated, even if they have no deleterious effects if added to the culture medium of nonelectroporated cells.

Dulbecco's modification of Eagle's medium (DMEM) is from ICN (Cat. No. 10-331-22). Fetal calf serum (Cat. No. 2406000AJ) and phosphate-buffered saline (PBS, Cat. No. 20012-027) are from Gibco Life Technologies. Calf serum is from ICN (Cat. No. 29-131-54). EGF is from Intergen (Cat. No. 4110-80). HEPES (Cat. No. H-9136), Lucifer yellow CH dilithium salt (Cat. No. L-0259), trypsin (Cat. No. T-0646), $MgCl_2$ (Cat. No. M-8266), Triton (Cat. No. T-6878), deoxycholate (Cat. No. D-5760), EDTA (Cat. No. E-5134), EGTA (Cat. No. E-3889), phenylmethylsulfonyl fluoride (PMSF, P-7626), aprotinin (Cat. No. A-6279), leupeptin (Cat. No. L-2023), benzamidine (Cat. No. B-6506), dithiothreitol (DTT, Cat. No. D-9779), $CaCl_2$ (Cat. No. C-4901), Tris (Cat. No. T-6791), paraformaldehyde (Cat. No. P-6148), bovine serum albumin (BSA, Cat. No. A-4503), SigmaFast DAB kit (Cat. No. D-9167 and Cat. No. U-5005), and sodium orthovanadate (Cat. No. S-6508) are from Sigma. Extran-300 detergent (Cat. No. B80002), SDS (Cat. No. 44244), NP-40 (Cat. No. 56009), glycerol (Cat. No. B10118), and peroxide (B 80017) are from BDH. CelTak™ is from BD Biosciences (Cat. No. 354240). The cell staining kit, including goat serum, secondary antibody, and avidin–biotin complex, was from Vector Labs (Vectastain kit Cat. No. PK-6101). Rabbit antipeptide antibodies against the double threonine and tyrosine phosphorylated (activated) Erk1/2 kinase are from Biosource International (Cat. No. 44-680). When stored frozen in aliquots, they are stable for more than 5 years. They are used at 1:500 for immunostaining and 1:10,000 for Western blotting. Three- and 6-cm tissue culture dishes are from Corning or Sarstedt. Please note that only these brands of plates fit the electroporation stand.

The Grb2-SH2 binding peptide is based on the sequence flanking the Y^{1068} of the EGF receptor (PVPE-Pmp-INQS, MW 1123). To enhance stability of the phosphate group, the phosphotyrosine analog, phosphono-methylphenylalanine (Pmp), which cannot be cleaved by phosphotyrosine phosphatases yet binds to SH2 domains with high affinity and specificity (Otaka *et al.*, 1994), is incorporated at the position of phosphotyrosine. The Pmp monomer is custom synthesized by *Color your enzyme Inc.*, (Kingston, Ontario, Canada). As control, we used the same peptide containing phenylalanine at the position of Pmp. Peptides are synthesized by the Queen's University Core Facility using standard Fmoc chemistry.

The system for electroporation *in situ* (Epizap Model EZ-16) can be purchased from Ask Science Products Inc. (Kingston, Ontario, Canada, phone: 613 545-3794). The inverted, phase-contrast and fluorescence microscope, equipped with filters for Lucifer yellow and fluorescein, was from Olympus (Model IX70).

III. PROCEDURES

The technique of *in situ* electroporation can be used equally effectively for large-scale biochemical experiments (Giorgetti-Peraldi *et al.*, 1997; Boccaccio *et al.*, 1998; Bardelli *et al.*, 1998) or for the detection of biochemical or morphological changes *in situ* (Raptis *et al.*, 2000a). Cells are grown on glass slides coated with conductive and transparent indium-tin oxide (ITO). The cell growth area is defined by a "window" formed with an electrically insulating frame made of Teflon as shown. A stainless-steel electrode is placed on top of the cells resting on the frame and an electrical pulse of the appropriate strength is applied, as illustrated in Fig. 15.1 (see also chapter 16 by Raptis *et al.*). The technique can be applied to a large variety of adherent cell types (Brownell *et al.*, 1996). Cells that do not adhere well can be grown and electroporated on the same conductive slides coated with CelTak™ or poly-l-lysine used according to the manufacturer's instructions.

A. Electroporation of Peptides into Large Numbers of Cells for Large-Scale Biochemical Experiments. Use of Fully Conductive Slides

To study the effect of protein interactions *in vivo* on cellular functions, such complexes can be disrupted through the introduction of peptides corresponding to the proteins' point(s) of contact. An example of this approach is described here.

Growth factors such as the epidermal growth factor (EGF) stimulate cell proliferation by binding to and activating membrane receptors with cytoplasmic tyrosine kinase domains. *In vitro*

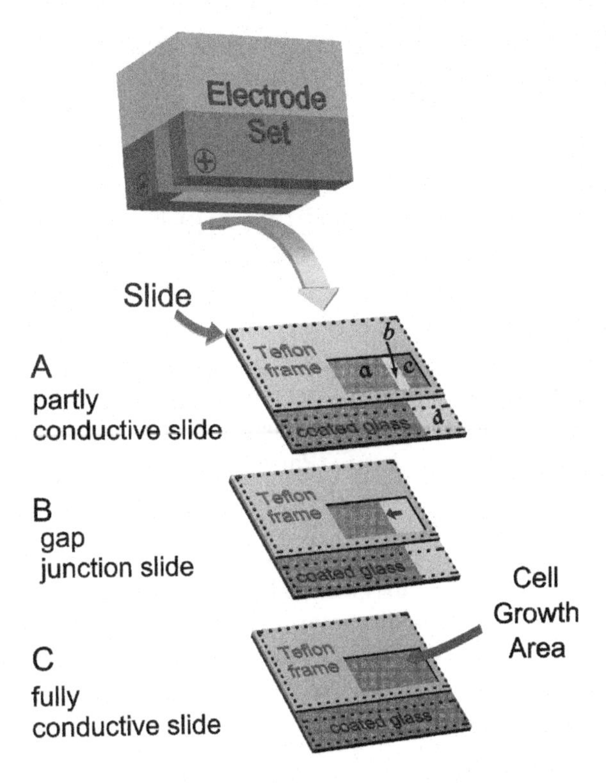

FIGURE 15.1 Electroporation electrode and slide assembly. Cells are grown on glass slides coated with conductive and transparent ITO within a "window" cut into a Teflon frame as shown. The window can be different sizes, depending on the cell growth area required. The peptide solution is added to the cells and introduced by an electrical pulse delivered through the electrode set, which is placed directly on the frame. Dotted lines point to the positions of negative and positive electrodes during the pulse. Three slide configurations are described. (A) Partly conductive slide assembly, with electroporated (*a*) and non-electroporated (*c*) cells growing on the same type of ITO-coated surface. (*b*) area where the conductive coating has been stripped, exposing the non-conductive glass underneath. Cells growing in areas *b* and *c* are not electroporated (Fig. 15.3). (B) Partly conductive slide assembly for use in the examination of gap junctional, intercellular communication. Arrow points to the transition line between conductive and non-conductive areas (Fig. 15.4). (C) Fully conductive slide assembly for use in biochemical experiments. In the setup shown, cell growth area can be up to 7 × 15 mm, but larger slides and electrodes offer larger areas, up to 32 × 10 mm (Fig. 15.2).

binding and receptor mutagenesis studies have shown that ligand engagement induces receptor autophosphorylation at distinct tyrosine residues, which constitute docking sites for a number of effector molecules, such as the growth factor receptor-binding protein 2 (Grb2), which are recruited to specific receptors through modules termed Src-homology 2 (SH2) domains (reviewed in *Schlessinger*, 2000). Grb2 binds to the receptors for PDGF and EGF at a number of sites, an event activating the Sos/Ras/Raf/Erk pathway, which is central to the mitogenic response stimulated by many growth factors. Previous results indicated that a synthetic phosphopeptide corresponding to the Grb2-binding site of the EGF receptor (EGFR, flanking the EGFR tyr^{1068}, PVPE-pY-INQS), when made in tandem with peptides that allow for translocation across the cell membrane, could inhibit EGF-mediated mitogenesis and Erk activation in newt myoblasts induced by 1 ng/ml EGF, but was less effective at 10 ng/ml (Williams *et al.*, 1997). To better determine the functional consequences of disrupting the association of Grb2 *per se* with different receptors *in vivo* in mammalian cells, we delivered large quantities of this peptide into intact, living NIH3T3 fibroblasts by *in situ* electroporation.

Solutions

Spent medium: Grow cells to confluence in DMEM with 10% calf serum. Seven days postconfluence, collect the culture supernatant and dilute 1:1 with fresh DMEM. Growth-arrest cells by incubating in spent medium prepared from the same line.

Lucifer yellow solution, 5 mg/ml: To make 10 ml, dissolve 50 mg Lucifer yellow in 10 ml calcium-free DMEM. Stable at 4°C for at least a month.

Peptides: The peptide concentration required varies with the strength of the signal to be inhibited. For the inhibition of the EGF-mediated Erk activation, prepare a solution of 5–10 mg/ml (~5–10 m*M*) of the Grb2-SH2-blocking peptide (PVPE-pmp-INQS, MW 1123 Da) in calcium-free DMEM (see *Comment* A).

Epidermal growth factor: To make a 10,000× stock solution, dissolve 100 μg of lyophilised EGF in 100 μl sterile water and freeze in 5 μl aliquots. Just prior to the experiment, add 1 μl stock solution to 10 ml calcium-free DMEM (final concentration, 100 ng/ml). The stock solution is stable at −20° or −70°C for up to 2 months.

Lysis buffer: 50 m*M* HEPES, pH 7.4, 150 m*M* NaCl, 10 m*M* EDTA, 10 m*M* $Na_4P_2O_7$, 100 m*M* NaF, 2 m*M* vanadate, 0.5 m*M* PMSF, 10 μg/ml aprotinin, 10 μg/ml leupeptin, 1% Triton X-100.

Steps

1. **Choice of slides**. For Western blotting experiments on cell extracts following electroporation, use fully conductive slides (Fig. 15.1C). Since the custom-made peptide is usually the most expensive reagent in this application, to avoid waste, choose the smallest possible cell growth area which provides a sufficient number of cells. Cell growth areas of 32 × 10 mm are generally sufficient to detect Erk1/2 activity inhibition by the Grb2-SH2 blocking peptide in EGF-stimulated, mouse NIH3T3 fibroblasts. In this case, the volume of the solution under the electrode is ~140 μl and will contain approximately 700–1,400 μg peptide in calcium-free DMEM. If fewer cells suffice, then slides with a cell growth area of 7 × 15 mm can be used, requiring ~40 μl of peptide solution. However, for the determination of [^{3}H]thymidine uptake, cell growth areas of 7 × 4 mm are preferred and they require only ~14 μl of solution (see *Comment* B).
2. **Plate the cells**. Uniform spreading of the cells is very important, as the optimal voltage depends in part on the degree of cell contact with the conductive surface (see *Comment* 3). Add a sufficient amount of medium (DMEM containing 10% calf serum) to cover the slide (approximately 9 ml for a 6 cm dish). Pipette the cell suspension in the window cut in the Teflon frame (Fig. 15.1) and place the petris in a tissue-culture incubator until confluent.
3. Prior to the experiment, **starve** the cells overnight in DMEM without serum. Alternatively, cells can be incubated in spent medium for 48 h; this treatment offers wider margins of voltage tolerance (see *Comment* C).
4. Prior to pulse application, **remove the growth medium** and wash the cells gently once with calcium-free DMEM.
5. Carefully **wipe** the Teflon frame with a folded Kleenex tissue to create a dry area on which a meniscus can form (see *Pitfall* 1).
6. **Add the peptide** solution to the cells with a micropipettor in calcium-free DMEM.
7. Carefully **place the electrode** on top of the cells and clamp it in place. To ensure electrical contact, a sufficient amount of growth medium or PBS should be present under the positive contact bar. Make sure there are no air bubbles under the negative electrode. If necessary, the electrode can be sterilized with 80% ethanol before the pulse and the procedure carried out in a laminar flow hood, using sterile solutions.
8. Apply three to six **pulses** of the appropriate voltage and capacitance (see *Comment* C).
9. **Remove the electrode set**. Since usually only a small fraction of the material enters the cells, the peptide solution may be carefully aspirated and used again. However, care must be exercised so that the cells do not dry (see *Pitfall* 1).
10. **Add serum-free growth medium** and incubate the cells for 2–5 min at 37°C to recover.
11. **Add EGF** to the medium to a final concentration of 100 ng/ml for 5 min.

Controls receive the same volume of calcium-free DMEM.

12. **Extract** the cells with 50 μl extraction buffer for a cell growth area of 32 × 10 mm. For smaller cell growth areas, the voltage can be adjusted accordingly.
13. To **detect activated Erk1/2**, load 100 μg of total cell extract protein on an acrylamide–SDS gel and analyse by Western immunoblotting using the antibody directed against the dually phosphorylated, i.e., activated, form of Erk1/2.
14. For examination of the effect of the peptide upon [^{3}H]thymidine incorporation into DNA, serum-starve 50% confluent, NIH3T3 cells as described earlier and electroporate in the presence or absence of peptide. Incubate in medium with or without EGF for 12 h at 37°C, followed by a 2 h incubation with 50 μCi/ml [^{3}H]thymidine. Wash the cells with PBS and measure acid-precipitable counts. Growth areas of 4 × 7 mm are sufficient for this experiment, and [^{3}H]thymidine can be added to the window only, in a volume of ~50 μl which is held in place by surface tension.

As shown in Fig. 15.2A, electroporation of the Grb2-SH2 blocking peptide caused a dramatic reduction in EGF-mediated Erk activation in mouse NIH3T3 cells at growth factor concentrations permitting full receptor stimulation (compare lanes 2 and 3 with lane 4). In addition, electroporation of this peptide reduced EGF-mediated [^{3}H]thymidine uptake (Fig. 15.2B). In contrast, the same peptide had only limited or no effect on Erk activation triggered by HGF, although it could inhibit PDGF signalling (Raptis *et al.*, 2000a). These findings demonstrate that the *in situ* electroporation approach described can very effectively inhibit growth factor-stimulated mitogenesis and thereby detect the differential specificity in the coupling of activated receptor tyrosine kinases to the Erk cascade.

B. Electroporation for the Study of Morphological Effects or Biochemical Changes *In Situ*: Use of Partly Conductive Slides

Assessment of Erk activity by Western blotting following electroporation of the Grb2-SH2 blocking peptide can reveal the involvement of this domain in growth factor-mediated Erk activation. However, to ensure that the treatment itself does not cause cell stress, examination of cellular morphology, in conjunction with measurement of gene product activity by immunocytochemistry, offers a distinct advantage. This approach can demonstrate the specificity of action of the Grb2-SH2 binding peptide, as well as examine the distribution of signal inhibition across the cell layer. An added advantage is that it requires a small number of cells, hence a substantially smaller volume of the peptides (~14 μl in the setup shown in Fig. 15.1A), compared to Western blotting (~140 μl for a cell growth area of 32 × 10 mm), which could be a significant consideration given their production costs. To precisely assess small background changes in morphology or gene expression levels, the presence of non-electroporated cells side by side with electroporated ones can offer a valuable control and this can be achieved by growing the cells on a conductive slide where part of the coating has been stripped by etching with acids, thus exposing the non-conductive glass underneath. However, as shown in Fig. 15.3, area *a* vs *b*, the slight tinge of the glass combined with the more effective staining of cells growing on ITO (possibly due to a chemical attraction of different reagents to the coating) can create problems in the interpretation of results. In addition, it was found that a number of cell lines grow slightly better on the conductive, ITO-coated glass than the nonconductive area, possibly due to the fact that the ITO-coated surface is not as smooth as glass, thus providing a better anchorage for the growth of adherent cells (Folkman and Moscona, 1978). As a result,

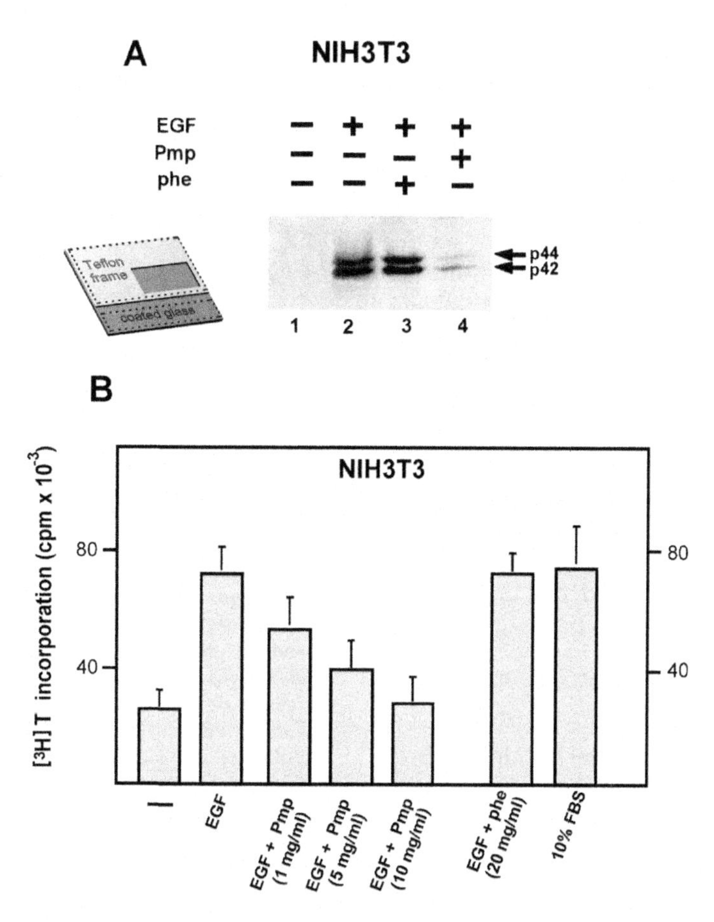

FIGURE 15.2 **(A) The Grb2-SH2 blocking peptide inhibits EGF-mediated Erk activation in living cells; detection by Western blotting.** The Grb2-SH2 blocking peptide was electroporated into NIH3T3 cells growing on fully conductive slides (inset, Fig. 15.1C, cell growth area 32 × 10mm) and growth-arrested by serum starvation. After a 5min incubation in DMEM, cells were stimulated with 100 ng/ml EGF (lanes 2–4) for 5min. Proteins in detergent cell lysates were resolved by polyacrylamide gel electrophoresis and analysed by Western blotting using the antibody against the dually phosphorylated, active Erk enzymes. Lane 1, control, unstimulated cells; lane 2, control non-electroporated, EGF-treated cells; lane 3, cells electroporated with the control, phenylalanine-containing peptide and EGF stimulated; and lane 4, cells electroporated with the Grb2-SH2-binding peptide and EGF stimulated. From Raptis *et al.* (2000), reprinted with permission. **(B) The Grb2-SH2 blocking peptide inhibits EGF-mediated DNA synthesis.** The Grb2-SH2 blocking peptide (Pmp) or its phenylalaline-containing counterpart (phe) were electroporated at the indicated concentrations into NIH3T3 cells growing on fully conductive slides (Fig. 15.1C, cell growth area, 4 × 7mm) and growth-arrested by serum starvation. Following incubation at 37°C and stimulation with EGF or 10% calf serum for 12h, cells were labeled for 2h with 50μCi/ml [^{3}H]thymidine and acid-precipitable radioactivity was determined. Numbers represent the mean ± SE from three experiments. From Raptis *et al.*, (2000), reprinted with permission.

cell density may be higher on the conductive than the etched side, which could have important implications if cell growth effects are being studied. It follows that, to assess the effect of the peptide, it is important to compare the staining and morphology of electroporated cells with non-electroporated ones while both are growing on the same type of surface. This was achieved by plating the cells on a slide where the conductive coating was removed in the pattern shown in Fig. 15.1A (Firth *et al.*, 1997). A thin line of plain glass separates the electroporated and control areas while etching extends to area Fig. 15.1A,d so that there is no electrical contact between the positive contact bar and area Fig. 15.1A,c. Application of the pulse results in electroporation of the cells growing in area Fig. 15.1A,a exclusively, while cells growing in area Fig. 15.1A,b or c do not receive any pulse. In this configuration, electroporated cells are being compared to nonelectroporated ones, while both are growing on ITO-coated glass. Because the coating is only ~1,600Å thick, this transition zone does not alter the growth of cells across it and is clearly visible microscopically, even under a cell monolayer (Figs. 15.3 and 15.4).

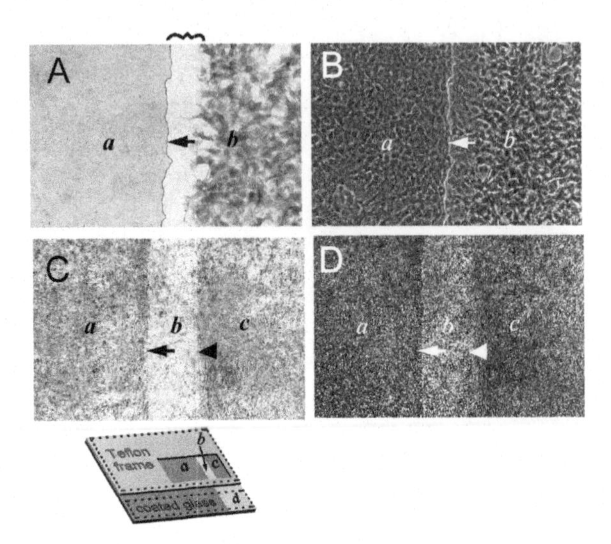

FIGURE 15.3 **The Grb2-SH2 blocking peptide inhibits EGF-mediated Erk activation in living cells; detection by immunocytochemistry.** The Grb2-SH2 blocking peptide (A and B) or its control, phenylalanine-containing counterpart (C and D) were introduced by *in situ* electroporation into NIH3T3 cells growing on partly conductive slides (inset, Fig. 15.1A) and growth arrested in spent medium. Five minutes after pulse application, cells were stimulated with EGF for 5 min, fixed, and probed for activated Erk1/2, and cells from the same field were photographed under bright-field (A and C) or phase-contrast (B and D) illumination. Magnification: A and B, 240×, C and D, 40×. Arrow points to the transition line between stripped (*b*) and electroporated (*a*) areas, while arrowhead points to the line between control ITO-coated (*c*) and stripped (*b*) areas (Fig. 15.1A). Cells growing on the left side (*a*) are electroporated, whereas cells on the stripped zone (*b*) or right side (*c*) of the slide do not receive any pulse. Note that the Grb2-SH2 blocking peptide dramatically reduced the EGF signal (A, *a*), whereas the degree of Erk activation is the same on both sides of the slide (*a* or *c*) for cells electroporated with the control, phenylalanine-containing peptide (C). In A, inhibition of the signal extends into approximatily three to four rows of adjacent cells in the non-electroporated area (squiggly bracket in *b*), probably due to movement of the peptide through gap junctions (Raptis *et al.*, 1994). At the same time, there is no detectable effect on cell morphology as shown by phase contrast (B and D). From Raptis *et al.* (2000), reprinted with permission.

Solutions

Peptide solution: 5–10 mg/ml in calcium-free DMEM. See Section III,A and *Comment* A

Lucifer yellow solution: 5 mg/ml in calcium-free DMEM. See Section III,A

Epidermal growth factor: 100 ng/ml in calcium-free DMEM. See Section III,A.

Steps

1. Choice of slides. Use partly conductive slides where the coating has been removed in a line as shown in Fig. 15.1A.
2. Plate the cells as described earlier and starve them from serum.
3. Aspirate the medium and wash the cells once with calcium-free DMEM.
4. Add the peptide solution as described previously.
5. Apply a pulse of the appropriate strength (see *Comment* C).
6. Add serum-free growth medium and place the cells in a 37°C incubator for the pores to close (2–5 min).
7. Treat the cells with EGF for 5 min as in Section III,A.
8. Fix the cells with 4% paraformaldehyde and probe with the anti-active Erk

antibody according to the manufacturer's instructions.

As shown in Fig. 15.3, electroporation of the Grb2-SH2 blocking peptide totally inhibited EGF-induced Erk activation (panel A, area "*a*"), while the control, phenylalanine-containing peptide had no effect (panel C, area "*a*"). This inhibition was uniform across the cell layer, in agreement with previous results indicating that *in situ* electroporation can introduce the material into essentially 100% of the treated cells. It is especially noteworthy that this inhibition ***extends into three to four rows of the adjacent, nonelectroporated cells*** growing on the non-conductive part of the slide (panel A, area "*b*", squiggly bracket), probably due to movement of the 1123 Da peptide through gap junctions (Raptis *et al.*, 1994). This finding constitutes compelling evidence that the observed inhibition must be due to the peptide rather than an artifact of electroporation. At the same time, as shown by phase-contrast microscopy (panel B), there was no alteration in the morphology of the electroporated cells under these conditions, suggesting that the observed effect is a result of a specific inhibition rather than toxic action. EGF stimulation for up to 30 min after peptide electroporation did not result in lower levels of Erk signal inhibition, indicating that the binding of the peptide to Grb2 is stable during this period of time. As expected, the phenylalanine-containing, control peptide (panels C and D) had no effect on Erk activation. In contrast, the Grb2-SH2 binding peptide had little effect in inhibiting Erk activity triggered by the hepatocyte growth factor (HGF) in NIH3T3 cells expressing the HGF receptor through transfection or in human A549 cells that naturally express this receptor (Raptis *et al.*, 2000a).

The introduction of peptides to interrupt signaling pathways using the modification of *in situ* electroporation described is a powerful approach for the *in vivo* assessment of the relevance of *in vitro* interactions. Results presented in Fig. 15.2 and 15.3 clearly demonstrate that an essentially complete and specific inhibition of EGF-dependent Erk activation can be achieved through peptide electroporation. The stepwise dissection of signaling cascades is essential for the understanding of normal proliferative pathways, which could lead to the development of drugs for the rational treatment of neoplasia.

C. Electroporation on a Partly Conductive Slide for the Assessment of Gap Junctional, Intercellular Communication

One of the targets of a variety of signals stemming from growth factors or oncogenes may be membrane channels, which serve as conduits for the passage of small molecules between the interiors of cells. Oncogene expression and neoplasia invariably result in a decrease in gap junctional, intercellular communication (GJIC) (Goodenough *et al.*, 1996). The investigation of junctional permeability is often conducted through microinjection of a fluorescent dye such as Lucifer yellow, followed by observation of its migration into neighboring cells. This is a time-consuming approach, requiring expensive equipment, while the mechanical manipulation of the cells may disturb cell-to-cell contact areas, interrupt gap junctions, and cause artefactual uncoupling. These problems can be overcome using a setup where cells are grown on a glass slide, half of which is coated with electrically conductive, optically transparent, indium-tin oxide. An electric pulse is applied in the presence of Lucifer yellow, causing its penetration into cells growing on the conductive part of the slide, and migration of the dye to non-electroporated cells growing on the non-conductive area is observed microscopically under fluorescence illumination.

The technique can be applied to a large variety of adherent cell types, including primary human lung carcinoma cells (Tomai *et al.*, 1998;

Raptis *et al.*, 1994; Brownell *et al.*, 1996; Vultur *et al.*, 2003).

Solutions

Lucifer yellow solution: 5 mg/ml in calcium-free DMEM or other growth medium. See Section III.A

Calcium-free growth medium with or without 5% dialysed serum

Steps

1. Plate the cells on partly conductive slides in 3 cm petris. Electroporated areas can be 4 × 4 mm and non-electroporated ones 4 × 3 mm (Fig. 15.1B). Other slide configurations are also available (Raptis *et al.*, 2000b).
2. Aspirate the medium. Wash the cells with calcium-free DMEM.
3. Add the Lucifer yellow solution.
4. Apply a pulse of the appropriate strength so that cells growing on the conductive coating at the border with the non-conductive area are electroporated without being damaged. As described in *Comment* C, this area receives slightly larger amounts of current than the rest of the conductive growth surface.
5. Add calcium-free DMEM containing 5% dialysed serum, remove the electrode, and incubate the cells for 3–5 min in a 37°C, CO_2 incubator. The inclusion of dialysed serum at this point helps pore closure.
6. Wash the unincorporated dye with calcium-free growth medium.
7. Microscopically examine under fluorescence and phase-contrast illumination (Fig. 15.4).
8. **Quantitate intercellular communication.** Photograph the cells with a 20× objective under fluorescence and phase contrast illumination (Figs. 15.4A and 15.4B). Identify and mark electroporated cells at the border with the non-conductive area (black stars) and fluorescing cells on the nonconductive side (white circles) where the dye has transferred through gap junctions. Divide the total number of fluorescing cells on the non-conductive area by the number of electroporated cells along the border

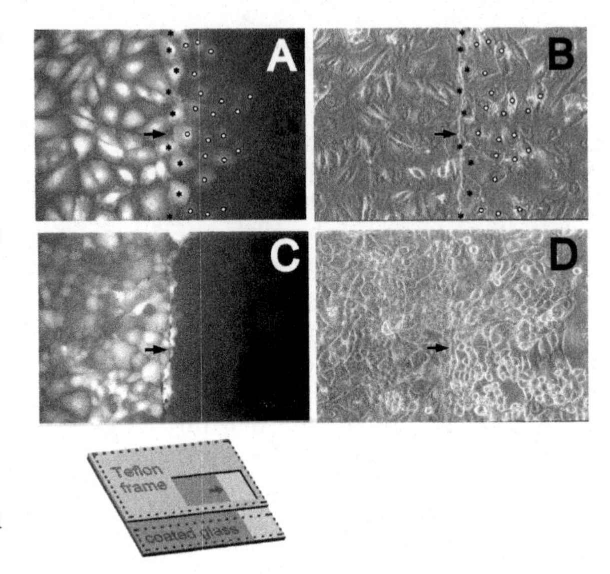

FIGURE 15.4 ***In situ* electroporation on a partly conductive slide for the measurement of intercellular, junctional communication.** (**A and B**) An established, mouse lung epithelial type II line (E10) was plated on partly conductive slides (inset, Fig. 15.1B) and at confluence was electroporated in the presence of 5 mg/ml Lucifer yellow. After washing away any unincorporated dye, cells from the same field were photographed under fluorescence (A) or phase-contrast (B) illumination (Raptis *et al.*, 1994). Note the gradient of fluorescence indicating dye transfer through gap junctions. To quantitate intercellular communication, the number of cells into which the dye transferred through gap junctions per electroporated border cell was calculated by dividing the total number of fluorescing cells on the non-conductive side (white circles) by the number of cells growing at the border with the conductive coating (black stars). (**C and D**) A spontaneous transformant of the E10 line (line E9), was plated on partly conductive slides, electroporated, washed, and photographed as described earlier. Fluorescence (C) and phase-contrast (D) illumination photograph of the same field. Note the absence of dye transfer through gap junctions. In all photographs, the left side is conductive. Arrows on the conductive side point to the interphase between conductive and non-conductive areas. Magnification: 200×. From Vultur *et al.* (2003), reprinted with permission.

with the etched side. The transfer from at least 200 contiguous electroporated border cells is calculated for each experiment (Raptis *et al.*, 1994). A careful kinetic analysis of dye transfer from 30 s to 2 h showed that the observed transfer is essentially complete by 5 min for all lines tested, while fluorescence is eliminated from the cells within approximately 60 min. After the transfer is complete, cells can be fixed with formaldehyde, in which case fluorescence is retained for approximately 2 h.

IV. COMMENTS

A. Peptides

The concentration of peptide required varies with the strength of the signal to be inhibited. For example, for the inhibition of the HGF-mediated Stat3 activation in MDCK cells, a concentration of 1 μg/ml of a peptide blocking the SH2 domain of Stat3 (^{P}YVNV) is sufficient (Boccaccio *et al.*, 1998), whereas for the inhibition of the EGF-mediated Erk activation in a variety of fibroblasts or epithelial cells, a concentration of 5–10 mg/ml (~5–10 m*M*) of the Grb2-SH2 blocking peptide is necessary (Raptis *et al.*, 2000a). The purity of the material is of utmost importance. Peptides must be HPLC-purified because impurities can cause cell death or give unexpected results. The pH of the peptide solution must be neutral, as indicated by the color of the DMEM medium where the peptide is dissolved. If it is too acidic, then it must be carefully neutralised with NaOH. In this case, the salt concentration of the no-peptide controls (DMEM without calcium) must be adjusted to the same level with NaCl because a change in conductivity may affect the optimal voltage required (see *Comment* C).

Any peptide which is soluble in DMEM or other aqueous buffer can be very effectively electroporated. Good solubility is especially important because the concentration needed for effective signal inhibition can be as high as 10 mg/ml. It was nevertheless found that, at least for certain applications, the inclusion of DMSO in the electroporation solution at a concentration of up to 5%, which might aid peptide solubility, did not affect results significantly. However, a number of peptides, e.g., peptides made as fusions with the homeobox domain or other membrane translocation sequences, are usually not sufficiently soluble for this application.

B. Slides

As described in detail in Chapter 43 by Raptis *et al.*, to obtain a uniform electrical field intensity over the entire area below the negative electrode, despite the fact that the conductive coating exhibits a significant amount of electrical resistance, the bottom surface of the negative electrode must be inclined relative to the glass surface in a manner proportional to the resistance of the coating. For electroporation of peptides, to minimise the volume of custom-made peptide used, slides with a conductivity of 2 Ω/sq, the most conductive commercial grade available, are used. The slides and electrodes come in different sizes, with the biggest cell growth area in this configuration being 32 × 10 mm. Depending on the experiment, if larger numbers of cells are required, extracts from two to three slides may be pooled. In this case, the peptide solution can be aspirated and used again. Alternatively, an electrode configuration with two positive contact bars can be employed, as described in Chapter 43 by Raptis *et al.*

The slides come with the apparatus, individually wrapped and sterile. However, they can be reused many times after washing with Extran-300 detergent while scrubbing with a toothbrush. In this case they must be sterilized with 80% ethanol for 20 min and the ethanol rinsed with sterile distilled water prior to plating the cells. Alternatively, the slides can be gas-sterilized. Do not autoclave. The glass can withstand high temperatures, but autoclaving would damage the Teflon frames.

C. Determination of the Optimal Voltage and Capacitance

Electrical field strength has been shown to be a critical parameter for cell permeation, as well as viability (Chang *et al.*, 1992). It is generally easier to select a discrete capacitance value for a given electroporated area and space between the conductive coating and the negative electrode and then precisely control the voltage. Both parameters depend upon the size of the electroporated area; larger conductive growth areas necessitate higher voltages and/or higher capacitances for optimal permeation. For the 32 × 10 mm cell growth area, some damage to the coating may be noted at the higher voltages necessary if a single pulse is employed. However, using higher capacitance values and multiple pulses with lower voltage settings can yield efficient cell permeation with no damage to the coating, and this treatment is also better tolerated by the cells. For greater growth areas, the dual positive contact bar design described in Chapter 43 by Raptis *et al.* can be employed.

The optimal pulse strength depends on the strain and metabolic state of the cells, as well as on the degree of cell contact with the conductive surface. Densely growing, transformed cells or cells in a clump require higher voltages for optimum permeation than sparse, subconfluent cells, possibly due to the larger amounts of current passing through an extended cell. Similarly, cells that have been detached from their growth surface by vigorous pipetting prior to electroporation require substantially higher voltages. It is especially striking that cells in mitosis remain intact under conditions where most cells in other phases of the cycle are permeated (Raptis and Firth, 1990). In addition, cells growing and electroporated on CelTak™-coated slides require substantially higher voltages than cells growing directly on the slide.

The margins of voltage tolerance depend on the size and electrical charge of the molecules to be introduced (Fig. 15.5). For the introduction of small, uncharged molecules such as Lucifer yellow or peptides, a wider range of field strengths permits effective permeation with minimal damage to the cells than the introduction of antibodies or DNA (Raptis and Firth, 1990; Brownell *et al.*, 1997). For a number of experiments involving cell growth, it may be necessary to electroporate serum-arrested cells. Voltages required are lower, and especially the margins of voltage tolerance were found to be substantially narrower for serum-starved cells compared to their counterparts growing in 10% calf serum (Brownell *et al.*, 1997). Also, it is important to keep all solutions at 37°C, which facilitates pore closure and efficient electroporation. If the material is applied in a medium with a lower salt concentration than DMEM, then the voltages required are lower, presumably due to the hypotonic shock to the cells and to the longer duration of the pulse because of the lower conductivity of the medium. Conversely, electroporation in a hypertonic solution requires higher voltages for optimum permeation.

Cell damage is manifested microscopically by the appearance of dark nuclei under phase-contrast illumination. For most lines this is most pronounced 5–10 min after the pulse. Such cells do not retain Lucifer yellow and fluoresce very weakly, if at all (Fig. 15.6). Despite the fact that every effort is made to make the electric field uniform over the whole cell growth area, the current flow along the border with the etched side is greater than the rest of the conductive surface. For this reason, as the voltage is progressively increasing, damaged cells will appear on this edge first (Fig. 15.6). Another area receiving a slightly higher current is corners of the window. This slight irregularity has to be taken into account when determining the optimal voltage.

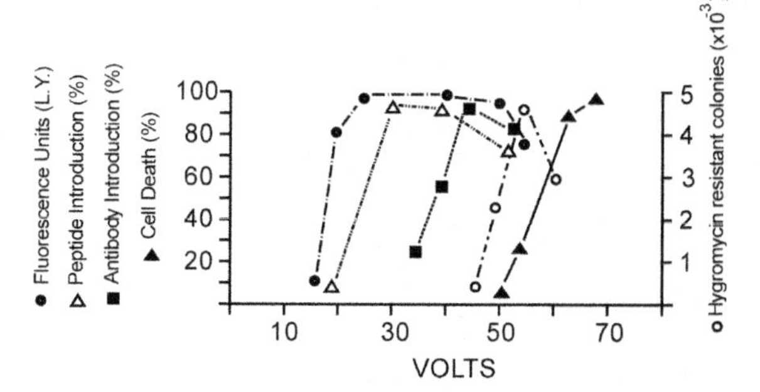

FIGURE 15.5 Effect of field strength on the introduction of different molecules. Three pulses of increasing voltage were applied to confluent rat F111 fibroblasts growing on a conductive surface of 32 × 10 mm from a 32 μF capacitor in the presence of 5 mg/ml Lucifer yellow (●), 5 mg/ml of the Grb2-SH2 blocking peptide (u), 5 mg/ml chicken IgG (■), or 100 μg/ml pY3 plasmid DNA, coding for resistance to hygromycin (Raptis and Firth, 1990) (s). Cells were lysed and Lucifer yellow fluorescence was measured using a Model 204A fluorescence spectrophotometer (●), probed with the anti-active Erk antibody (u), probed for incorporated IgG (■), or selected for hygromycin resistance (s) (Raptis and Firth, 1990). Introduction of chicken IgG was quantitated from the percentage of cells staining positive with their respective antibodies. Cell killing (▲) was assessed by calculating the plating efficiency of the cells after the pulse. Note that a wider range of voltages (20–50 V) permits efficient introduction of Lucifer yellow with no detectable loss in cell viability than the introduction of IgG or DNA. Points represent averages of at least three separate experiments. L.Y., Lucifer yellow.

D. Example of the Determination of Optimal Voltage for the Introduction of Peptides

Prepare a series of slides with cells plated uniformly in a 4 × 7 mm window on a partly conductive slide (Figs. 15.1A or 15.1B). Set the apparatus at 0.5 μF capacitance. Prepare a solution of 5 mg/ml Lucifer yellow and electroporate at different voltages (0.5 μF, three pulses) to determine the upper limits where a small fraction of the cells at the border with the etched side (probably the more extended ones) are killed by the pulse, as determined by visual examination under phase-contrast and fluorescence illumination (Fig. 15.6). Depending on the cells, this voltage can vary from 20 to 40 V. Repeat the electroporation using the peptide solution at different voltages starting at 10 V below the upper limit and at 2 V increments. The Lucifer yellow dye offers an easy way to test for cell permeation.

Results of a typical experiment of electroporation of cells growing on a 32 × 10 mm surface are shown in Fig. 15.5. The application of three exponentially decaying pulses of an initial strength of 25 V from a 32 μF capacitor to rat F111 cells growing on a conductive growth area of 32 × 10 mm, resulted in essentially 100% of the cells containing the introduced Lucifer yellow, whereas introduction of a nine amino acid peptide required 30 V and the stable expression of DNA 45 V, respectively, for maximum signal. If a 20 μF capacitor is used, the corresponding voltages are ~80–180. Electroporation on 7 × 15 mm slides requires a 2 μF capacitor and voltages of 25–50. For etched slides with a conductive area of 4 × 4 mm, if a 0.5 μF capacitor and three pulses are used, the voltages are 20–40. However, if a 0.1 μF capacitor and six pulses are used, the voltages will be ~30–50.

Under the appropriate conditions, electroporation was shown not to affect the activity of Erk1/2 or the stress-activated kinases JNK/SAPK and p38^{hog}. This was shown by probing with antibodies specific for the activated forms of these kinases (Fig. 15.7); no activation of JNK/SAPK or p38^{hog} was found under conditions of up to 70 V (Figs. 15.7C and 7E). These kinases were, however, slightly activated at voltages higher than 85 V, when more than 60% of the cells were killed by the pulse (not shown).

V. PITFALLS

1. Care must be taken so that cells do not dry during the procedure, especially during wiping of the frame with a tightly folded

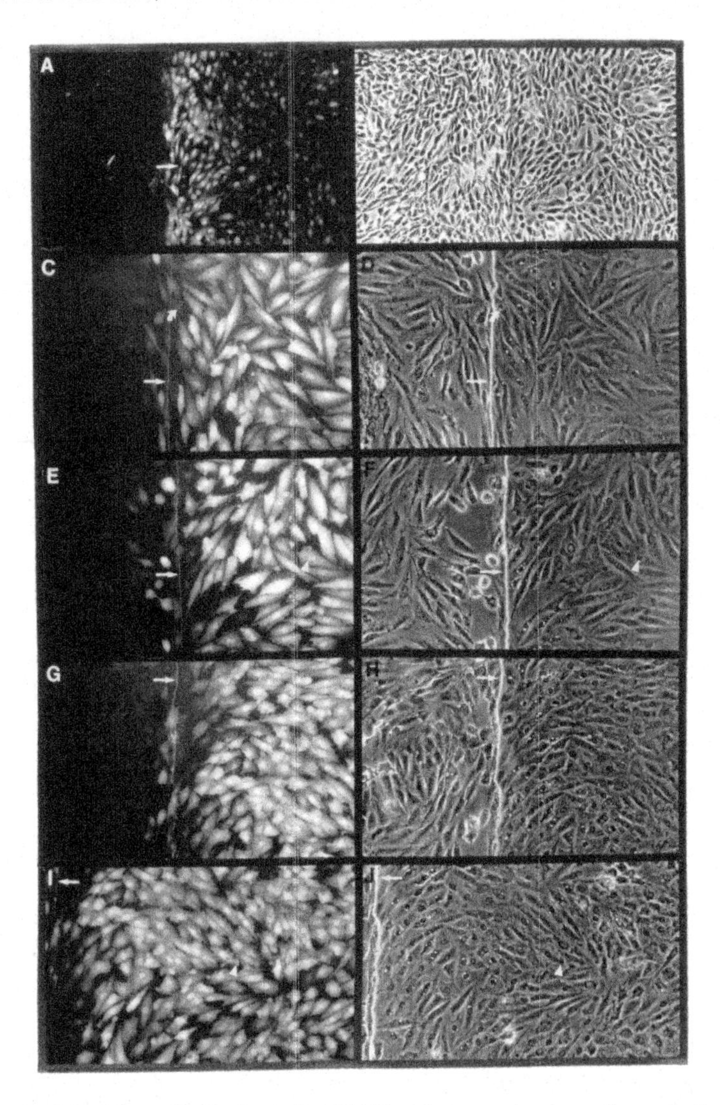

FIGURE 15.6 Determination of the optimal voltage. Rat F111 fibroblasts growing on partly conductive slides (Fig. 15.1B, conductive area, 4 × 4mm) were electroporated in the presence of 5mg/ml Lucifer yellow using three pulses of increasing voltage delivered from a 0.2μF capacitor. (A and B) 18V, (C and D) 26V, (E and F) 28V, (G and H) 32V, and (I and J) 40V. After washing of the unincorporated dye, cells were photographed under fluorescence (A, C, E, G, I) or phase-contrast (B, D, F, H and J) illumination. Straight arrow points to the interphase between conductive (right) and non-conductive (left) areas. Curved arrows in C and D point to a cell which has been killed by the pulse. Note the dark, pycnotic and prominent nucleus under phase contrast and the flat, nonrefractile appearance. Such cells do not retain any electroporated material as shown by the absence of fluorescence (C). Note that the number of such cells along the border with the non-conductive area increases with voltage. Arrowheads in E and F point to a cell that has a prominent nucleus under phase contrast (F) but has retained the dye (E). Such cells rapidly recover their normal morphology, indistinguishable from their non-electroporated counterparts. White arrowheads in I and J point to membrane blebs which tend to enclose Lucifer yellow and fluoresce strongly. Such membrane blebbing tends to be more prominent under higher voltages. Note that if the determination of intercellular communication is desired, then the voltage must be such that cells at the border with the non-conductive area are electroporated without being damaged (e.g., 18V, A and B), whereas for all other applications, voltages of approximately 26–32V would be preferred (C to H). Magnification: A and B, 120X; C–J, 240×.

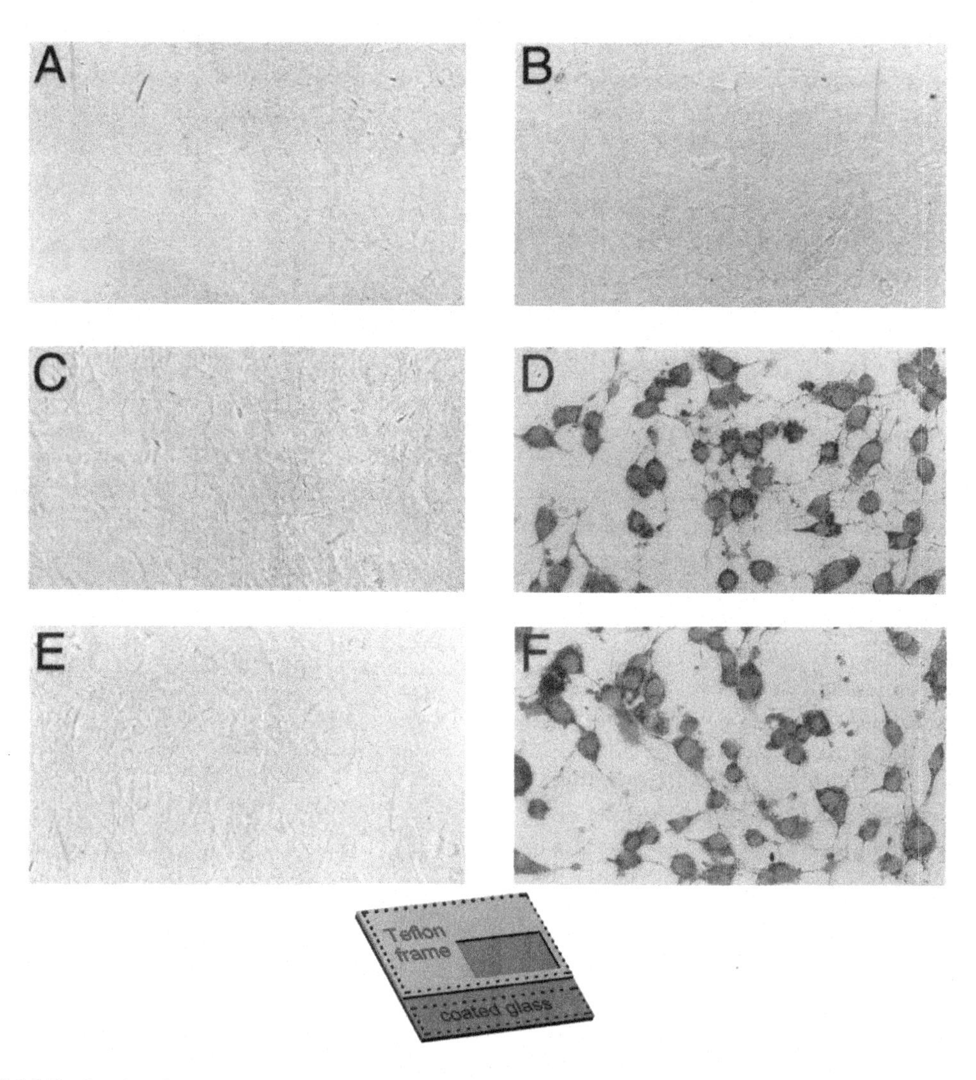

FIGURE 15.7 ***In situ*** **electroporation does not affect ERK activity or the stress pathway.** (**A, C, and E**) NIH3T3 cells were plated on fully conductive slides (inset, Fig. 15.1C, conductive growth area 4 × 8 mm), growth arrested in 50% spent medium, and electroporated in the presence of PBS containing 0.025% DMSO (0.2 μF, 70 V, four pulses). Ten minutes after the pulse, cells were fixed and stained for activated ERK (A), activated JNK/SAPK (C), or activated $p38^{hog}$ (E), respectively. Electroporated cells were photographed under bright-field illumination. (**B, D, and F**) NIH3T3 cells were plated on conductive slides, treated with UV light for 10 min, fixed, and stained for activated ERK (B), activated JNK/SAPK (D), or activated $p38^{hog}$ (F), respectively. From Brownell *et al.* (1998), reprinted with permission. Magnification: 240×.

Kleenex. It was found that serum-starved cells were more susceptible than their counterparts grown in medium containing serum. The morphology of cells that have been killed by drying is very similar to cells that have been killed by the pulse (Fig. 15.6). Slightly dried cells may incorporate Lucifer yellow and appear almost normal under phase contrast, whereas cells that have dried to a great extent display dark nuclei and may not retain Lucifer yellow. It was also found that the combination of even slight drying with electroporation may have undesirable effects on gene expression (e.g., induction of *fos* by serum; unpublished observations). In the case of electroporation on a partly conductive slide (Figs. 15.1A or 15.1B), drying of the cells is immediately suspected if cells growing on the non-electroporated area exhibit Lucifer yellow fluorescence.

2. Accurate determination of the optimal voltage is very important. The limits of voltage tolerance are narrower for serum-starved cells (Brownell *et al.*, 1997) or if the introduction of larger molecules is attempted.
3. For the determination of GJIC, it is important to wash the dye using a calcium-free solution (growth medium or PBS). If calcium-containing growth medium is used instead, the values obtained may be reduced, presumably because of the calcium influx, which was shown to interrupt junctional communication.

Acknowledgments

The financial assistance of the Canadian Institutes of Health Research, the Canadian Breast Cancer Research Initiative, the Natural Sciences and Engineering Research Council of Canada (NSERC), and the Cancer Research Society Inc. is gratefully acknowledged. AV is the recipient of NSERC and Ontario Graduate studentships, a Queen's University graduate award, and a Queen's University travel grant. HB was the recipient of a studentship from the Medical Research Council of Canada and a Microbix Inc. travel award. We are grateful to Dr. Erik Schaefer of Biosource Int. for numerous suggestions and valuable discussions.

References

Bardelli, A., Longati, P., Gramaglia, D., Basilico, C., Tamagnone, L., Giordano, S., Ballinari, D., Michieli, P., and Comoglio, P. M. (1998). Uncoupling signal transducers from oncogenic MET mutants abrogates cell transformation and inhibits invasive growth. *Proc. Natl. Acad. Sci. USA* **95**, 14379–14383.

Boccaccio, C., Ando, M., Tamagnone, L., Bardelli, A., Michielli, P., Battistini, C., and Comoglio, P. M. (1998). Induction of epithelial tubules by growth factor HGF depends on the STAT pathway. *Nature* **391**, 285–288.

Boussiotis, V. A., Freeman, G. J., Berezovskaya, A., Barber, D. L., and Nadler, L. M. (1997). Maintenance of human T cell anergy: Blocking of IL-2 gene transcription by activated Rap1. *Science* **278**, 124–128.

Brownell, H. L., Firth, K. L., Kawauchi, K., Delovitch, T. L., and Raptis, L. (1997). A novel technique for the study of Ras activation; electroporation of [α^{32}P]GTP. *DNA Cell Biol.* **16**, 103–110.

Brownell, H. L., Lydon, N., Schaefer, E., Roberts, T. M., and Raptis, L. (1998). Inhibition of epidermal growth factor-mediated ERK1/2 activation by *in situ* electroporation of nonpermeant [(alkylamino)methyl]acrylophenone derivatives. *DNA Cell Biol.* **17**, 265–274.

Brownell, H. L., Narsimhan, R., Corbley, M. J., Mann, V. M., Whitfield, J. F., and Raptis, L. (1996). Ras is involved in gap junction closure in mouse fibroblasts or preadipocytes but not in differentiated adipocytes. *DNA Cell Biol.* **15**, 443–451.

Chang, D. C., Chassy, B. M., Saunders, J. A., and Sowers, A. E. (1992). "Guide to Electroporation and Electrofusion." Academic Press, New York.

Firth, K. L., Brownell, H. L., and Raptis, L. (1997). Improved procedure for electroporation of peptides into adherent cells *in situ*. *Biotechniques* **23**, 644–645.

Folkman, J., and Moscona, A. (1978). Role of cell shape in growth control. *Nature* **273**, 345–349.

Gambarotta, G., Boccaccio, C., Giordano, S., Ando, M., Stella, M. C., and Comoglio, P. M. (1996). Ets up-regulates met transcription. *Oncogene* **13**, 1911–1917.

Giorgetti-Peraldi, S., Ottinger, E., Wolf, G., Ye, B., Burke, T. R., and Shoelson, S. E. (1997). Cellular effects of phosphotyrosine-binding domain inhibitors on insulin receptor signalling and trafficking. *Mol. Cell. Biol.* **17**, 1180–1188.

Goodenough, D. A., Goliger, J. A., and Paul, D. L. (1996). Connexins, connexons, and intercellular communication. *Annu. Rev. Biochem.* **65**, 475–502.

Kwee, S., Nielsen, H. V., and Celis, J. E. (1990). Electropermeabilization of human cultured cells grown in monolayers: Incorporation of monoclonal antibodies. *Bioelectrochem. Bioenerg.* **23**, 65–80.

Marais, R., Spooner, R. A., Stribbling, S. M., Light, Y., Martin, J., and Springer, C. J. (1997). A cell surface tethered enzyme improves efficiency in gene-directed enzyme prodrug therapy. *Nature Biotechnol.* **15**, 1373–1377.

Matsumura, T., Konishi, R., and Nagai, Y. (1982). Culture substrate dependence of mouse fibroblasts survival at 4°C. *In Vitro* **18**, 510–514.

Nakashima, N., Rose, D., Xiao, S., Egawa, K., Martin, S., Haruta, T., Saltiel, A. R., and Olefsky, J. M. (1999). The functional role of crk II in actin cytoskeleton organization and mitogenesis. *J. Biol. Chem.* **274**, 3001–3008.

Neumann, E., Kakorin, S., and Toensing, K. (2000). Principles of membrane electroporation and transport of macromolecules. *In "Electrochemotherapy, Electrogenetherapy and Transdermal Drug Delivery"* (M. J. Jaroszeski, R. Heller, and R. Gilbert, eds.), pp. 1–35. Humana Press, Clifton, NJ.

Otaka, A., Nomizu, M., Smyth, M. S., Shoelson, S. E., Case, R. D., Burke, T. R., and Roller, P. P. (1994). Synthesis and structure-activity studies of SH2-binding peptides containing hydrolytically stable analogs of O-phosphotyrosine. *In "Peptides; Chemistry, Structure and Biology"* (R. S. Hodges and J. A. Smith, eds.), pp. 631–633. Escom, Leiden.

Raptis, L., Brownell, H. L., Firth, K. L., and MacKenzie, L. W. (1994). A novel technique for the study of intercellular, junctional communication; electroporation of adherent cells on a partly conductive slide. *DNA Cell Biol.* **13**, 963–975.

Raptis, L., Brownell, H. L., Vultur, A. M., Ross, G., Tremblay, E., and Elliott, B. E. (2000a). Specific inhibition of growth factor-stimulated ERK1/2 activation in intact cells by electroporation of a Grb2-SH2 binding peptide. *Cell Growth Differ.* **11**, 293–303.

Raptis, L., and Firth, K. L. (1990). Electroporation of adherent cells *in situ*. *DNA Cell Biol.* **9**, 615–621.

Raptis, L., Tomai, E., and Firth, K. L. (2000b). Improved procedure for examination of gap junctional, intercellular communication by *in situ* electroporation on a partly conductive slide. *Biotechniques* **29**, 222–226.

Robinson, M. J., and Cobb, M. H. (1997). Mitogen-activated protein kinase pathways. *Curr. Opin. Cell Biol.* **9**, 180–186.

Schlessinger, J. (2000). Cell signaling by receptor tyrosine kinases. *Cell* **103**, 211–225.

Tomai, E., Brownell, H. L., Tufescu, T., Reid, K., Raptis, S., Campling, B. G., and Raptis, L. (1998). A functional assay for intercellular, junctional communication in cultured human lung carcinoma cells. *Lab. Invest.* **78**, 639–640.

Vultur, A., Tomai, E., Peebles, K., Malkinson, A. M., Grammatikakis, N., Forkert, P. G., and Raptis, L. (2003). Gap junctional, intercellular communication in cells from urethane-induced tumors in A/J mice. *DNA Cell Biol.* **22**, 33–40.

Williams, E. J., Dunican, D. J., Green, P. J., Howell, F. V., Derossi, D., Walsh, F. S., and Doherty, P. (1997). Selective inhibition of growth factor-stimulated mitogenesis by a cell-permeable Grb2-binding peptide. *J. Biol. Chem.* **272**, 22349–22354.

Yang, T. A., Heiser, W. C., and Sedivy, J. M. (1995). Efficient *in situ* electroporation of mammalian cells grown on microporous membranes. *Nucleic Acids. Res.* **23**, 2803–2810.

CHAPTER

16

In Situ Electroporation of Radioactive Nucleotides: Assessment of Ras Activity or ^{32}P Labeling of Cellular Proteins

Leda Raptis, Adina Vultur, Evi Tomai, Heather L. Brownell, and Kevin L. Firth

OUTLINE

I. INTRODUCTION

Binding to nucleotide(s) can determine a state of activation of a protein; a number of GTP-binding proteins such as Ras exist in two distinct, guanine nucleotide-bound conformations, the active Ras-GTP state and the inactive Ras-GDP form, so that the fraction of Ras bound to GTP (percentage Ras-GTP/GTP + GDP) can determine its state of activation (Lowy and Willumsen, 1993). Several indirect assays are in existence for measurement of Ras activity (Scheele *et al.*, 1995; Taylor *et al.*, 2001). However, in a number of instances, a direct measurement of Ras-GTP binding is necessary (Egawa *et al.*, 1999) and is commonly performed through the addition of [^{32}P]orthophosphate to the growth medium followed by Ras immunoprecipitation and guanine nucleotide elution (Downward, 1995). This approach is relatively inefficient due to the fact that the isotope is incorporated into all phosphate-containing cellular components. To circumvent this problem, [α^{32}P]GTP has been introduced into the cell and its breakdown into [α^{32}P]GDP after Ras binding monitored as described earlier (Downward, 1995). Because, contrary to free bases or nucleosides, most nucleotides do not cross the cell membrane, [α^{32}P]GTP has to be introduced into intact cells after cell membrane permeabilization.

Protein phosphorylation is a ubiquitous regulator of a large variety of cellular functions. The *in vivo* radiolabeling and detection of phosphoproteins is usually conducted through the addition of [^{32}P]orthophosphate to the culture medium, followed by immunoprecipitation and electrophoretic separation of precipitated proteins. Just as in the case of Ras activity measurement, this method is relatively inefficient, hence ATP, the common immediate phosphate donor nucleotide, may be used for *in vivo* protein labeling, which must be introduced into intact cells through membrane permeabilization.

This article describes a technique where the introduction of nucleotides into adherent cells is performed through *in situ* electroporation. Cells are grown on a glass surface coated with electrically conductive, optically transparent indium-tin oxide at the time of pulse delivery, a coating that promotes excellent cell adhesion and growth. Unlike other techniques of cell membrane permeabilization, such as streptolysin-O (SLO) treatment, *in situ* electroporation does not detectably affect cellular metabolism, presumably because the pores reseal rapidly so that the cellular interior is restored to its original state; under the appropriate conditions there is no increase in the activity of the extracellular signal-regulated kinase (Erk1/2) or two stress-activated kinases, JNK/SAPK or p38^{hog} (Brownell *et al.*, 1998). Results show that Ras activity measurement through electroporation of [α^{32}P]GTP could be performed using approximately 50–100 times lower amounts of radioactivity; although the ^{32}P is in the form of [α^{32}P]GTP exclusively, this technique offers higher specificity compared to labeling through the addition of [^{32}P]orthophosphate to the culture medium. In addition, labeling of two viral phosphoproteins, the large tumor antigen of simian virus 40 and adenovirus E1A, by *in situ* electroporation of [γ^{32}P]ATP requires a fraction of the amount of radioactive phosphorus, while offering enhanced specificity.

II. MATERIALS AND INSTRUMENTATION

Dulbecco's modification of Eagle's medium (DMEM) is from ICN (Cat. No. 10-331-22). Phosphate-free DMEM (Cat. No. D-3916) is from Sigma. Fetal calf serum (Cat. No. 2406000AJ) is from Life Technologies Inc. Calf serum is from ICN (Cat. No. 29-131-54). The following reagents are from Sigma: insulin (Cat. No. I-6643), NaCl (Cat. No. S-7653), HEPES (Cat. No. H-9136), Lucifer yellow CH dilithium salt (Cat. No. L-0259), trypsin (Cat. No. T-0646), $MgCl_2$ (Cat. No. M-8266), Triton X-100 (Cat. No. T-6878), deoxycholate (Cat. No. D-5760), EDTA (Cat. No. E-5134), EGTA (Cat. No. E-3889), phenylmethylsulfonyl fluoride (PMSF, P-7626), aprotinin (Cat. No. A-6279), leupeptin (Cat. No. L-2023), benzamidine (Cat. No. B-6506), dithiothreitol (DTT, Cat. No. D-9779), $CaCl_2$ (Cat. No. C-4901), Tris-base (Cat. No. T-6791), sodium orthovanadate (S-6508), and LiCl (Cat. No. L-4408). The following reagents are from BDH: Extran-300 detergent (Cat. No. S6036 39), SDS (Cat. No. 44244), TLC silica gel plates containing a fluorescence indicator (Cat. No. M05735-01), isopropanol (Cat. No. ACS720), concentrated ammonia solution (Cat.

No. ACS033-74), Nonidet P-40 (Cat. No. 56009), and glycerol (Cat. No. B10118). GTP (Cat. No. 106 372) and GDP (Cat. No. 106 208) are from Boehringer Mannheim. The monoclonal anti-Ras antibody is from Oncogene Science (Pan-Ras Ab2, Cat. No. OP22), the monoclonal antibody to simian virus 40 large tumor antigen is from Pharmingen (p108, Cat. No. 14121A), and the monoclonal antibody to adenovirus E1A is from Calbiochem (M73, Cat. No. DP11). Staph. A Sepharose beads (17-0780-03) for immunoprecipitation are from Pharmacia. X-ray film is from Kodak (X-OMAT AR, Cat. No. 165 1454). CelTak™ (Cat. No. 354240) is from BD Biosciences. Tissue culture petri dishes (6 cm diameter) are from Corning or Sarstedt.

The purity of the material to be electroporated is of paramount importance. [α^{32}P]GTP (Cat. No. NEG 006H) and [γ^{32}P]ATP (Cat. No. NEG 002A) are from Dupont NEN Research Products (HPLC purified).

The apparatus for electroporation *in situ* (Epizap model EZ-16) is available from Ask Science Products Inc. (487 Victoria St. Kingston, Ontario Canada). The inverted, phase-contrast and fluorescence microscope, equipped with a filter for Lucifer yellow (excitation: 435, emission: 530), is from Olympus (Model IX70).

The technique can be applied to a large variety of adherent cell types. We have used a number of lines, such as the Fisher rat fibroblast F111 and its polyoma or simian virus 40 virus-transformed derivatives, mouse fibroblast NIH 3T3, mouse Balb/c 3T3, mouse NIH 3T6, mouse C3H10T½ fibroblast derivatives expressing a *ras*-antimessage [e.g., lines R14 and 25B8 (Raptis *et al.*, 1997)], Ras^{leu61}-transformed 10T½, and rat liver epithelial T51B, as well as a variety of differentiated adipocytes (Brownell *et al.*, 1996). All cells can be grown in plastic petri dishes in DMEM supplemented with 5% calf serum in a humidified 5% CO_2 incubator with the exception of R14 and 25B8, which are grown in DMEM supplemented with 10% fetal calf serum. Cells that do not adhere can be grown and electroporated on the same conductive slides coated with CelTak™, poly-l-lysine or collagen.

III. PROCEDURE

The apparatus for *in situ* electroporation is described in Fig. 16.1. Cells are grown on conductive and transparent glass slides, which are placed in a petri dish to maintain sterility. The cell growth area is defined by a "window" formed with an electrically insulating frame made of Teflon. The pulse is transmitted through a stainless-steel negative electrode, which is slightly larger than the cell growth area and is placed on top of the cells, resting on the Teflon frame. Another stainless-steel block is used as a positive contact bar. A complete circuit is formed by placing the electrode set on top of the slide as shown in Figs. 16.1A and 16.1B. The frame creates a gap between the conductive coating and the negative electrode so that current can only flow through the electroporation fluid and cells growing in the window. In order to obtain a uniform electric field strength over the entire area below the negative electrode, despite the fact that the conductive coating exhibits a significant amount of electrical resistance, the bottom surface of the negative electrode must be inclined relative to the glass surface, rising in the direction of the positive contact bar, in a manner proportional to the resistance of the coating (Raptis and Firth, 1990); glass with a surface resistivity of $2\,\Omega/sq$ requires an angle of 1.5°, whereas glass of $20\,\Omega/sq$ requires an angle of 4.4°. The procedure described is for glass with a surface resistivity of $20\,\Omega/sq$, which is readily available and relatively inexpensive, hence it can be discarded after use to limit exposure to radioactivity (see *Comment* 1). Similarly, the electrode can be made out of inexpensive aluminum for a single use.

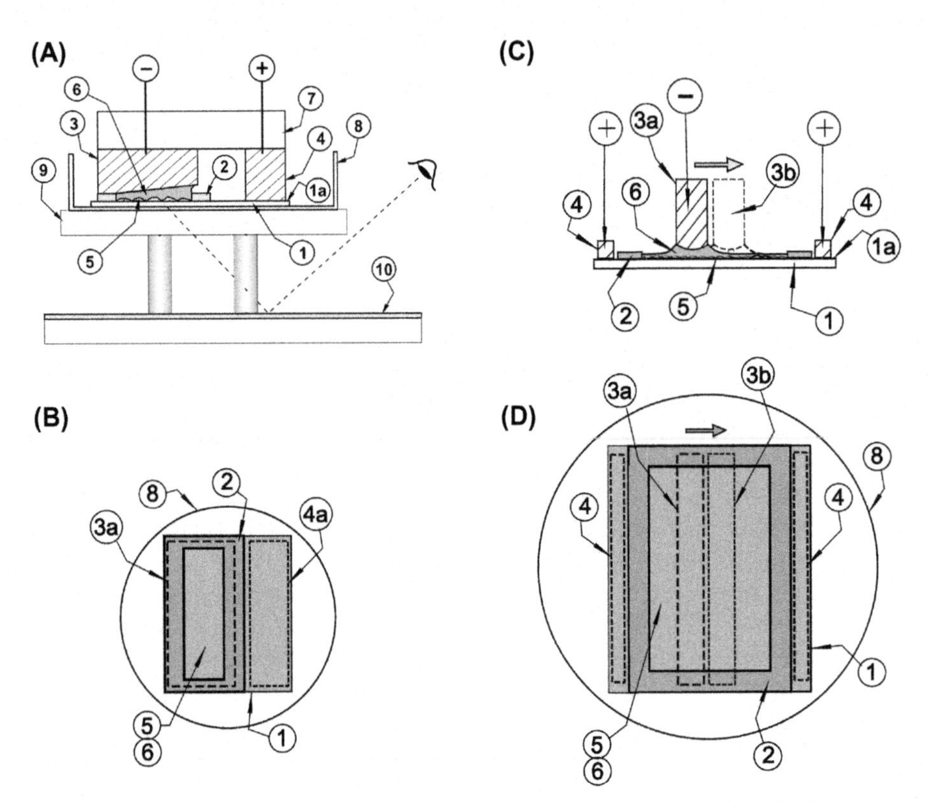

FIGURE 16.1 **Electroporation electrode assembly. (A) Side view.** A Delrin carrier (7) holds the negative electrode (3) and the positive contact bar (4) so that they form one unit, which can be placed on top of the slide (1) with its frame (2) and cells (5) in place. Negative (−) and positive (+) signs indicate the electrical connecting points via which the pulse of electricity is delivered to the electrodes from the pulse generator. The underside of the negative electrode (3) is machined to a slight angle, which compensates for the surface resistivity of the conductive slide to provide uniform electroporation of the whole cell growth area. The negative electrode rests on a Teflon frame (2), which insulates it from the conductive surface. The fluid containing the material to be electroporated just fills the cavity below the negative electrode. When a capacitor is discharged, current passes through the electroporation fluid (6) and cells (5) attached to the conductive glass slide (1) to the positive contact bar (4) and back to the pulse source. Note that the angle of the negative electrode has been exaggerated to better illustrate the meniscus of the radioactive electroporation solution (6). The slide and electrode fit in a 6 cm petri dish (8) that is locked in place on a stand (9). The top plate supports the petri dish and is made of transparent acrylic so that the operator can look in the mirror (10) to ensure that the liquid (6) is properly filling the cavity without air bubbles. **(B) Top view.** The outline of the conductive slide (1) with a Teflon frame (2) in place to define the area of cell growth and electroporation are indicated [from Brownell *et al.* (1997), reprinted with permission]. **Upscaling. (C) Side view.** Cells to be electroporated (5) are grown on a glass slide (1), coated with ITO (1a). The negative electrode (3a and 3b) is a narrow steel bar mounted across the width of the slide, resting on the Teflon frame (2), which is moved across the surface of the slide as shown by the arrow, by an insulated carrier. The underside of the negative electrode is curved in both directions, such as to optimise the uniformity of electrical field. Note that only the area of cells immediately below the electrode is electroporated by a given pulse. The curvature of the negative electrode has been exaggerated to better illustrate its contour. Note that due to the narrow shape of the electrode, air bubbles do not get trapped easily under it, hence a stand with a mirror may not be necessary. **(D) Top view.** The outline of the conductive slide with a Teflon frame (2) in place to define the area of cell growth and electroporation and the counterelectrodes (4) are indicated. The assembly is placed in a 10 cm petri dish (8) [from Raptis *et al.* (2003) and Tomai *et al.* (2003), reprinted with permission].

Upscaling

A large number of signal transducers are present in small amounts in the cell so that a large number of cells may be required to obtain a strong signal. Uniform electroporation of a cell growth area of 32 × 10 mm using 20 Ω/sq glass (i.e., an angle of 4.4°) and a Teflon frame with a thickness of 0.279 mm using the assembly in Figs. 16.1A and 16.1B requires a volume of ~280 μl. Simple scale-up of this assembly, e.g., to a cell growth area of 50 × 30 mm, which can be accommodated in a standard, 10 cm petri dish, is faced with the problem of burning the ITO coating that occurs at the higher voltage and capacitance settings required to electroporate this larger area because of the resistance generated by the greater distance the current has to travel. In addition, the volume required, ~1.7 ml, cannot be held in place by surface tension, while the cost of purchase and disposal of the isotope can be prohibitive for certain experiments. These problems can be solved by using an assembly with a narrow, moveable electrode that electroporates a "strip" of cells at a time (Figs. 16.1C and 16.1D); in this configuration, only cells immediately below the negative electrode are electroporated by a given pulse of electricity. After electroporation of the first strip of cells, the electrode is translocated laterally, dragging the solution under it by surface tension so that a new strip of cells is electroporated using mostly the same solution (Figs. 16.1C and 16.1D). The electric circuit formed during pulse delivery starts at the negative electrode, passes through the electroporation fluid, the cells and the conductive slide surface, to the two positive contact bars, one on each side of the slide. The two positive contact bars form parallel circuit paths, both carrying current from the conductive surface. To compensate for the resistance of the coating, the bottom surface of the negative electrode must be inclined toward each of the positive contact bars; a 25 mm radius on the bottom of the electrode produces successive strips of even electroporation over the entire cell growth area. Using this assembly, an area of 32 × 10 mm can be electroporated using less than 50 μl of solution with a 2.5 mm-wide electrode, whereas an area of 50 × 30 mm can be electroporated effectively in four, 8 mm-wide strips, with a total of 200 μl of solution.

A. Electroporation of [α^{32}P]GTP for Measurement of Ras Activity

Solutions

1. ***Lucifer yellow solution, 5 mg/ml***: To make 1 ml, add 5 mg Lucifer yellow to 1 ml phosphate-free DMEM.
3. ***[α ^{32}P]GTP***: Prepare a solution of 500–2000 μCi/ml in phosphate-free DMEM (*see Comment* 2).
4. ***Ras extraction buffer***: 50 m*M* HEPES, pH 7.4, 150 m*M* NaCl, 5 m*M* $MgCl_2$, 1% Triton, 0.5% deoxycholate, 0.05% SDS, 1 m*M* EGTA, 1 m*M* PMSF, 10 μg/ml aprotinin, 10 μg/ml leupeptin, 10 m*M* benzamidine, and 1 m*M* vanadate. The stock solutions can be made ahead of time. To make 50 ml, add 2.5 ml of HEPES stock, 1.5 ml of NaCl stock, 0.25 ml of $MgCl_2$ stock, 0.5 ml Triton, 0.05 g deoxycholate, 0.25 ml of SDS stock, and 0.5 ml of EGTA stock and then the protease inhibitors, 0.5 ml of PMSF stock solution, 0.05 ml of aprotinin stock, 0.05 ml of leupeptin stock, 0.5 ml of benzamidine stock, and 0.05 ml of vanadate stock and bring the volume to 50 ml with distilled H_2O on the day of the experiment.

   10% sodium dodecyl sulfate stock solution: Dissolve 100 g in 1 liter H_2O. Store at room temperature. Stable for more than a year.
   5 M NaCl stock solution: Dissolve 292.2 g NaCl in 1 liter distilled H_2O. Autoclave and store at room temperature. Stable for more than a year.

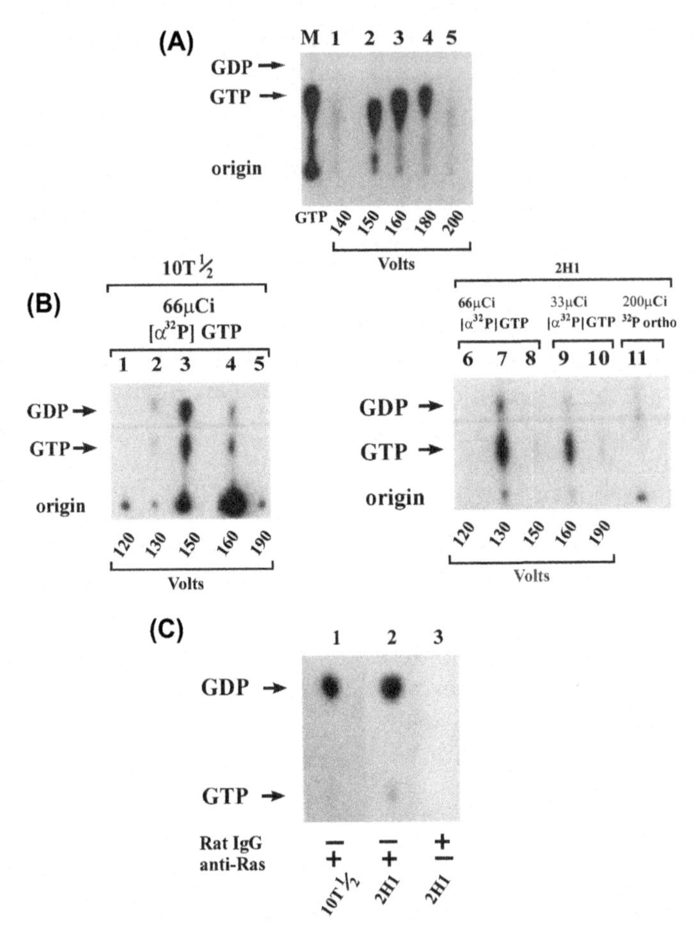

FIGURE 16.2 **Assessment of Ras activity through electroporation of [α^{32}P]GTP.** Cells were grown on conductive glass (cell growth area, 32 × 10mm, Figs. 16.1A and 16.1B) and starved from serum and phosphates. A solution containing [α^{32}P]GTP was added to the cells and introduced with three pulses delivered from a 20μF capacitor at different voltages as indicated. Cells were subsequently placed in a humidified 37°C, CO_2 incubator for 3h. Ras was extracted and precipitated with the pan-ras Ab2 monoclonal antibody, the bound GTP and GDP eluted and separated by thin-layer chromatography (see text). The plate was exposed for 15h to Kodak XAR-5 film with an intensifying screen. In all panels, arrows point to the positions of cold GTP and GDP standards, respectively. **(A) Electroporation does not induce a rapid breakdown of intracellular GTP.** Lanes 1–5: Mouse 10T½ fibroblasts were electroporated using voltages of 140–200V as indicated in the presence of 5μCi [α^{32}P]GTP. Three hours later, nucleotides in a 2μl aliquot of each clarified lysate were separated as described earlier without Ras immunoprecipitation. Lane M: As a marker, an aliquot of the [α^{32}P]GTP used in this experiment was run in parallel. **(B) Assessment of Ras activity through electroporation of [α^{32}P]GTP in normal 10T½ and their *ras*-transformed counterparts.** Lanes 1–10: 10T½ cells (lanes 1–5) or their *ras*^{val12}-transformed counterparts, 2H1 (lanes 6–10), were electroporated using voltages of 120–190V as indicated in the presence of 66 or 33μCi [α^{32}P]GTP, respectively. Proteins were extracted, and the Ras-bound GTP and GDP were separated as described earlier. As a control (lane 11), 2H1 cells growing in a 3cm petri to were metabolically labeled with 200μCi [α^{32}P]orthophosphate and processed as described previously. **(C) Assessment of Ras activity in normal 10T½ fibroblasts and their *ras*-transformed counterparts, 2H1, using the standard SLO permeabilization assay.** This assay was performed as described (Brownell *et al.*, 1997) and is shown here as a comparison. Lane 1, 10T½; lane 2, *ras*^{val12}-transformed, 10T½-derived line 2H1; and lane 3, 2H1 precipitated with control rat IgG instead of anti-Ras antibodies

1 M $MgCl_2$ stock solution: Dissolve 203.3 g $MgCl_2 \cdot 6H_2O$ in 1 liter H_2O. Autoclave and store at room temperature. Stable for more than a year.
100 mM EGTA stock solution: Dissolve 38.04 g in 800 ml distilled H_2O. Adjust pH to 8.0 with NaOH and complete volume to 1 liter with distilled H_2O. Stable for more than a year at room temperature.
100 mM PMSF stock solution: Dissolve 17.4 mg PMSF in 10 ml isopropanol and store in aliquots at −20°C. Stable for several months.
10 mg/ml aprotinin stock solution: Dissolve 100 mg in 10 ml of 0.01 *M* HEPES, pH 8.0. Aliquot and store at −20°C. Stable for several months.
10 mg/ml leupeptin stock solution: Dissolve 100 mg in 10 ml distilled H_2O. Aliquot and store at −20°C. Stable for several months.
1 M benzamidine stock solution: Dissolve 1.56 g in 10 ml distilled H_2O. Aliquot and store at −20°C. Stable for several months.
1 M vanadate stock solution: Dissolve 1.84 g in 10 ml distilled H_2O. Aliquot and store at −20°C. Stable for several months.

5. ***Guanine nucleotide elution buffer***: 2 m*M* EDTA, 2 m*M* DTT, 0.2% SDS, 0.5 m*M* GTP, and 0.5 m*M* GDP.

1 M dithiothreitol stock solution: Dissolve 3.09 g DTT in 20 ml of 0.01 sodium acetate (pH 5.2). Store in 1 ml aliquots at −20°C. Stable for more than a year.
0.5 M EDTA, pH 8.0 stock solution: Add 186.1 g disodium ethylenediaminetetraacetate · $2H_2O$ to 800 ml distilled H_2O. To dissolve, adjust the pH to 8.0 with NaOH pellets while stirring. Bring volume to 1 liter. Stable for more than a year at room temperature.
10 mM GTP stock solution: Dissolve 52.3 mg in 10 ml distilled H_2O and store at −20°C. Stable for a month.
10 mM GDP stock solution: Dissolve 44.3 mg in 10 ml distilled H_2O and store at −20°C. Stable for a month.
See earlier list for other required stock solutions: To make 500 µl of elution buffer, add 10 µl of 10% SDS stock solution, 2 µl of EDTA stock solution, 1 µl of DTT stock solution, 25 µl each of GTP and GDP stock solutions, and bring volume up to 500 µl with H_2O.

6. ***Thin-layer chromatography (TLC) running buffer***: 66% isopropanol and 1% concentrated ammonia. To make 100 ml, mix 66 ml isopropanol, 1 ml concentrated ammonia solution, and 33 ml distilled H_2O in a fume hood. Make fresh the day of the experiment. Depending on the size of the chromatography tank, this volume may need to be increased.

Steps

The appropriate institutional regulations for isotope use must be followed for all experiments.

1. **Choice of slides.** Cell growth areas of 32 × 12 mm are sufficient for most cases. However, for some lines with low Ras levels, a larger number of cells may be required to obtain an adequate signal. In this case, the assembly in Figs. 16.1C and 16.1D (cell growth area, 50 × 30 mm) may be used. Make sure the glass is clean and free of fingerprints (see *Comment* 5).
2. **Plate the cells.** Uniform spreading of the cells is very important, as the optimal voltage depends in part on the degree of cell contact with the conductive surface (*see Comment* 3). Place the sterile glass slide inside a 6 cm (for 32 × 10 mm) or 10 cm (for 50 × 30 mm) petri dish. Add a sufficient amount of medium to cover the slide (approximately 9 ml for a 6 cm dish). Pipette the cell suspension in the window (Fig. 16.1) and place the petris in a tissue-culture incubator.
3. **Starve the cells from phosphates** by placing them in phosphate-free DMEM and the required amounts of dialysed serum

for 2–3 h or overnight, depending on the experiment.

4. Prior to pulse application, **remove the growth medium** and wash the cells gently with phosphate-free DMEM.
5. Carefully **wipe** the Teflon frame with a folded Kleenex tissue to create a dry area on which a meniscus can form (see *Pitfall* 1).
6. **Add the [α^{32}P]GTP solution.** The volume of the solution under the electrode varies with the electroporation assembly and cell growth area. For the setup in Fig. 16.1A and a cell growth area of 32 × 10 mm, the volume is ~280 μl, whereas the assembly in Fig. 16.1C requires ~200 μl for a cell growth area of 50 × 30 mm. Depending on the exact concentration, this volume will contain ~200 μCi [α^{32}P]GTP in phosphate-free DMEM.
7. Carefully **place the electrode** on top of the cells. Make sure there is a sufficient amount of electroporation buffer under the positive contact bar to ensure electrical contact. Make sure there are no air bubbles between the negative electrode and the cells by looking in the mirror. If necessary, the electrode can be sterilized with 70% ethanol before the pulse, and the procedure carried out in a laminar-flow hood, using sterile solutions.
8. Apply three to six **pulses** of the appropriate strength (40–200 V, see *Comment* 3) from a 10- or 20-μF capacitor, depending on the apparatus used (Fig. 16.1A vs Fig. 16.1C).
9. **Remove the electrode set.** Because usually only a small fraction of the material penetrates into the cells, the [α^{32}P]GTP solution can be carefully aspirated and used again.
10. **Add phosphate-free medium** containing dialysed serum if permitted by the experimental protocol and incubate the cells for the desired length of time (see *Comment* 4).
11. **Remove the unincorporated material**: wash the cells twice with phosphate-free medium lacking serum.
12. **Extract the proteins.** Add 1 ml of extraction buffer to the window area of the slide. Scrape the cells using a rubber policeman into a 15 ml tube and rock the tubes on ice for 20 min. Centrifuge for 30 min at 1,000 rpm in a Beckman J-6 centrifuge to clarify. Preclear the lysates by adding 100 μl packed Staph. A–Sepharose beads, incubating on ice for 1 h, and centrifuging for 5 min at 1,000 rpm in a Beckman J-6 centrifuge.
13. **Immunoprecipitate Ras.** Incubate the precleared supernatant overnight with pan ras Ab2 antibody bound to Staph. A Sepharose beads while rocking on ice.
14. **Wash the immunoprecipitate** four times with 1 ml of extraction buffer lacking the inhibitors. Use a Hamilton syringe to completely remove all traces of wash solution.
16. **Elute** GTP and GDP off the beads by adding 10–20 μl elution buffer to the beads and incubating at 68°C for 20 min.
17. **Spot** the eluate containing the labeled nucleotides on a silica gel TLC plate containing a fluorescence indicator. Spot 1 μl each of the stock GTP and GDP solutions to serve as cold standards, easily visible under UV light. Develop the plate using a solution of 1% ammonia–66% isopropanol for about 3–4 h.
18. Dry the TLC plate, **expose** to Kodak X-OMAT AR film, and excise the spots for liquid scintillation counting or submit to phosphorimager analysis (see Fig. 16.2).

B. Electroporation of [γ^{32}P]ATP for Labeling of Cellular Proteins: Labeling of the Simian Virus 40 Large Tumor Antigen or Adenovirus E1A

Solutions

1. *[γ^{32}P]ATP*: 600–1000 μCi/ml in phosphate-free DMEM.

2. ***Simian virus large tumor antigen (SVLT) or adenovirus E1A extraction buffer***: 122 m*M* NaCl, 18 m*M* Tris-base, pH 9.0, 0.8 m*M* $CaCl_2$, 0.43 m*M* $MgCl_2$, 10% glycerol, 1% NP-40, 10 μg/ml aprotinin, 10 μg/ml leupeptin, 1 m*M* PMSF, and 1 m*M* vanadate. Add 8.0 g NaCl, 2.42 g Tris-base, 0.1 g $CaCl_2$, 0.04 g $MgCl_2$, 10 ml NP-40, and 100 ml glycerol to 800 ml H_2O. Adjust pH to 9.0 and store at −20°C. On the day of the experiment, add 0.1 ml aprotinin stock, 0.1 ml leupeptin stock, 1 ml PMSF stock, and 0.1 ml vanadate stock solutions to 100 ml of this solution.
3. ***Tris–LiCl solution for washing immunoprecipitates***: 0.1 *M* Tris-base and 0.5 *M* LiCl. Make a stock solution of 10X (4 litres). Add 484 g of Tris-base and 848 g of LiCl in 4 litres of H_2O. Adjust pH to 7.0 using concentrated HCl.
4. ***SDS gel-loading buffer***: 15 ml 10% SDS, 1.5 ml mercaptoethanol, 6 ml glycerol, 3 ml 1.25 *M* Tris, pH 6.8, 0.75 ml 0.3% bromphenol blue, and water to 30 ml.
5. *10% SDS stock solution*
6. *1.25 M Tris–HCl (pH 6.8) stock solution*: Add 151.3 g Tris-base to 800 ml distilled water. Adjust the pH to 6.8 using concentrated HCl and the volume to 1 liter with distilled water.

Steps

1. **Plate, starve from phosphates, and wash** the cells with phosphate-free DMEM as in Section III,A.
2. **Add the [γ^{32}P]ATP solution.**
3. Apply three to six **pulses** of the appropriate strength (40–200 V, see *Comment* 3) from a 10- or 20-μF capacitor, depending on the apparatus used (Fig. 16.1A vs Fig. 16.1C). If the setup in Fig. 16.1 A is employed, then the solution can be carefully aspirated and reused once more after the pulse.
4. **Add phosphate-free medium** and incubate the cells for 2–3 h in a tissue-culture incubator.
5. **Extract** the proteins by adding 1 ml SVLT extraction buffer to the window, scraping into a 15 ml tube, and rocking on ice. Clarify by spinning for 30 min at 1,000 rpm in a Beckman J-6 centrifuge (2,000 *g*).
6. **Precipitate** with the pAb108 (SVLT) or M73 (E1A) monoclonal antibody and wash three times with PBS, twice with the Tris–LiCl solution, and once with H_2O. Elute labeled proteins from the beads with SDS gel-loading buffer and resolve by acrylamide gel electrophoresis (Fig. 16.3).

IV. COMMENTS

The technique of *in situ* electroporation is very versatile. A large variety of molecules, such as peptides (Boccaccio *et al.*, 1998; Raptis *et al.*, 2000), nucleotides (Brownell *et al.*, 1997), antibodies (Raptis and Firth, 1990), or drugs (Marais *et al.*, 1997), can be introduced, alone or in combination, at the same or different times, in continuously growing or growth arrested cells or cells at different stages of thefir division cycle (see Chapter 16 by Raptis *et al.*).

1. The slides can normally be washed with Extran-300 and reused a number of times. However, in the case of introduction of radioactive material, exposure of personnel to irradiation is an important consideration. The removal of ^{32}P-labeled nucleotides from the ITO-coated glass is difficult because the phosphate group is attracted to this coating (Tomai *et al.*, 2000), hence the use of inexpensive slides and electrodes that can be discarded after use is highly desirable. Slides with a conductivity of 20 Ω/sq are sufficiently inexpensive that they can be discarded after use or stored for the ^{32}P to decay before washing. The use of less conductive ITO-coated glass (100 Ω/sq) would reduce the cost of the slides further. However, in our experience,

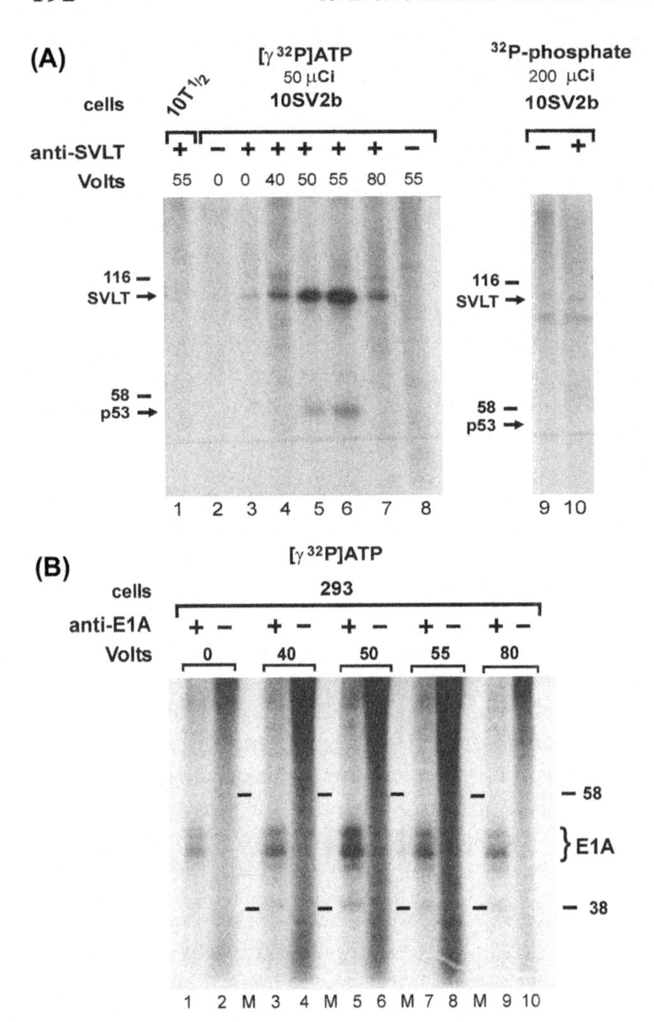

FIGURE 16.3 Labeling of the simian virus 40 large tumor antigen or adenovirus E1A through *in situ* electroporation of [γ^{32}P]ATP. (A) Mouse 10T½ cells (lane 1) or their SVLT-transformed derivatives (line 10SV2b, lanes 2–8) were grown on 50 × 30 mm conductive areas (Figs. 16.1C and 16.1D). A solution containing 50 μCi [γ^{32}P]ATP in phosphate-free DMEM was added to the cells, and six capacitor-discharge pulses of 10 μF, 40–80 V were applied as indicated. Cells were placed in a humidified incubator for 30 min. For a comparison (lanes 9 and 10), the same SVLT-transformed cells were labeled *in vivo* with the indicated amounts of [^{32}P]orthophosphate. SVLT was precipitated from detergent extracts with the pAb108, anti-SVLT antibody (lanes 1, 3–7, and 10), or normal mouse IgG (lanes 2, 8, and 9) and labeled proteins were resolved by acrylamide gel electrophoresis. Dried gels were exposed for 1 h to Kodak XAR-5 film with an intensifying screen. Note the intense and specific labeling of SVLT and the associated phosphoprotein, p53, by *in situ* electroporation (lanes 5 and 6) compared to cells labeled *in vivo* with 200 μCi [^{32}P]orthophosphate (lanes 9 and 10). **(B)** Human 293 cells transformed with adenovirus DNA were labeled as described earlier and extracts were preadsorbed with normal mouse IgG (lanes 2, 4, 6, 8, and 10) or immunoprecipitated using the M73, anti-E1A antibody (lanes 1, 3, 5, 7, and 9). Bracket points to the position of the phosphorylated E1A bands. M, molecular weight marker lanes.

the conductivity of this grade of glass is not sufficiently consistent for electroporation experiments due to problems related to uniformity of thickness encountered with the thinner coating of the less conductive, commercially available surface. On the other hand, the use of more conductive glass (2 Ω/sq) requires a smaller electrode angle, which reduces the cost of the material substantially. However, this glass is not a regular production item, hence it is more expensive.

2. The [α^{32}P]GTP must be of the highest purity. Because a number of lots were found to contain varying amounts of [α^{32}P]GDP, it is wise to test the preparation by thin-layer chromatography before use.
3. **Determination of the optimal voltage and capacitance.** Electrical field strength has been shown to be a critical parameter for cell permeation, as well as viability (Chang *et al.*, 1992). It is generally easier to select a discrete capacitance value and then control the voltage precisely. The optimal voltage

depends on the strain and metabolic state of the cells, as well as the degree of cell contact with the conductive surface, possibly due to the larger amounts of current passing through an extended cell (Yang *et al.*, 1995; Raptis and Firth, 1990). Densely growing, transformed cells or cells in a clump require higher voltages for optimum permeation than sparse, subconfluent cells. Similarly, cells that have been detached from their growth surface by vigorous pipetting prior to electroporation require substantially higher voltages. In addition, cells growing and electroporated on collagen, poly-l-lysine, or CelTak™-coated slides require substantially higher voltages than cells growing directly on the slide.

The margins of voltage tolerance depend on the size and electrical charge of the molecules to be introduced. For the introduction of small, uncharged molecules such as Lucifer yellow or nucleotides, a wider range of field strengths permits effective permeation with minimal damage to the cells than the introduction of antibodies or DNA (Raptis and Firth, 1990; Brownell *et al.*, 1997). For all cells tested, the application of multiple pulses at a lower voltage can achieve a better permeation and is better tolerated than a single pulse. This is especially important for the electroporation of serum-starved cells where the margins of voltage tolerance were found to be substantially narrower compared to their counterparts growing in 10% calf serum (Brownell *et al.*, 1997). Results of a typical experiment are shown in Fig. 16.4. Rat F111 cells were grown on slides with a 50 × 30 mm cell growth area (Figs. 16.1C and 16.1D) and six pulses were delivered from a 10 μF capacitor. Following electroporation of the first strip of cells, the negative electrode was translocated laterally (Fig. 16.1C, arrow) so that the whole area was electroporated in four strips. The application of six exponentially decaying pulses of an initial strength of ~40–55 V resulted in essentially 100% of the cells containing the introduced dye, Lucifer yellow, whereas [α^{32}P]GTP required ~50 V for a maximum signal.

Cell damage is microscopically manifested by the appearance of dark nuclei under phase-contrast illumination. For most lines, this is most prominent 5–10 min after the pulse. Such cells do not retain Lucifer yellow and fluoresce very weakly, if at all (Fig. 16.5). It was also noted that the current flow along the corners of the window is slightly greater than the rest of the conductive area. For this reason, as the voltage is progressively increasing, damaged cells will appear on this area first. This slight irregularity has to be taken into account when determining the optimal voltage.

Example of determination of the optimal voltage. Prepare a series of slides with cells plated uniformly in a 32 × 10 mm window (Figs. 16.1A and 16.1B). Set the apparatus at 20 μF capacitance. Prepare a solution of 5 mg/ml Lucifer yellow in phosphate-free DMEM and a solution of [α^{32}P]GTP or [γ^{32}P]ATP containing 5 mg/ml Lucifer yellow in phosphate-free DMEM. Electroporate the Lucifer yellow solution at different voltages, 20 μF, three pulses, to determine the upper limits where a small fraction of the cells at the corners of the window (usually the more extended ones) are killed by the pulse, as determined by visual examination under phase-contrast and fluorescence illumination 5–10 min after the pulse (Fig. 16.5). Depending on the cells and growth conditions, this voltage can vary from 130 to 190 V. Repeat the electroporation using the radioactive nucleotide solution at different voltages starting at 20 V below the upper limit and at 5 V increments. The Lucifer yellow offers a convenient marker for cell permeation and it was found not to affect the results.

4. Serum was shown to facilitate pore closure (Bahnson and Boggs, 1990).
5. The slides come with the apparatus and are sterile. If sterility is compromised or if the slides have been washed to reuse, then place

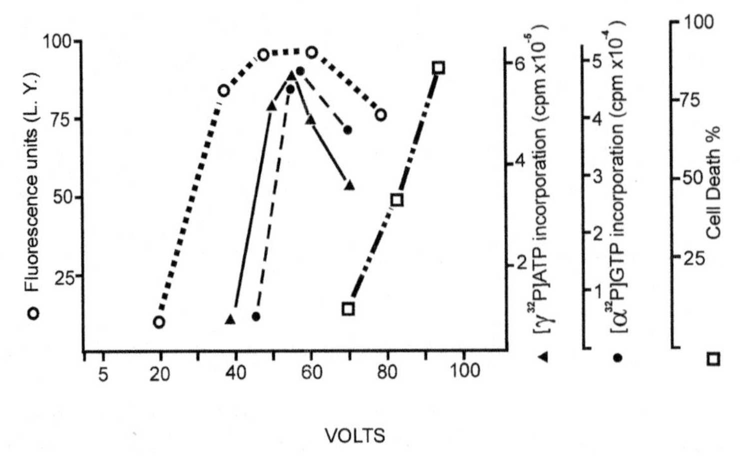

FIGURE 16.4 Effect of field strength on the introduction of nucleotides. Six pulses of increasing voltage were applied from a 10μF capacitor to serum-starved 10T½ cells growing on a conductive surface of 50 × 30mm in the presence of 10μCi [α^{32}P]GTP (●) or 10μCi [γ^{32}P]ATP (▲). Total protein labeling was quantitated using a nitrocellulose filter-binding assay (Buday and Downward, 1993). Numbers refer to cpm per 100μg of protein in clarified extracts. As a control, cells were electroporated with 5mg/ml Lucifer yellow (s), and its introduction was assessed by fluorescence measurement of cell lysates using a Perkin–Elmer Model 204A fluorescence spectrophotometer. Cell killing (u) was calculated from the plating efficiency of the cells 2h after the pulse [from Raptis *et al.* (2003) and Tomai *et al.* (2003), reprinted with permission].

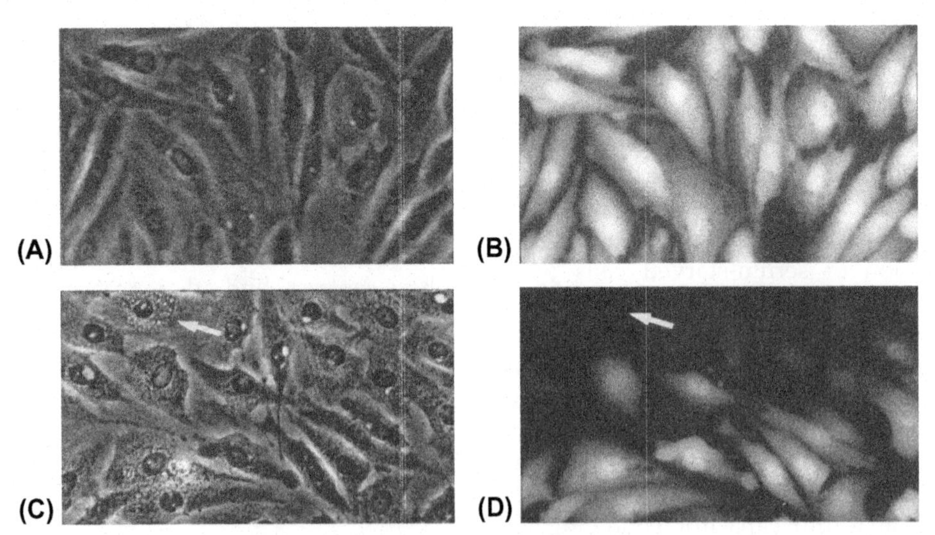

FIGURE 16.5 Determination of optimal voltage. Rat F111 fibroblasts growing on conductive slides (Figs. 16.1A and 16.1B, cell growth area, 32 × 10mm) were electroporated in the presence of 5mg/ml Lucifer yellow using three pulses of 155 (A and B) or 190 (C and D) volts delivered from a 20μF capacitor. After washing of the unincorporated dye, cells were photographed under phase-contrast (A and C) or fluorescence (B and D) illumination. Arrows in C and D point to a cell that has been killed by the pulse. Note the dark, pycnotic and prominent nucleus under phase contrast and the flat, nonrefractile appearance. Such cells do not retain any electroporated material, as shown by the absence of fluorescence (C). It is especially striking that cells at the top of frames C and D, situated at the edge of the electroporated area, received a larger amount of current and have been killed by the pulse. Magnification: 240 times.

in petris and sterilize by adding 80% ethanol for 20 min and then removing the ethanol by rinsing with sterile distilled water.

Other characteristics of the technique. Under the appropriate conditions, *in situ* electroporation does not affect cell morphology or the length of the G1 phase of serum-stimulated cells and does not induce c-*fos* (Raptis and Firth, 1990; Brownell *et al.*, 1997). In addition, it does not affect activity of the extracellular signal-regulated kinase (Erk1/2) or two kinases commonly activated by a number of stress-related stimuli, JNK/SAPK and $p38^{hog}$ (Robinson and Cobb, 1997), presumably because the pores reseal rapidly so that the cell interior is restored to its original state (Brownell *et al.*, 1998).

Measurement of steady-state Ras activity is possible using this method because, contrary to other methods of cell permeabilization, such as streptolysin-O (Buday and Downward, 1993), the cells are not detectably affected by the procedure so that they can be incubated for long periods of time before extraction. In addition, electroporation does not appear to induce a rapid breakdown of intracellular GTP in any of the lines tested, even under conditions where a substantial fraction of the cells are killed by the pulse (Fig. 16.2). As a result, determination of the Ras-bound, GTP/GTP + GDP ratio is made easier by the fact that although the optimal voltage must be determined empirically as in all electroporation experiments, excessively high voltages, despite the fact that they may kill a substantial proportion of the cells, do not alter the ratios obtained (Fig. 16.2B), presumably because such cells rapidly lyse, without affecting the results.

V. PITFALLS

1. Care must be taken so that cells do not dry during the procedure, especially during wiping of the frame with a tightly folded Kleenex. It was found that serum-starved cells were especially susceptible. The morphology of cells that have been killed by drying is very similar to cells that have been killed by the pulse (Fig. 16.5).
2. Accurate determination of the optimal voltage is very important. For most nucleotide introduction applications, optimal labeling was observed in the range of 140–160 V, whereas these margins were found to be narrower for serum-starved cells. Nevertheless, the Ras-bound GTP/GTP + GDP ratios or the profile of SVLT or E1A labeling obtained was found to be the same even when a substantial proportion of the cells were killed by the pulse, in which case merely ^{32}P incorporation is reduced.

Acknowledgments

The financial assistance of the Canadian Institutes of Health Research, the Natural Sciences and Engineering Research Council of Canada, and the Cancer Research Society Inc. to LR is gratefully acknowledged. AV is the recipient of NSERC and Ontario Graduate studentships, a Queen's University graduate award, and a Queen's University travel grant. ET is the recipient of a Queen's University graduate award, an NSERC studentship, and awards from the Thoracic Society and the Lemos foundation. HB was the recipient of a studentship from the Medical Research Council of Canada and Queen's University graduate school and Microbix Biosystems Inc. travel awards.

References

Bahnson, A. B., and Boggs, S. S. (1990). Addition of serum to electroporated cells enhances survival and transfection efficiency. *Biochem. Biophys. Res. Commun.* **171**, 752–757.

Boccaccio, C., Ando, M., Tamagnone, L., Bardelli, A., Michielli, P., Battistini, C., and Comoglio, P. M. (1998). Induction of epithelial tubules by growth factor HGF depends on the STAT pathway. *Nature* **391**, 285–288.

Brownell, H. L., Firth, K. L., Kawauchi, K., Delovitch, T. L., and Raptis, L. (1997). A novel technique for the study of Ras activation; electroporation of [α^{32}P]GTP. *DNA Cell Biol.* **16**, 103–110.

Brownell, H. L., Lydon, N., Schaefer, E., Roberts, T. M., and Raptis, L. (1998). Inhibition of epidermal growth factor-mediated ERK1/2 activation by *in situ* electroporation of nonpermeant [(alkylamino)methyl]acrylophenone derivatives. *DNA Cell Biol.* **17**, 265–274.

Brownell, H. L., Narsimhan, R., Corbley, M. J., Mann, V. M., Whitfield, J. F., and Raptis, L. (1996). Ras is involved in gap junction closure in mouse fibroblasts or preadipocytes but not in differentiated adipocytes. *DNA Cell Biol.* **15**, 443–451.

Buday, L., and Downward, J. (1993). Epidermal growth factor regulates the exchange rate of guanine nucleotides on p21ras in fibroblasts. *Mol. Cell. Biol.* **13**, 1903–1910.

Chang, D. C., Chassy, B. M., Saunders, J. A., and Sowers, A. E. (1992). Guide to Electroporation and Electrofusion. Academic Press, New York.

Downward, J. (1995). Measurement of nucleotide exchange and hydrolysis activities in immunoprecipitates. *Methods Enzymol.* **255**, 110–117.

Egawa, K., Sharma, P. M., Nakashima, M., Huang, Y., Huver, E., Boss, G. R., and Olefsky, J. M. (1999). Membrane-targeted phosphatidylinositol kinase mimics insulin actions and induces a state of cellular insulin resistance. *J. Biol. Chem.* **274**, 14306–14314.

Lowy, D. R., and Willumsen, B. M. (1993). Function and regulation of ras. *Annu. Rev. Biochem.* **62**, 851–891.

Marais, R., Spooner, R. A., Stribbling, S. M., Light, Y., Martin, J., and Springer, C. J. (1997). A cell surface tethered enzyme improves efficiency in gene-directed enzyme prodrug therapy. *Nature Biotechnol.* **15**, 1373–1377.

Raptis, L., Brownell, H. L., Vultur, A. M., Ross, G., Tremblay, E., and Elliott, B. E. (2000). Specific inhibition of growth factor-stimulated ERK1/2 activation in intact cells by electroporation of a Grb2-SH2 binding peptide. *Cell Growth Differ.* **11**, 293–303.

Raptis, L., Brownell, H. L., Wood, K., Corbley, M., Wang, D., and Haliotis, T. (1997). Cellular ras gene activity is required for full neoplastic transformation by simian virus 40. *Cell Growth Differ.* **8**, 891–901.

Raptis, L., and Firth, K. L. (1990). Electroporation of adherent cells *in situ*. *DNA Cell Biol.* **9**, 615–621.

Raptis, L., Vultur, A., Balboa, V., Hsu, T., Turkson, J., Jove, R., and Firth, K. L. (2003). *In situ* electroporation of large numbers of cells using minimal volumes of material. *Anal. Biochemi.*

Robinson, M. J., and Cobb, M. H. (1997). Mitogen-activated protein kinase pathways. *Curr. Opini. Cell Biol.* **9**, 180–186.

Scheele, J. S., Rhee, J. M., and Boss, G. R. (1995). Determination of absolute amounts of GDP and GTP bound to Ras in mammalian cells: Comparison of parental and Ras-overproducing NIH 3T3 fibroblasts. *Proc. Natl. Acad. Sci. USA* **92**, 1097–1100.

Taylor, S. J., Resnick, R. J., and Shalloway, D. (2001). Nonradioactive determination of Ras-GTP levels using activated ras interaction assay. *Methods Enzymol.* **333**, 333–342.

Tomai, E., Klein, S., Firth, K. L., and Raptis, L. (2000). Growth on indium-tin oxide-coated glass enhances ^{32}P-phosphate uptake and protein labelling of adherent cells. *Prep. Biochem. Biotechnol.* **30**, 313–320.

Tomai, E., Vultur, A., Balboa, V., Hsu, T., Brownell, H. L., Firth, K. L., and Raptis, L. (2003). *In situ* electroporation of radioactive compounds into adherent cells.

Yang, T. A., Heiser, W. C., and Sedivy, J. M. (1995). Efficient in situ electroporation of mammalian cells grown on microporous membranes. *Nucleic Acids. Res.* **23**, 2803–2810.

CHAPTER

17

Tracking Individual Chromosomes with Integrated Arrays of *lac*^{op} Sites and GFP-*lac*^{i} Repressor: Analyzing Position and Dynamics of Chromosomal Loci in *Saccharomyces cerevisiae*

Frank R. Neumann, Florence Hediger, Angela Taddei, and Susan M. Gasser

OUTLINE

I. INTRODUCTION

The visualisation of specific DNA sequences in living cells, achieved through the integration of *lac* operator arrays (*lac*^{op}) and expression of a GFP-lac repressor fusion, has provided new tools to examine how the nucleus is organised and how basic events such as sister chromatid separation occur (Straight *et al.*, 1996; Belmont, 2001). In contrast to other methods, such as fluorescence *in situ* hybridisation, the lac^{op}/GFP-*lac* repressor (GFP-lac^{i}) technique is noninvasive and therefore interferes minimally with nuclear structure and function. In addition, it facilitates analysis of the rapid dynamics of specific DNA loci (Gasser, 2002). Although this technique has been adapted to organisms from bacteria to humans, the ease with which GFP fusions can be targeted to specific chromosomal sites depends on the ability of the organism to carry out homologous recombination. This process is very efficient in budding yeast, allowing pairs of chromosomal loci to be analysed at the same time through the use of two bacterial repressors (lac^{i} and *tetR*) fused to different GFP variants. Given the relatively advanced state of the art in budding yeast, this article presents protocols optimised for this organism. These provide a starting point for adapting multilocus tagging to other species. Moreover, the techniques described here for the quantitative analyses of locus dynamics are universally applicable.

II. MATERIALS AND INSTRUMENTATION

Yeast minimal and rich media (SD, YPD) are described in Guthrie *et al.* (1991). Cells can be mounted on a depression slide (Milian SA, Cat. No. CAV-1, Fig. 17.2A) upon 1.4% agarose (Eurobio Cat. No. 018645) containing SD medium with 4% glucose (Fluka). Aliquots of this can be kept at 4°C for several months. Alternatively, cells can be immobilised on a 18-mm coverslip treated with concanavalin A (Con A, Sigma, Cat. No. C-0412) in a cell observation chamber (Ludin chamber, Life Imaging Services, Fig. 17.2B). Con A dissolved to 1 mg/ml in H_2O is stable at −20°C for months. Wide-field microscopy is performed on a Metamorph-driven Olympus IX 70 inverted microscope with Olympus Planapo 60×/NA = 1.4 or Zeiss Planapo 100×/NA = 1.4 objectives on a **piezoelectric translator** (**PIFOC;** Physik Instrumente), illuminating with a PolychromeII monochromator (T.I.L.L. Photonics). Also needed is a CoolSNAP-HQ digital camera (Roper Scientific) or equivalent, and both the FITC filter set for detecting GFP (Chroma, Ref. 41001) and the CFP/YFP filter set (e.g., Chroma, Ref. 51017). Confocal microscopy can be performed on a Zeiss LSM510 Axiovert 200M, equipped with a Zeiss Plan-Apochromat 100×/NA = 1.4 oil immersion or a Plan-Fluar 100×/NA = 1.45 oil immersion objective. The stage is equipped with a hyperfine motor HRZ 200. Temperature is stabilised using a temperature-regulated box surrounding the microscope (The Box, Life Imaging Services). Software used for analysis is (a) Excel (Microsoft), (b) ImageJ public domain software (Rasband), (c) Imaris v 3.3 (Bitplane), (d) Mathematica 4.1 (Wolfram Research), and (e) Metamorph v 4.6r6 **(Universal Imaging Corp.)**.

III. PROCEDURES

A. Preparations

1. Plasmids and Strains

Yeast transformation and growth are as described (Guthrie *et al.*, 1991). The lac^{op}/GFP-lac^{i} system for site recognition exploits the high affinity and specificity of the bacterial *lac* repressor for its recognition sequence (lac^{op}). All procedures are performed analogously for the *tetR*/tet^{op} system (Michaelis *et al.*, 1997).

1. Plasmids or integrations of repetitive arrays are difficult to propagate in both bacteria and yeast due to recombination induced excision events. To avoid this, bacteria should be grown at 30°C in a recombination-deficient strain [STBL2 (Invitrogen Life Technologies) or SURE (Stratagene)].

2. Integrate a copy of *lac* repressor fused in frame to sequences encoding the S65T V163A, S175G derivative of GFP and a nuclear localisation signal, e.g., pAFS144 into the yeast strain. This red-shifted GFP derivative has a higher emission intensity and longer fluorescence time than natural GFP (Straight *et al.*, 1998). The *lac*^{i} later helps to stabilise the *lac*^{op} array in yeast.
3. Insert a multimerised *lac*^{op} array (usually 256 copies or ~10 kb) into the chromosome by standard transformation using a linearised construct that integrates by homologous recombination. Integration is directed to a genomic locus by a unique cleavage within a polymerase chain reaction (PCR)-generated genomic sequence >200 bp inserted into the host plasmid (e.g., pAFS52 integration is selected by growth on SD-trp; Straight *et al.*, 1996; Heun *et al.*, 2001a; Hediger *et al.*, 2002). In yeast as few as 24 contiguous *lac*^{op} sites can be detected readily.
4. Check the proper insertion by standard colony PCR and/or Southern blotting (Guthrie *et al.*, 1991). Binding of *lac*^{i}-GFP to the *lac*^{op} array results in a bright focal spot, detected readily by fluorescence microscopy within the nucleoplasm. Confirmed transformants with bright signals should be frozen and stored immediately as individual colony isolates. When strains are recovered from frozen stocks, they should be grown on selective medium to avoid further excision events.

Note: Other GFP fusions, optimised forms of CFP or YFP (or ECFP and EYFP), have also been used successfully in yeast (Lisby *et al.*, 2003). The lac repressor used is also modified to prevent tetramerisation, thus minimising artefactual higher order interactions between *lac*^{op} sites (Straight *et al.*, 1996).

5. Double tagging. If the position or mobility of two genomic loci is to be compared, one should avoid tagging both with the same repeats. It has been shown that identical arrays can undergo a pairing event that, at least in the case of the *tet* system, depends on the expression of the repressor (*tetR;* Fuchs *et al.*, 2002). By using *tet*^{op} for one site, and *lac*^{op} for the second, the risk of spurious pairing is eliminated. Useful pairs of GFP derivatives are CFP and YFP, or GFP and the new monomeric mRFP (Campbell *et al.*, 2002).
6. In contrast to the *lac*^{i}–GFP fusion (Figs. 17.1A and 17.1B), the *tetR*–GFP gives a high and generally diffuse nucleoplasmic background in yeast, both in the presence and in the absence of *tet*^{op} repeats (Figs. 17.1C and 17.1D).
7. Dynamics. If movement analysis is to be pursued, it is important to differentiate the movement of the nucleus itself or that induced by mechanical vibrations from the dynamics of the chromosome. Nuclear movement must be determined and then subtracted from that of a specifically tagged site, using any of the following methods.
   - **a.** Visualisation of the nuclear envelope with Nup49–GFP (Belgareh *et al.*, 1997; Heun *et al.*, 2001a). In this case the nuclear center can be interpolated from the oval or circular pore signal in an automated fashion by software such as ImageJ or Metamorph (Figs. 17.1A and 17.1B). The DNA locus position is then determined relative to the nuclear center for each frame.
   - **b.** Diffuse nucleoplasmic signal of *tetR*–GFP (Figs. 17.1C and 17.1D). The center of the nucleus is defined by interpolation frame by frame and locus movement is calculated relative to this.
   - **c.** By comparing the motion of two tagged loci, one can calculate average movement without concern for nuclear drift. The fact that both loci are moving has to be taken into account for movement quantitation (see later).

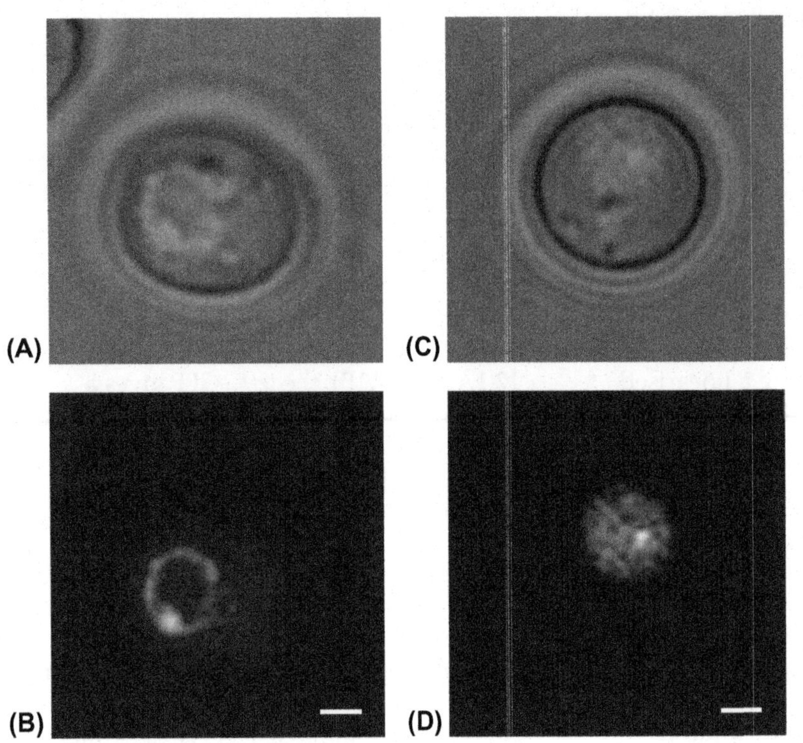

FIGURE 17.1 (**A** and **C**) An overlay of the phase image and the fluorescence image of a GFP-tagged yeast cell in G1 phase. (**B** and **D**) The corresponding fluorescence image. The lac^{op} array is integrated at the *LYS2* locus; the nucleus is visualised by the tagged nuclear pore component Nup49-GFP (**A,B**) or by using the diffuse staining of nucleoplasm by *tetR*-GFP (**C,D**). Bar: 1 μm.

2. *Growth and Cell Preparation*

1. All yeast strains to be analysed should be cultured identically and preferably to an early exponential phase of growth ($<0.5 \times 10^7$ cells/ml) in synthetic or YPD medium, starting from a fresh overnight culture. Depletion of glucose or growth on alternative carbon sources can alter chromatin dynamics. Wash cells once before observation to avoid YPD autofluorescence. We recommend two mounting techniques for living cell visualisation.

2a. *SD–agarose-filled slides* (Fig. 17.2A): Immobilised cells between an agarose patch on a depression slide and a coverslip to avoid flattening or distortion of the yeast by coverslip pressure on a normal glass slide. Cells sealed in this way are in a closed environment in which the depletion of O_2 and production of CO_2 bubbles can influence growth and impair visualization. Optimally this technique is used for imaging periods limited to <60 min.

i. Melt an aliquot of SD/agarose at 95°C until the agarose has completely melted, but not longer.
ii. Vortex briefly and transfer 150 μl into the well of a depression slide that is preheated either by a heating block or by passage through the flame of a Bunsen burner.
iii. Immediately pass a normal microscope slide over the depression to remove excess agarose as depicted in Fig. 17.2A.
iv. While the agarose solidifies, recover the cells from 1 ml of culture by centrifugation for 1 min at >10,000 *g*.
v. Resuspend the cells in ~20 μl of appropriate medium.
vi. Once the agarose has solidified, remove the upper slide by sliding along the depression

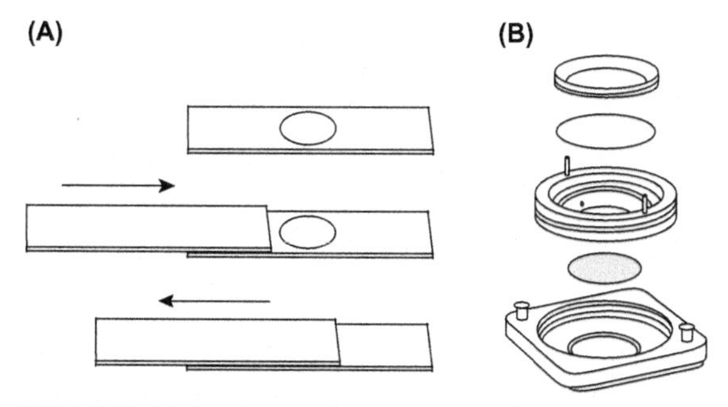

FIGURE 17.2 Yeast cells can be immobilised for imaging either using an agarose patch on a depression slide (**A**) or using a cell observation chamber (e.g., Ludin chamber; **B**).

slide surface and place ~5μl of concentrated cells on the agarose patch.

vii. Close with a coverslip, eliminate eventual air bubbles, and seal with nail polish.

Note: Monitor bud emergence and cell division carefully, as some brands of nail polish contain solvents that influence yeast cell physiology negatively.

2b. *Cell observation chamber* (Ludin chamber, Fig. 17.2B): The second technique uses a Ludin chamber in which cells are attached noncovalently to a coverslip by a lectin. The medium-filled chamber is assembled as shown in Fig. 17.2B. A flow of fresh medium can be applied.

i. Coat 18-mm coverslips with 10μl Con A (1mg/ml in H_2O) and let them air dry for >20min. Coated slides can be kept for weeks at room temperature.

ii. Adhere cells to the Con A-coated coverslip by sedimenting 1ml of the culture at 1*g* for 3min at room temperature.

iii. Remove excess culture and add ~1ml fresh preheated medium before closing the chamber.

3. *Temperature Control*

In order to have a stable condition for microscopic observation, the temperature of the microscope and room should be controlled carefully (±2°C). Two mechanisms are used standardly. The first is to enclose the entire imaging part of the microscope in a commercially available temperature-regulated box (e.g., Life Imaging Services or Zeiss). A second, less precise method is to regulate the temperature of the slide through a heated stage.

B. Image Acquisition

1. *General*

The choice of imaging technique depends on the question being asked. To derive quantitative information on the position of a given locus relative to a fixed structure (e.g., the spindle pole body, nucleolus, or nuclear envelope), three-dimensional (3D) stacks and detection of different wavelengths may be necessary. An analysis of fine movement and chromatin dynamics, however, requires the rapid and extended capture of one or more fluorochromes. Bleaching of the signal is often a major limiting factor in time-lapse imaging. One should note that chromatin movement is very fast [movements >0.5μm in less than 10s (Heun *et al.*, 2001a)], making it necessary to have rapid image acquisition with a minimal interval between sequential images. To optimise acquisition, parameters such as image resolution, the number of *z* frames, intervals between frames, light intensity, and exposure time can be varied. In all cases, it is of utmost importance to minimise and monitor laser- or light-induced damage to the organism during imaging, in part by comparing the time required for one division cycle in imaged and nonimaged cells.

Cell Cycle Determination

As position and mobility of a chromosomal locus can vary with stages of the cell cycle, it is crucial to determine precisely what stage each imaged cell is in. This is done by monitoring bud presence and bud size, as well as the shape and position of the nucleus, as visualised by the Nup49-GFP fusion and a transmission or phase image. Figure 17.3 summarises the morphologies that characterise each stage of the cell cycle.

2. *Wide-Field Microscopy and Deconvolution*

For the imaging of large fields of cells, best results are obtained with a wide-field microscope equipped with **a PIFOC**, Xenon light source, and monochromator that allows a broad and continuous range of incident light wavelengths, as well as rapid switching between these values. Images are acquired by a high-speed monochrome CCD camera run by a rapid imaging software, such as Metamorph. The limiting step is often the speed of signal transfer from the CCD chip to the RAM and/or hard disk of your computer.

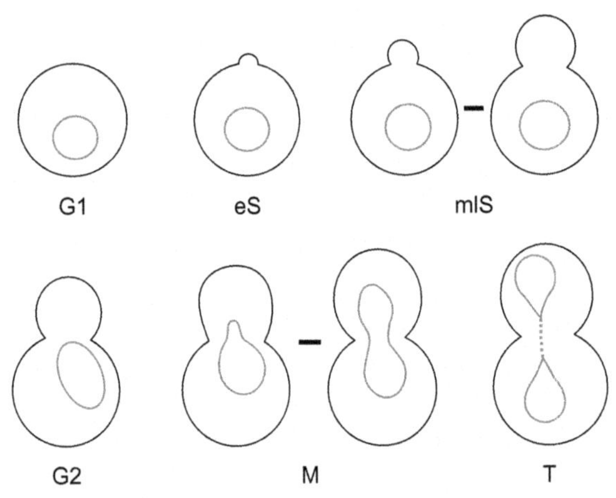

FIGURE 17.3 Diagrams of a budding yeast cell at different characteristic points in the cell division cycle. The following criteria are used to identify the indicated stage. *G1 phase*, unbudded cells with round nuclei or attached pairs of posttelophase cells that have two round, clearly separated nuclei; *early S*, with initial bud emergence, cells are in early S; *mid-to-late S*, cells with a bud big enough to form a ring at the bud neck, in which nuclei are still round and centred in the mother cell; *G2 phase*, large budded cells (bud ≥ two-thirds of mother) with the nucleus at the bud neck; *mitosis (M)*, large budded cells in which the nucleus extends into the daughter cell due to spindle extension; *telophase (T)*, two globular cells with two distinct nuclei that remain connected by residual NE structures.

z-Stacks

Wide-field microscopy is well adapted to experiments in which a large number of cells (200–300) need to be scored, e.g., when determining the subnuclear position of a given locus relative to the nuclear envelope or another tagged locus or landmark (e.g., spindle pole body or nucleolus). The reference point should optimally be tagged with a different fluorescent protein. If two loci bind the same fluorescent fusion proteins, then their intensities should be significantly different. Rapid through-focus stacks of images using the full chip capacity of the camera are taken of cells growing on agar or in a Ludin chamber (such that 20–30 individual cells are resolved per field). Optimal parameters for GFP are as follows: exposure time, 100–200 ms; *z* spacing of 200 nm for 18 focal planes, excitation wavelength 475 nm. For dual-wavelength capture, images of both wavelengths (CFP: 432 nm, ~300 ms; YFP: 514 nm, ~150 ms) must be acquired before the focal plane changes. A phase image is taken after every stack of fluorescence images. Wide-field images have out-of-focus haze and deconvolution of the *z* stack is often necessary to reassign blurred intensities back to their original source. Use Metamorph software or other available deconvolution packages.

Three-Dimensional Time Lapse

The conditions for capturing 3D time-lapse series are as follows: 5–11 optical *z* slices taken every 1 to 4 min, *z* sections are 200 to 400 nm in depth, and the exposure time is ~50 ms. Using these settings, up to 300 stacks of five sections each (1500 frames) at 1-min intervals can be captured without affecting cell cycle progression. More rapid sampling with this system, however, leads to bleaching and potential cellular damage. Until this can be remedied by more rapid and more sensitive CCD cameras, wide-field microscopy is recommended for less rapid time-lapse imaging (intervals ⩾60s) on larger fields and confocal microscopy (see later) for very rapid time-lapse imaging (intervals ⩽2s) on small regions of interest (typically one yeast nucleus).

For very long imaging times (>1h), stray light should be suppressed by inserting an additional shutter. Deconvolution is performed using the Metamorph fast algorithm with five iterations, a sigma parameter of 0.7, and a frequency of 4.

3. *Confocal Microscopy*

To follow chromatin dynamics in individual cells with rapid time-lapse microscopy, the Zeiss LSM510 scanning confocal microscope is particularly well adapted, although the laser and acousto-optic tuneable filter (AOTF) system is limited in activation wavelengths. Its positive attributes are an ability to limit scanhead motion to a minimal region of interest (ROI), rapid and well-regulated scanning speeds, and the possibility to adjust pinhole aperture and laser intensities to very low levels, while maintaining maximal sensitivity.

General Settings

To reduce the risk of damage by illumination, the laser transmission is kept as low as possible, and the cells are imaged as rapidly as possible within a minimal ROI. Useful settings for the Zeiss LSM510 are as follows.

Laser: argon/2 458, 488, or 514nm tube current 4,7 amp. Output 25%.

Filters: Channel 1: Lp 505 for GFP alone; channel 1 Lp 530, channel 3 Bp 470–500 for YFP/CFP single track acquisition.

Channel setting: Pinhole 1–1.2 airy unit (corresponding to optical slice of 700 to 900nm); detector gain: 930 to 999; amplifier gain: 1–1.5; amplifier offset: 0.2–0.1V; laser transmission AOTF = 0.1–1% for GFP alone, 1–15% for YFP, and 10–50% for CFP in single track acquisition. In order to use minimal laser transmission the pinhole must be aligned regularly.

Scan setting: Speed 10 (0.88μs/pixel); 8 bits one scan direction; 4 average/mean/line; zoom 1.8 (pixel size: 100 × 100nm)

Imaging intervals: 1.5s

Note: If CFP and YFP signals are very weak, images can be acquired sequentially using the more sensitive LSM 510 channel 1 in multi-track mode. This allows the use of broader filters: long-pass filter Lp 475 for CFP and Lp 530 for YFP. Alternatively, and to avoid any cross talk, recover the YFP signal as before and use Bp 470–500 on channel 3 for CFP. These latter parameters will slow the imaging process.

Two- or Three-Dimensional Time Lapse

If maximal capture speed is desired, only one image per time point can be taken, as long as the GFP spot stays in the imaged plane of focus (called 2D time lapse). Often the plane of focus has to be changed manually to follow the spot. Image acquisition in 3D has two main advantages. (1) The GFP spot does not have to be followed manually as it is always present in one of the focal planes. A subsequent maximal projection along the *z* axis produces a complete 2D time sequence without loss of focus on the GFP spot. (2) After image reconstruction, one can visualise the nucleus and calculate distances in 3D. Such measurements are nonetheless compromised by the reduced optical resolution in *z* (⩾0.5μm for 488-nm light).

Specific 3D time-lapse settings are as follows: six to eight optical slices in *z*, 300- to 450-nm spacing in *z* with Hyperfine HRZ 200 motor using a ROI of 3 × 3 to 4 × 4μm and time intervals of 1.5s. A 12-min time-lapse series at 0.2% laser transmission did not influence cell cycle progression.

C. Image Analysis

1. *z Stacks*

Determination of the subnuclear position of a GFP-tagged locus is monitored relative to the centre of the Nup49-GFP ring. Nuclei in which the tagged locus is at the very top or bottom of the nucleus are not scored because the pore signal no longer forms a ring but a surface and a peripheral spot will appear internal.

1. Measure the distance from the centre of intensity of the GFP spot to the nearest pore signal along the nuclear diameter, as well as the nuclear diameter itself, using the middle of the GFP-Nup49 ring as the periphery (Fig. 17.4). Several programs can export coordinates of points of interest, and the publicly available point picker plug in for ImageJ (Rasband) is useful.
2. Calculate the distances/diameter ratio, e.g., using Excel. Determine the precise relative radial position by dividing the distance between pore and the spot by half of the calculated diameter, thus normalising distances.
3. Classify the position of each spot with respect to three concentric zones of equal surface (Fig. 17.4). The peripheral zone (zone I) is a ring of width = 0.184 × the nuclear radius (r). Zone II lies between 0.184 and 0.422r and zone III is the core of the nucleus with radius = 0.578r. The three zones are of equal surface no matter where the nuclear cross section is taken.
4. Compare the measured distribution to another (e.g., other cell cycle phase, another condition or a random distribution) with a χ^2 analysis. If only percentages of one zone (e.g., the outermost zone) are compared for different conditions (or to a random distribution), a proportional test should be applied. Statistical significance is determined using a 95% confidence interval.

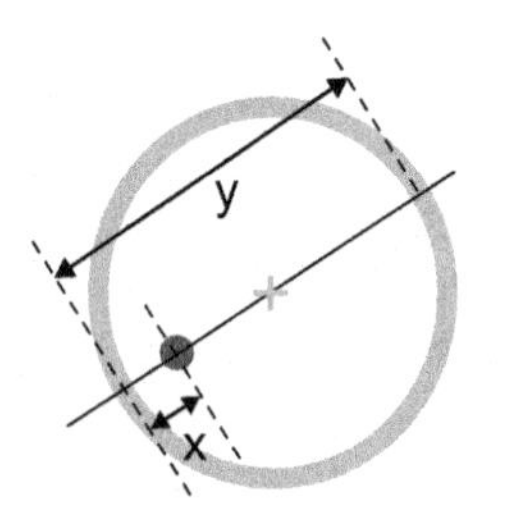

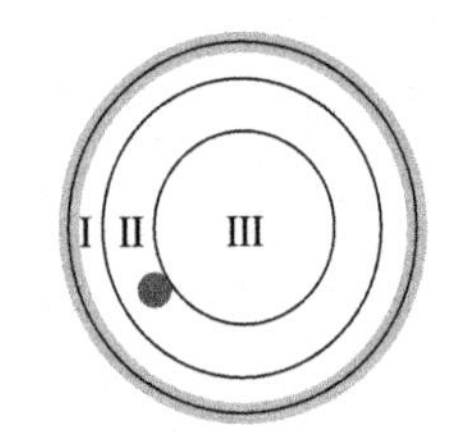

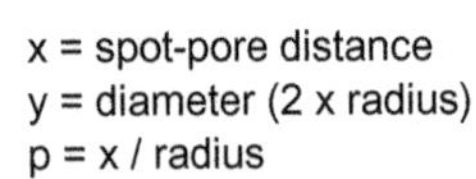

FIGURE 17.4 Analysis of DNA locus position. Relative locus position is calculated by normalising measured distance x by the radius (0.5 × measured distance y). The relative radial distances can then be classified and attributed to three groups of equal surface. The peripheral zone (zone I) is a ring of width = 0.184 × the nuclear radius (r). Zone II lies between 0.184 and 0.422r, and zone III is the centre of the nucleus with radius = 0.578r. In a predicted random distribution every group would contain one-third of the cells.

2. *Three-Dimensional Time Lapse*

Locus Tracking

A prerequisite for the precise description of chromatin movement is the knowledge of the coordinates of the locus and of the nuclear centre for each frame of a time-lapse movie. In collaboration with D. Sage and M. Unser (Swiss Federal Institute of Technology, Lausanne), a best-fit algorithm has been developed that reliably tracks a moving spot in 2D time-lapse movies or in maximal projections of z stacks in 3D time lapse using nuclei carrying Nup49-GFP or expressing *tetR*-GFP to detect the nucleoplasmic signal. This system is complete and dramatically improves reproducibility and the speed of analysis, while allowing user intervention at several stages. The algorithm has been implemented as a Java plug in for the public domain ImageJ software (Rasband; Sage *et al.*, 2003). The spatiotemporal trajectory is exported as *x,y* coordinates for each time point in a spreadsheet. An implementation for 3D image stacks over time is also available (Sage *et al.*, 2005). Automated image analysis requires three steps.

a. *Alignment phase.* The first step is an alignment module that compensates for the translational movement of the nucleus, cell, or microscope stage. This is achieved by a modifiable threshold on the image. The extracted points are then fitted within an ellipse using the least-squares method. Finally, each image is realigned automatically with respect to the centre of the ellipse.
b. *Preprocessing phase.* To facilitate the detection of the tagged locus, the images are convolved with a Mexican-hat filter. This preprocessing compensates for background variations and enhances small spot-like structures.
c. *Tracking phase.* The final step is the tracking algorithm. Using dynamic programming, which takes advantage of the strong dependency of the spot position in one frame on its position in the next, the optimal trajectory over the entire period of the movie is determined. The following three criteria influence spot recognition: (1) maximum intensity (i.e., the tagged DNA is usually brighter than the pore signal), (2) smoothness of trajectory, and (3) position relative to the nuclear centre. This latter criterion is necessary because Nup49-GFP staining can be confused with a weak perinuclear locus. All three parameters can be modulated individually in order to optimise the tracking for different situations (loci that are more mobile, more peripheral, of variable intensity, etc). Most importantly, the program has the option of further constraining the optimisation by forcing the trajectory to pass through a manually defined pixel. In that way mistracked spots can again be added manually to the correct trajectory, which is recalculated quasi-instantaneously. This tracking method proves to be extremely robust and reproducible due to its global approach.

Note: Some commercially available software are also able to track objects [e.g., Imaris (Bitplane), Volocity (Improvision)], although tracking efficiency is variable and usually requires uniformly high-quality images. The algorithms are mostly based on threshold principles, which are rarely modifiable or interactive, and which are ill-suited for noisy images.

Characterisation of Movement

Because each time-lapse series represents a single cell, it is indispensable to average 8–10 movies over a total time >40 min for a given strain or condition. Subtle differences require a larger data source. Useful parameters for quantitative analysis include the following.

a. *Track length.* The projected track of the tagged locus can be visualised using LSM software, ImageJ, Excel, or other programs (Fig. 17.5A). The sum of all 1.5-s step lengths within a time-lapse series yields the total track length of that movie. From this, average track length and velocity (μm/min) can be calculated, but often this parameter is not very revealing.
b. *Step size.* A histogram of step size distribution describes the nature of the movement more precisely. Statistical parameters such as mean, median, and standard deviation of individual and groups of movies can be calculated and compared with statistical tests (e.g., ANOVA). Even small but reproducible significant differences can be documented due to the large number of measurements.

c. *Large movements.* Often differences in mobility are not obvious by comparing average speed, yet the frequency of large steps >500 nm will vary significantly. These indicate transient high velocity movements. We generally score for steps larger than 500 nm during seven frames (10.5s), an interval that has proven useful for distinguishing patterns of mobility between different physiological states and stages of the cell cycle (Heun *et al.*, 2001b). These are reported as the number of large steps per 10 min, averaged over at least 50 min of time-lapse imaging. Although a 500 nm is a meaningful cutoff, any threshold over 300 nm can be used.
d. *Mean square displacement (MSD).* Observing the movement of a DNA locus over time not only gives information about its velocity, but also about the subvolume of the nucleus that it occupies during a given period of time. It has been shown for several chromosomal loci that chromosomal domains are able to move apparently randomly in a given subvolume (Gasser, 2002). This constraint can be quantified by MSD analysis, assuming that the movement of the spot follows a random walk. Ideally it describes a linear relationship between different time intervals and the square of the distance travelled by a particle during this period of time (MSD or $\langle \Delta d^2 \rangle$, where $\Delta d^2 = \{d(t) - d(t + \Delta t)\}^2$ (Berg, 1993; Marshall *et al.*, 1997; Vazquez *et al.*, 2001). In order to get the numbers, one must calculate the distances traveled by the spot for each time interval (1.5, 3, 4.5s...) and plot the square of the mean against increasing time intervals. These calculations and the corresponding graphs can be performed easily in Excel (Microsoft) or Mathematica (Wolfram Research). A representative MSD graph is shown in Fig. 17.5B. In these curves, the slope reflects the diffusion coefficient of the particle, and the linearity of the curve is usually lost at larger time intervals due to spatial constraint on the freedom of movement of the locus, i.e., the random walk of the particle is obstructed by the nuclear envelope or other subnuclear constraints, leading to a plateau (horizontal dashed line in Fig. 17.5B). The height of this plateau is related to the volume in which the particle is restricted. The slope of the MSD relation is directly correlated with diffusion coefficient. As explained earlier, in enclosed systems, the diffusion coefficient decreases with increasing Δt due to space

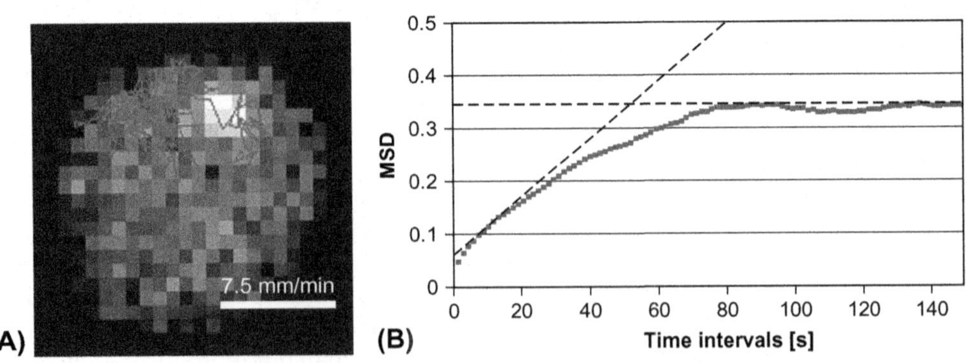

FIGURE 17.5 Analysis of DNA locus dynamics. (**A**) The projected trace of 300 images of a movie of the *LYS2* locus. The average track length in 5 min is 37.4 μm. Bar: 1 μm. (**B**) A mean square displacement (MSD, $\langle \Delta d^2 \rangle$ in [μm^2]) analysis on an average of 8 movies of the *LYS2* locus. All cells were observed in G1 phase.

constraints exerted on the particle dynamics. Nevertheless, the maximal diffusion coefficient can be calculated for very short time intervals and reflects the intrinsic mobility of particles (see sloping dashed line in Fig. 17.5B). For chromosomal loci in yeast, we observed a maximal diffusion coefficient in the range of 1×10^{-4} to $1 \times 10^{-3} \mu m^2/s$ based on short time intervals. If distances are measured between two separate moving loci, $\langle \Delta d^2 \rangle$ reflects two times the MSD of an individual spot or locus moving relative to a fixed point (Vazquez *et al.*, 2001). A more theoretical discussion of these parameters is found in Berg (1993).

IV. COMMENTS

It is very difficult to accurately quantify the intensity of a small, mobile GFP-lac^i focus. Even in deconvolved images it can differ by twofold in sequential images.

This protocol shows the optimal method for the described microscope setups. For different microscopes, the values and methods of this protocol are simply a starting point for further optimisation. As improvements in technology (e.g., more sensitive and rapid CCD cameras) and reagents (e.g., more stable or more intense GFP variants) evolve, future adjustments of this protocol will be indispensable.

The method described here can also be applied to *Schizosaccharomyces pombe* with a few changes, one being immobilisation on a coverslip with isolectin B (1 mg/ml) (Williams *et al.*, 2002) or lectin from *Bandeiraea simplicifolia* (lyophilized powder, Sigma Cat. No. L2380).

V. PITFALLS

1. To ensure that DNA movements are not the result of nuclear rotation, fluorescence recovery after photobleaching of GFP nuclear pore components should be performed over the same time intervals used to monitor DNA movement.
2. To increase oxygen concentration and to prevent massive production of CO_2 under the cover slide, vortex the agarose/medium before making the patch.
3. Growth conditions must be standardised thoroughly, because both choice of carbon source and its concentration significantly influence subnuclear position and dynamics of tagged loci.
4. Cells grown in minimal medium may not pellet as well as cells grown in YPD. Concentrate cells by centrifuging 2 volumes of culture in the same 1.5-ml tube.
5. In the Ludin chamber yeast cells often bud upwards into the medium (i.e., parallel to the optical axis). Thus it is important to scan the entire cell in transmission mode not to miss the presence of a bud.
6. Observations made on individual cells are often not representative of entire populations. It is crucial to verify observed differences with the appropriate statistical tests.

References

Belgareh, N., and Doye, V. (1997). Dynamics of nuclear pore distribution in nucleoporin mutant yeast cells. *J. Cell Biol.* **136**(4), 747–759.

Belmont, A. S. (2001). Visualizing chromosome dynamics with GFP. *Trends Cell Biol.* **11**(6), 250–257.

Berg, H. C. (1993). *"Random Walks in Biology."* Princeton Univ. Press, Princeton, NJ.

Campbell, R. E., Tour, O., Palmer, A. E., Steinbach, P. A., Baird, G. S., Zacharias, D. A., and Tsien, R. Y. (2002). A monomeric red fluorescent protein. *Proc. Natl. Acad. Sci. USA* **99**(12), 7877–7882.

Fuchs, J., Lorenz, A., and Loidl, J. (2002). Chromosome associations in budding yeast caused by integrated tandemly repeated transgenes. *J. Cell Sci.* **115**(Pt 6), 1213–1220.

Gasser, S. M. (2002). Visualizing chromatin dynamics in interphase nuclei. *Science* **296**(5572), 1412–1416.

Guthrie, C., and Fink, G. R. (1991). *"Guide to Yeast Genetics and Molecular Biology."* Academic Press, San Diego.

Hediger, F., Neumann, F. R., Van Houwe, G., Dubrana, K., and Gasser, S. M. (2002). Live Imaging of Telomeres: yKu and Sir Proteins Define Redundant Telomere-Anchoring Pathways in Yeast. *Curr. Biol.* **12**(24), 2076–2089.

Heun, P., Laroche T., Raghuraman, M. K., and Gasser, S. M. (2001a). The positioning and dynamics of origins of replication in the budding yeast nucleus. *J. Cell Biol.* **152**, 385–400.

Heun, P., Laroche, T., Shimada, K., Furrer, P., and Gasser, S. M. (2001b). Chromosome dynamics in the yeast interphase nucleus. *Science* **294**(5549), 2181–2186.

Lisby, M., Mortensen, U. H., and Rothstein, R. (2003). Colocalization of multiple DNA double-strand breaks at a single Rad52 repair centre. *Nature Cell Biol.* **5**(6), 572–577.

Marshall, W. F., Straight, A., Marko, J. F., Swedlow, J., Dernburg, A., Belmont, A., Murray, A. W., Agard, D. A., and Sedat, J. W. (1997). Interphase chromosomes undergo constrained diffusional motion in living cells. *Curr. Biol.* **7**(12), 930–939.

Michaelis, C., Ciosk, R., and Nasmyth, K. (1997). Cohesins: Chromosomal proteins that prevent premature separation of sister chromatids. *Cell* **91**(1), 35–45.

Rasband, W. ImageJ. National Institute of Health, Bethesda, MD.

Sage, D., Neumann, F. R., Hediger, F., Gasser, S. M., and Unser, M. (2005). *"Automatic Tracking of Individual Fluorescent Particles: Application to Chromatin Dynamics."* in IEEE Transactions on Image Processing., in press.

Straight, A. F., Belmont, A. S., Robinett, C. C., and Murray, A. W. (1996). GFP tagging of budding yeast chromosomes reveals that protein-protein interactions can mediate sister chromatid cohesion. *Curr. Biol.* **6**(12), 1599–1608.

Straight, A. F., Sedat, J. W., and Murray, A. W. (1998). Time-lapse microscopy reveals unique roles for kinesins during anaphase in budding yeast. *J. Cell Biol.* **143**(3), 687–694.

Vazquez, J., Belmont, A. S., and Sedat, J. W. (2001). Multiple regimes of constrained chromosome motion are regulated in the interphase Drosophila nucleus. *Curr. Biol.* **11**(16), 1227–1239.

Williams, D. R., and McIntosh, J. R. (2002). mcl1+, the *Schizosaccharomyces pombe* homologue of CTF4, is important for chromosome replication, cohesion, and segregation. *Eukaryot Cell* **1**(5), 758–773.

SECTION VI

CYTOSKELETON ASSAYS

CHAPTER

18

Microtubule Motility Assays

N. J. Carter, and Robert A. Cross

OUTLINE

I. INTRODUCTION

As the pivotal role of microtubule (MT) motors in the cell cycle becomes more widely recognized, so the demand for motility assays will increase. This article presents a robust and straightforward set of protocols based on experience gained in our laboratory. Readers seeking to set up more specialised assays should refer to one of the excellent methodological compendia that are available (Inoue and Spring, 1997; Cross and Kendrick Jones, 1991; Scholey, 1993).

II. MATERIALS AND INSTRUMENTATION

A. Hardware

1. Microscope

Unstained MTs are visualized most conveniently by video microscopy using computer-enhanced differential interference contrast (DIC). MTs assembled from fluorescently labeled tubulin can be visualized using epifluorescence. This article concentrates on these two contrast modes. It is also possible to visualise microtubules by dark field, which gives a high contrast image of unstained microtubules, or by phase contrast, but both modes are susceptible to dirt in the solutions and so may be inconvenient for routine work. Interference reflection produces higher contrast than DIC and gives information in the Z direction, but again is susceptible to dirt.

For all modes of contrast the main requirement of the microscope is that it should be as simple as possible. It should also be big and heavy to give it some vibration resistance. Uprights are less expensive, optically straightforward, and more convenient if the bathing solution is to be exchanged during observation (which is often the case). Inverted scopes are more stable and provide much better access to the specimen if other inputs (micromanipulators, flash-photolytic lasers, evanescence prisms) are envisioned.

Human eyes work in colour and have a higher dynamic range, better spatial resolution, and a bigger view field than a video camera. It is very useful to be able to use them to find focus. To do this the microscope needs to incorporate some sort of device to switch the light between the eyepieces and the camera.

2. Illumination

One hundred Watt Hg lamps generate large amounts of heat and it is necessary to remove this heat from the illuminating light. Glass heat filters may be used, but the best way to filter out heat is to use a hot mirror, a mirror that transmits infrared but reflects visible light.

DIC optics are optimised for visible wavelengths, and it is conventional to use a green interference filter to give narrow band illumination. In practice the improvement this brings is often too slight to be obvious by eye, and removing the green filter can be a convenient way to brighter illumination. The UV emitted by Hg lamps is removed substantially removed by optical (lead) glass, but for DIC an in-line UV filter is nonetheless a sensible precaution.

A useful improvement in both fluorescence and DIC image quality can be achieved by sending the illuminating light through a fibre-optic light scrambler (Technical Video). Light from the lamp is focussed into one end of the fibre (Fig. 18.1), and the other end emits a uniform disc of light (of Gaussian intensity profile), which is used to illuminate the microscope.

3. Optical Train

For maximum image quality in all contrast modes, it is advisable to reduce the number of optical components between the objective and the camera to a minimum. The most critical component is the objective. The light intensity transmitted by a lens is proportional to the square of its numerical aperture (NA), whilst resolution rises linearly with the NA. It is important therefore to use a 1.4 NA (therefore oil immersion) 60 or 100× planapo objective in order to maximise light-gathering power and resolution, particularly so in epifluorescence, where the objective doubles as the condensor.

We use a novel configuration that replaces the condensor on a Zeiss Axiovert with a home-built fitting that improves light collection from the Zeiss tungsten lamp and mounts a second objective in place of the condensor. This arrangement (Fig. 18.1) gives good DIC of microtubules with a field about 40 μm and

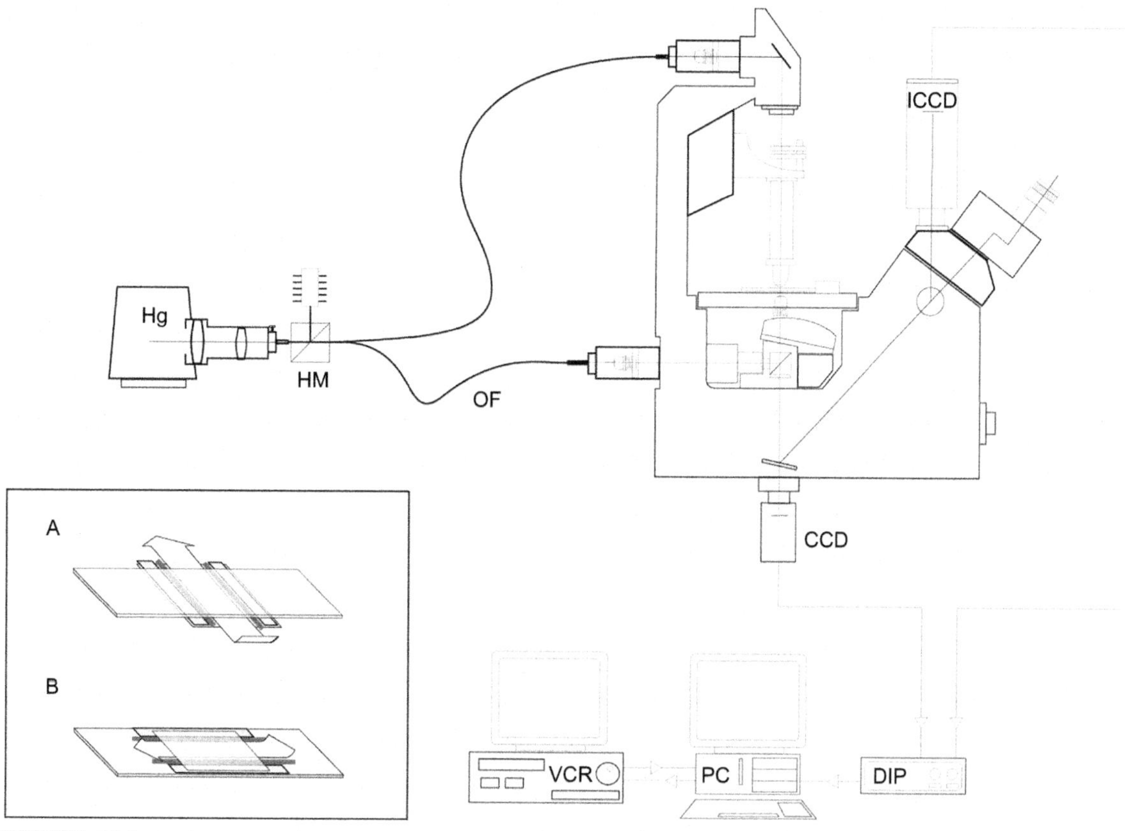

FIGURE 18.1 Video microscope. An inverted microscope set up for both fluorescence and DIC microscopy. Hg, mercury lamp; HM, hot mirror; OF, optical fibre; CCD charge-coupled device camera (for DIC); ICCD, intensified CCD (for fluorescence); DIP, digital image processor; PC, personal computer; VCR, video cassette recorder.

brings the considerable advantages of the tungsten lamp, which is much more stable than the Hg lamp. For smaller view fields, the higher radiant intensity of the Hg lamp is necessary.

4. *Antivibration Hardware*

Vibration will degrade the highly magnified image. Low frequencies (people walking across the floor) will cause the image to bounce around, whilst high frequencies will be averaged out by the video framing rate and cause the image to blur. The extent of this problem is, however, often overemphasized. Before purchasing an expensive and awkward vibration-damping equipment, try placing a few layers of bubble wrap under the microscope baseplate.

5. *Temperature Control Hardware*

Motility rates for several microtubules motors are extremely temperature sensitive in the range of room temperature, and temperature control is consequently important if measurements taken at different times are to be compared. The best way is to temperature clamp the entire microscope. If you have air conditioning this will already be happening. To warm specimens above ambient temperature, we use a homemade plexiglass box with a warm-air blower coupled to the side. Another option is to use water jackets for the objective and stage. These are readily fashioned by wrapping flexible, narrow-bore copper tubing around several times and connecting the ends to a water bath with a circulator (e.g., Techne) using silicon tubing. If cooling is necessary, we find it useful to wrap the microscope in a tent made of cling film, reducing condensation.

6. *Camera*

For fluorescence of moving objects, two types of low-light level / high-contrast / high-framing rate cameras are suitable: intensified charge-coupled device (ICCD) and intensified silicon-intensified tube (ISIT) cameras. ICCD cameras are better for current purposes because ISIT cameras, although more sensitive, introduce spatial and intensity distortions across the view field that are tedious to correct for. All the aforementioned produce an analogue video signal, e.g., ISIT Hamamatsu C2400-08 and ICCD Hamamatsu C2400-97E. An alternative approach that is becoming feasible uses a digital camera to record direct to computer hard disc in time lapse. A cooled CCD camera coupled to a generation 4 intensifier gives excellent sensitivity for fluorescence work, but is limited by the quantum efficiency of the intensifier (about 30%) and is currently very expensive and the digital data can be awkward to archive. If considering this route, be sure to test the software, which in our experience can place more severe limits on performance than the specifications of the hardware.

For DIC, a nonintensified scientific grade CCD is fine, e.g., grey-scale CCD camera Hamamatsu C2400 77e.

7. *Camera Coupling/Magnification*

The ideal magnification sets four or more camera pixels across the width of the MT. For a 512 × 512 pixels (2/3 in.) CCD, this corresponds to a square field with sides of 20–25 μm. Zoom couplings are wonderfully convenient but are not always a good idea because they absorb a lot of light. Magnification must be calibrated using a stage micrometer.

8. *Image Processor*

CCD cameras are often offered with a hardware box providing real time analogue enhancement of the video signal (any or all of gain, back-off and shading correction). A digital video processor is better, which is able to digitise incoming video frames, perform frame averaging, contrast enhancement, background subtraction, and caption overlay and then re-encode the image as an analogue video signal, all in real time. The Hamamatsu Argus 20 is so well thought out that it is virtually standard equipment for video microscopy laboratories. The best and most flexible arrangement is to adjust the gain and backoff on an Argus 20 or similar processor controls such that there are no areas in the image that are completely saturated and then apply further digital enhancement. The resulting signal is displayed on a monitor and is fed to the PC for direct grabbing of video clips and to the VCR for archiving.

9. *Monitor*

It is worth investing in a high-quality 14-in. multiformat monitor. Larger monitors look impressive but are only helpful if they have to be placed a long distance from the operator.

10. *Video Recording*

At the time of writing, the most convenient and practical way to store large amounts of video is still to use video tape. Recording to VCRs inevitably involves some degradation of the image (loss of spatial resolution, noise, contrast effects). For practical purposes the resolution loss is potentially the most serious problem. The effective resolution following recording can be visualised by recording and replaying a test card image having black and white lines at various spatial frequencies. Currently the best option is digital recording to video tape. Digital video tape recorders input and output an analogue signal, but encode data digitally to tape. The digitization involves some compression of the incoming video signal, but most of the compression is on the chrominance rather than the luminance, with the result that spatial information is relatively well preserved, particularly for grey-scale signals. Unlike other compression schemes such as DVD, there is no compression along the temporal axis. There are currently two sizes of tape: digital video (DV) and mini-DV. There are also two different formats: DVCAM (Sony) and DVCPRO (Panasonic). These two formats use the same compression scheme but DVCPRO spaces the tracks further apart on the tape, giving (arguably) better reliability and accuracy for editing, but with correspondingly less recording time on the tapes. Digital video recorders tend to be built for the pro market and are more robustly engineered than consumer machines, but are also more expensive.

We have evolved a video-recording strategy that offers maximum flexibility: Time-lapse digital recording of grabbed video frames to computer hard disc (with no resolution loss) and simultaneous real-time recording to digital VCR. The VCR runs uninterrupted in the background and generates an archive. The operator is free to go back to this archive at a later date and transfer interesting sequences to the computer for analysis. Captured digital sequences can be archived to external hard discs. We now have a fast digital capture and analysis application RETRAC II, which allows batch digitisation of video clips from tape for subsequent analysis. RETRAC II can be downloaded from our website (http://mc11.mcri.ac.uk/Retrac/index.html).

Analogue video recording is still in widespread use. Different video formats unfortunately operate in different countries. In the United States and Japan, NTSC format applies (525 lines per frame; 30 frames/s, typically captured at 640 × 480 pixels). European countries use PAL, which has higher spatial resolution but lower time resolution (625 lines per frame, 25 frames per sec, typically captured at 768 × 576 pixels).

11. Computing

The most convenient way to analyse motility is to capture a sequence of frames into computer memory and to track objects using a mouse-driven cursor. It helps to have a hard disc big enough to hold 2 day's work (more is dangerous because of the temptation not to back up) and enough hard memory to hold the stack of captured frames. A typical 20 frame stack uses about 8 Mb of application memory; if you want to use larger stacks then you need more memory. Processed stacks can conveniently be archived to removable discs. We use 100-Mb zip discs or 230-Mb magneto-optical discs. CD writers are getting less expensive and are worth considering if a permanent archive is required. Video compression protocols (JPEG, MPEG) are best avoided, as all involve some data loss. That said, Quicktime has become a standard for digital video and can use a variety of compressors, some of which are lossless.

12. Frame Grabber Card

Large numbers of video grabbing cards are available. Only a few are supported by Retrac and NIH Image, the freeware software packages that are recommend later. Because the situation is fluid, please check the software documentation for a list of supported cards.

13. Software

On the Mac, the best route for analysis is NIH Image, which can be customised using macros to track objects and output data in spreadsheet-compatible format. Macros for basic tracking through NIH Image stacks are available for download from our web page http://mc11.mcri.ac.uk/retrac.html. NIH Image runs on system 9 macs, with support for the Scion LG3 frame capture board, or in classic in OSX but without the Scion support. Wayne Rasband is continuing development of ImageJ, an NIH image—inspired java programme that runs in OSX and currently has partial support for the LG3 board (http://rsb.info.nih.gov/ij/). NIH Image and ImageJ are both available for the PC.

RETRAC 2 for Windows is purpose-written for the analysis of motility assay data. The latest version supports time-lapse frame grabbing from either VCR or live video, autofocus, autocontrast, tracking (including drift correction) spatial filtration, and magnification. The programme now incorporates a powerful file manager. Figure 18.2 shows a screenshot during tracking.

14. Glassware

The type of slide used does not matter. The type of coverslip does. The thickness of the coverslip should be matched to the objective. The objective will be marked appropriately [e.g., 60/planapo DIC 1.4 0.17/160 means a 60× objective selected as strain free for DIC, aplanatic (flat field); apochromatic (low chromatic aberration for blue yellow and green); optimised for cover glasses 0.17 mm thick and with a 160-mm focal length]. We use Chance 22 × 22-mm No.1.5 coverslips. In the past we have used these without any special cleaning treatment and rejected "bad" batches of coverslips that show poor binding of motor and/or a poor image because of surface contamination. This is still a workable approach, but we have begun to use a cleaning procedure that appears effective in removing contamination and making the coverslip reproducibly hydrophilic, as evidenced by the spreading of a drop of buffer placed on the surface so that it wets the entire surface. This coverslip cleaning procedure is based heavily on that given on the Technical

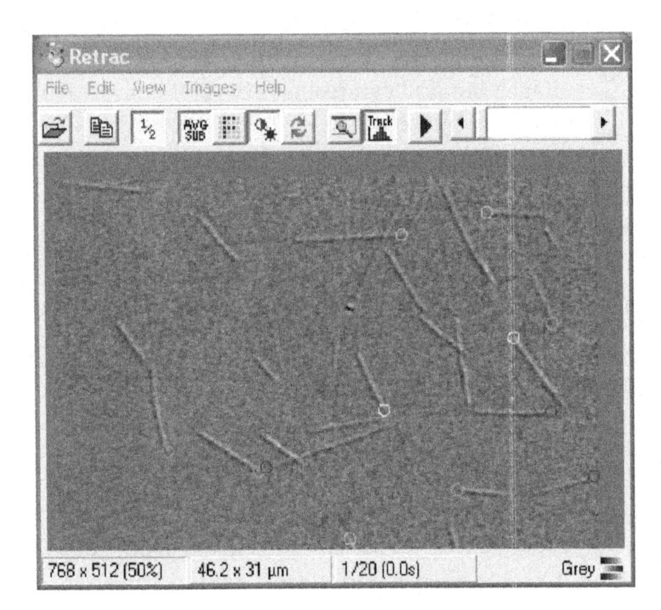

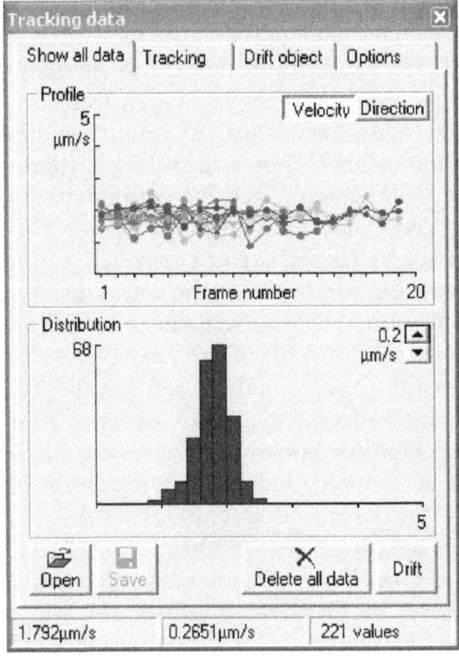

FIGURE 18.2 A screenshot from RETRAC 2.

Video website (http://www.technicalvideo.com/Products/CCP.html). Our localized variant is on our website (http://mc11.mcri.ac.uk/protocols.html).

III. PROCEDURES

A. Taxol-Stabilised Microtubules

Solutions

1. *1M K–PIPES:* PIPES dissolves around its isoelectric point of about pH 6.5. Take 500 ml water, add 65 g solid KOH, and then, after cooling if necessary, slowly add 302 g PIPES buffer (Sigma P-6757). Once everything is dissolved, monitor pH and roughly adjust by adding more KOH pellets as necessary. Allow the warm solution to cool and then fine-adjust pH using 5*M* KOH. Be careful not to overshoot, as there is no way back.
2. *100 mM NaGTP stock solution:* Because nucleoside triphosphates such as GTP and ATP undergo rapid hydrolysis at acidic pH, efforts should be made to control pH when dissolving and storing them. Dissolve 1 g NaGTP (Sigma G8877) in 15 ml 10 m*M* Na-PIPES, pH 6.9, monitoring pH. Rapidly reneutralise pH by titrating in 5*M* KOH. Fine adjust pH and then make volume up to 19.11 ml. Store frozen at −20°C in aliquots of 5–2000 μl. Do not add $MgCl_2$ to the stock solution (it precipitates).
3. *100 mM MgATP stock solution:* Dissolve 5.87 g NaATP (Sigma A7699 ATP ultra or Boehringer 519 987) in 60 ml 10 m*M* K-PIPES, pH 6.9, monitoring pH continuously and holding as close as possible to neutral using concentrated KOH. Once the ATP is dissolved, add 10 ml of 1*M* $MgCl_2$ and readjust pH to 6.9. Adjust volume to 100.0 ml and freeze in aliquots of 5–5000 μl.
4. *Taxol stock solution:* Wear gloves and work in the fume hood. Inject 2.93 ml anhydrous dimethyl sulfoxide (DMSO, Aldrich 27685-5) into a 25-mg bottle of taxol (Sigma T 7402). Dissolve by vortexing and store as 2- to 20-μl aliquots at −20°C. Taxol is stable in DMSO but unstable in water. It is insoluble in aqueous buffers above about 18 μ*M*. DMSO is explosive if it gets wet. Store small volumes at room temperature over beds of Sephadex G-50.
5. *0.2M NaEGTA:* Dissolve 15.2 g EGTA (Sigma E 4378) in 190 ml water. Adjust pH to neutral by adding concentrated NaOH and then make volume to 200.0 ml. Store at room temperature.
6. *1M* $MgCl_2$*:* 20.33 g $MgCl_2 \cdot 6H_2O$ to 100 ml water. Sterile filter and store at room temperature.
7. *BRB 80 (Brinkley reassembly buffer):* 80 m*M* K-PIPES, 1 m*M* $MgCl_2$, 1 m*M* EGTA, pH 6.9. Make up as a 10× stock, store at 4°C, and dilute freshly for use.

Purified tubulin at about 100 μ*M* (protocol for tubulin preparation on our web page) in BRB80 should be flash frozen in 10- to 25-μl aliquots in the presence of 30% glycerol by immersion in liquid nitrogen and stored either at −70°C or preferably in liquid nitrogen.

Steps

1. Thaw an aliquot of tubulin (typically 200 μ*M*) and add stock 100 m*M* NaGTP to 1 m*M* and $MgCl_2$ to 2 m*M*. Warm to 37°C and incubate for 20 min.
2. After 20 min, add taxol from a 10 m*M* stock in DMSO to 20 μ*M* final. Dilute microtubules 1000-fold for use using BRB80 buffer supplemented with 20 μ*M* taxol.

B. Preparation of Flow Cells

Steps

1. Apply single-sided Scotch tape to the long edges of a microscope slide such that the

strip of glass surface between the two pieces of tape is 8–10 mm wide. Trim away overhangs with a razor blade.
2. Extrude two parallel stripes of Apiezon M grease from a syringe with a squared-off wide-bore needle along the inner edges of the tape strips.
3. Press a clean coverslip onto the grease. The volume of the flow cell can be adjusted by spacing the grease strips apart and/or by placing spacers between the coverslip and the slide. Single-sided Scotch magic tape is about 50 μm thick, giving a flow cell of about 10 mm × 5 mm × 50 μm, or 25 μl. Thinner metal or cellophane foils can be used to make a shallower flow cell and conserve sample. It is helpful to make the flow cell shallow because the microtubules below the top surface scatter light and reduce contrast. For inverted microscopes, it is convenient to arrange flow crosswise. The inset to Fig. 18.1 illustrates flow cells for inverted (A) and upright (B) microscopes.

C. Surface Adsorption of Motor

Solutions

1. *Motility buffer:* BRB 80 plus 1 m*M* MgATP. For fluorescence work only, degas and add 1% of 100× antibleach mix (GOC), which is 100 mg/ml glucose oxidase (Sigma G7016), 18 mg/ml catalase (Sigma C100), and 300 mg/ml glucose (Sigma G7528) in BRB80 plus 50% glycerol. When aliquoting, fill tubes to exclude oxygen, cap, and store at −20°C.
2. *100× diluted MTs:* either motility buffer or motility buffer plus GOC.

Steps

1. Place the flow cell flat. Using a Gilson, inject into the cell 1 chamber volume of motor solution. The solution is drawn into the cell by capillarity. Incubate the slide in a moisture chamber for 2–5 min at 20°C to allow the motor to adsorb to the glass.
2. Wash the cell with 2 chamber volumes of assay buffer, applying the solution to one side of the chamber using a micropipette and drawing the solution gently through the cell using the capillary action of the torn edge of a strip of Whatman 3MM, placed at the exit of the chamber.
3. Flow in 1 volume of MTs in motility buffer + taxol and mount the slide on the microscope stage, oiling the condensor to the bottom of the slide (it may be possible to use a dry condensor for quick-and-dirty assays).
4. Extra for fluorescence work 1. Degas some BRB80. To 10 ml, add 100 μl of 100× GOC. Take another 3 ml and add MgATP to 2 m*M*. Fill and cap tubes to exclude oxygen and hold buffers on ice. Add taxol to 20 μ*M* freshly before use.

D. Microscope Setup

1. DIC

Before the day's work, align the microscope roughly using a test specimen (a slide made using a suspension of plastic beads provides a stable and realistic test specimen). Switch on the lamp and allow a few minutes for the arc to stabilise. Rack down the objective and oil it to the slide. Insert some neutral density filtration to protect your eyes from the intensely bright light, focus roughly on the top surface of the grease at the edge of the chamber, and then drive the stage to centre the sample below the objective. Find some beads attached to the undersurface of the coverslip. Open the condensor aperture and close the field aperture. Obtain Koehler illumination by focussing and centring the condensor so that a sharp image of the field diaphragm appears in the view field. Open the field diaphragm again and adjust DIC sliders close to extinction.

Focussing on MTs in the experimental flow cell is also best done using the grease surface as a guide. Focus as described earlier and then remove neutral density filters and switch in the video system. Adjust fine focus to image the surface. Adjust light intensity to almost saturate the camera (this is the point where signal to noise is maximal). With the contrast on the Argus set to maximum, microtubules should be visible without background subtraction. Defocus slightly, collect a background image, and subtract. Microtubules should now be clearly visible.

2. *Epifluorescence*

A test sample of multispectral fluorescent beads is very useful (Molecular Probes multispeck M-7900). Switch on the arc lamp and allow a few minutes for the arc to stabilise. Once the lamp is stable, align the microscope for epifluorescence: Remove an objective and place a piece of paper on the stage. Inset some neutral density filtration. Close the field diaphragm slightly and focus and centre the image of the lamp filament that appears on the paper. Replace the objective.

Focussing on microtubules in the experimental cell is much easier with dark-adapted eyes. Using the full intensity of the mercury lamp, rack the objective down until MTs are visible, first as a dim red glow, and then as sharply defined bright red lines on a black background. Immediately reduce the illumination intensity to protect against photobleaching, switch in the intensified camera, and start recording.

E. Recording Data

The most flexible arrangement for data recording is to set up time-lapse digital recording of video frames to a computer hard disc (with no resolution loss) and simultaneous recording to VCR. The VCR runs uninterupted in the background for 3h per tape and generates an archive. The operator is free to go back to this archive at a later date and recapture interesting sequences for analysis.

F. Analysing Data Calibration

Image a stage graticule, a slide with etched lines at 1- or 10-μm intervals (from microscope manufacturers). It is important to calibrate both in X and Y; simply rotate the camera 90°. Most systems will give a different number of pixels per micrometer in X and Y. Tracking software compensates for this effect.

The best way to track is to follow the tip of a moving microtubule: tracking the centroids, as common in cell tracking, for example, will give you the wrong answer as soon as the microtubule bends. For maximum accuracy, the time lapse between frames should be adjusted to minimise the effects of operator error when tracking using the mouse. In practice we try to collect 20 frames and adjust the time lapse so that the microtubules move across the full field (22 μm) during this time.

IV. COMMENTS

A. Archiving Data

It is very important to have a formal system for identifying every video frame on every tape. In this way there is no possibility of confusing data sets. The simplest way to do this is to time and date stamp the frames as they are generated, using the overlay feature of the Argus. As ever, keeping careful written notes also helps a lot. For complex experiments it can be useful to speak notes onto the audio track of the tape. Digital clips are archived most conveniently on external.

B. Imminent Technology

As computers get quicker, it is realistic to start recalculating images in real time. Autocontrast is one interesting possibility, whereby the pixels of each incoming frame are parsed and the look-up table is stretched to optimise contrast. It will be some time before we can dispense with the VCR. Real-time recording of uncompressed grey-scale video to disc is pushing the limits at present, but sufficiently fast sustained data transfer rates will soon be available. This is not the real problem, however. One frame of PAL video is 768×512 pixels, which, with 8 bit (256 greys) data, means that each frame is 384 kb. Real-time recording to hard disc fills the disc up at about 0.5 Gb per minute, and it soon becomes necessary to archive data to video tape.

C. Workstation Ergonomics

It is worth paying some attention to the ergonomics of your microscope workstation. Microscope focus, mouse, keyboard video, and contrast-adjustment electronics all need to be within easy reach of a seated operator. Screens should be visible with only a slight turn of the head. It is very helpful to have a foot switch to dim the room lights and blinds on windows.

D. Best Practice

Because of inherent uncertainties about the way a particular protein attaches to a particular glass, motility assays are at their strongest when used to measure the relative motility in different treatments of samples. It is commonly assumed that motility assays measure motor-driven microtubule sliding under zero load. It is probably more correct to assume that an unspecified, variable, (but low) load applies.

V. PITFALLS

A. Computerphilia

The most common fault in video microscopy is to overprocess an indifferent optical image. Too much processing can seriously degrade the amount of information in the image. A good primary image has high spatial resolution (sharpness), high contrast, and low background noise. Obtaining one is partly a function of specimen preparation and partly of microscope setup.

B. Lamp Intensity Fluctuates

In DIC, a troublesome problem is sudden variations in light intensity caused by the arc of the mercury lamp wandering. These are not noticeable in normal modes of microscopy, but with electronic amplification of contrast they become annoying. The only solution is to change the lamp. Cooling the lamp using a fan may help. Mercury lamps typically need changing after 100 h, because after that their intensity drops fairly rapidly.

C. Microtubules Fishtail or Do Not Move At All

Some motor proteins bind better to the glass surface than others. Erratic motility may be due to your protein denaturing on the glass or binding in such a way that its force-generating conformational change is inhibited. Areas of uncoated glass can also bind microtubules and inhibit sliding. Increase motor concentration if possible or try infusing the motor twice over and/or reducing or eliminating the wash step prior to infusing microtubules. Including casein at 0.1–1 mgml in the assay buffer efficiently protein coats glass. Because motor activity can also be sensitive to thiol oxidation, try including 5 mM DTT in your motility buffer.

References

Cross, R. A., and Kendrick, Jones J. (1991). Motor proteins. *J. Cell. Sci.* Suppl. 14.

Inoué, S., and Spring, K. R. (1997). *"Video Microscopy."* Plenum Press, New York.

Kron, S. J., Toyoshima, Y. Y., Uyeda, T. Q. R., and Spudich, J. A. (1991). *Assays for actin sliding movement over myosin-coated surfaces. Methods Enzymology* **196**, 399–416.

Scholey, J. M. (1993). *Motility assays for motor proteins. Methods Cell Biol* **39**.

CHAPTER

19

In Vitro Assays for Mitotic Spindle Assembly and Function

Celia Antonio, Rebecca Heald, and Isabelle Vernos

OUTLINE

I. INTRODUCTION

During the past two decades, cell biology has entered a phase in which technology is so powerful that fundamental questions concerning the morphogenesis and function of cellular organelles can be addressed. One essential and beautiful structure is the mitotic spindle. The work of Lohka and Maller (1985), followed by that of a few other laboratories (Murray and Kirschner, 1989; Sawin and Mitchison, 1991; Shamu and Murray, 1992), opened up a novel approach to studying such a complex and dynamic structure. The idea is not to purify individual spindle components and put them back together hoping that a spindle will assemble, it is rather to open up the cell and prepare a cytoplasmic extract as crude and concentrated as possible to keep the conditions close to the *in vivo* situation. At first sight, this approach seems uninformative: one merely mimics *in vitro* what happens *in vivo*. However, several methods have been developed to manipulate the system, such as the addition of reagents and depletion of proteins. Also the microscopy techniques have evolved and now allow the study of different aspects of spindle formation and function (Desai *et al.*, 1998; Kalab *et al.*, 2002). Thus, this system can be used both to analyze the mechanism of spindle assembly and function and to evaluate the role of individual molecules in the process. More methods are still being developed or improved that will increase the usage and the utility of these extracts even more.

II. MATERIALS AND INSTRUMENTATION

A. Preparation of *Xenopus laevis* Egg Extracts

Incubator at 16°C, clinical centrifuge, DuPont Sorvall RC-5 centrifuge, HB-4 or HB-6 rotor with rubber adaptors (Sorvall Cat. No. 00363), Beckman ultraclear SW50 tubes (Cat. No. 344057), Sarstedt 13-ml adaptor tubes (Cat. No. 55.518), 1-ml syringes, 18- and 27-gauge needles, glass pasteur pipettes.

Mature female frogs are from African Reptile Park. Pregnant mare serum gonadotropin (Intergonan) is from Intervet. Human chorionic gonadotropin (HCG, Cat. No. CG-10), cytochalasin D (Cat. No. C-8273), EGTA (Cat. No. E-4378), 4.9M $MgCl_2$ (Cat. No. 104.20), and ATP (Cat. No. A-2383) are from Sigma. $CaCl_2$ (Cat. No. 2383), KCl (Cat. No. 1.04936.1000), NaCl (Cat. No. 1.06404.1000), and l-cysteine (Cat. No. 1.02838.1000) are from Merck. Leupeptin (Cat. No. E18) and pepstatin (Cat. No. E19) are from Chemicon. Aprotinin (Cat. No. 981532) and creatine phosphate (Cat. No. 0621714) are from Roche. HEPES is from Biomol. Sucrose is from USB (Cat. No. 21938).

B. Preparation of DNA Beads

Biotin-21-dUTP (Cat. No. 5201-1) is from Clontech. Biotin-14-dATP (Cat. No. 19524-016) is from Invitrogen. Thio-dCTP (Cat. No. 27-7360-02), thio-dGTP (27-7370-04), and G-50 gel-filtration (NICK) columns (17-0855-01) are from Amersham Pharmacia Biotech. Restriction enzymes and Klenow (Cat. No. M02125) are from New England Biolabs. The magnetic particle concentrator (MPC) and Kilobase BINDER kit containing Dynabeads M-280 Streptavidin, washing, and binding solutions can be obtained from Dynal. Trizma base (Cat. No. T-1503) is purchased from Sigma, and EDTA (Cat. No. 1.08418.1000) is from Merck.

C. Spindle and Aster Assays

Freshly prepared CSF extract, sperm nuclei (3000/μl)(Murray, 1991), rhodamine-labelled tubulin (2–3 mg/ml) (Hyman *et al.*, 1991), 76 × 26-mm microscope slides, 22 × 22-mm coverslips, fluorescence microscope. PIPES (Cat. No. P-6757), Triton X-100 (Cat. No. T-8787), glutaraldehyde (Cat. No. G-5882), $NaHB_4$ (Cat. No. S-9125), GTP (Cat. No. G-8877), and Hoechst dye (bisbenzimide, Cat. No. H33342) are from Sigma. Glycerol (Cat. No. 1.04093.2500) and formaldehyde (Cat. No. 1.04003) are from Merck. Protein A dynabeads (Cat. No. 100.02) are from Dynal. Dimethyl sulfoxide (DMSO, Cat. No. 27,685-5) is from Sigma-Aldrich, and Paclitaxel (taxol equivalent) (Cat. No. P-3456) is from Molecular Probes. For sedimentation and immunofluorescence experiments, corex tubes (15 ml) are equipped with plastic adaptors (home-made, see Evans *et al.*, 1985) to support 12-mm round coverslips. The tubes are centrifuged in an HB-4 rotor containing rubber adaptors. pGEX expression vectors are from Amersham Pharmacia Biotech.

III. PROCEDURES

A. Preparation of *X. laevis* CSF Egg Extracts

Good protocols detailing the preparation of Xenopus egg extracts and sperm nuclei have already been published (Murray, 1991). We describe here only the preparation of metaphase cytostatic factor-arrested "CSF" extracts as optimized in our laboratory for spindle assembly reactions.

Solutions

1. *Pregnant mare serum gonadotropin (PMSG)*: Dissolve in sterile water to a final concentration of 200 units/ml. Store at 4°C
2. *Human chorionic gonadotropin (HCG)*: Dissolve in sterile water to a final concentration of 2000 units/ml. Store at 4°C.
3. *MMR*: 100 m*M* NaCl, 2 m*M* KCl, 1 m*M* $MgCl_2$, 2 m*M* $CaCl_2$, 0.1 m*M* EDTA, 5 m*M* HEPES, pH 7.8. Prepare a 20X MMR stock solution. Adjust pH with NaOH, autoclave, and store at room temperature. Before use make 1X MMR from the 20X stock solution and adjust pH again if necessary.
4. *Dejellying solution*: 2% l-cysteine, pH 7.8. Prepare in water and adjust pH with NaOH. Make fresh just before use.
5. *LP*: dissolve leupeptin and pepstatin in DMSO to a final concentration of 10 mg/ml. Store in 50-μl aliquots at −20°C.
6. *Aprotinin*: Dissolve in sterile water to a final concentration of 10 mg/ml. Store in 50-μl aliquots at −20°C.
7. *Cytochalasin D*: Dissolve in DMSO to a final concentration of 10 mg/ml. Store in 50-μl aliquots at −20°C.
8. *20X XB salts*: 2*M* KCl, 20 m*M* $MgCl_2$, 2 m*M* $CaCl_2$. Filter sterilize and store at 4°C.
9. *XB*: 1X XB salts containing 50 m*M* sucrose and 10 m*M* HEPES. Adjust pH to 7.7 with KOH. Prepare fresh before use.
10. *CSF-XB*: Prepare from XB buffer by adjusting $MgCl_2$ to 2 m*M* and adding 5 m*M* EGTA. To prepare CSF-XB with protease inhibitors (CSF-XB with PI), add LP stock solutions and aprotinin to a final concentration of 10 μg/ml.
11. *Energy Mix*: 20X stock contains 150 m*M* creatine phosphate, 20 m*M* ATP, 2 m*M* EGTA, 20 m*M* $MgCl_2$. Store in 100-μl aliquots at −20°C.

Steps

1. Inject four to six frogs subcutaneously with 0.5 ml (1000 units) PMSG each using 1-ml syringes and a 27-gauge needle at least 4 days before planning to make an extract. They should be used within 2 weeks after the priming injection. The number of frogs required depends on the quantity and quality of eggs. Five milliliters of eggs (one SW50 tube) gives approximately 1 ml of extract.
2. Sixteen to 18 h before use, inject frogs subcutaneously with 0.25–0.5 ml (500–1000 units) HCG. Place the frogs in individual boxes containing 500 ml MMR at 16°C.
3. Prepare all solutions before starting collecting the eggs: 2 liters MMR, 500 ml XB, 400 ml CSF-XB, 100 ml CSF-XB with PI, and 500 ml of dejellying solution. Rinse all glassware with distilled water (eggs stick to plastic dishes). Cut off the end of a glass pasteur pipette and fire polish it to make a wide-mouth pipette.
4. Collect laid eggs in MMR. Frogs can also be squeezed, which often gives the highest quality eggs. Keep eggs from different frogs in separate batches. Discard batches of eggs

containing more than 5% of lysed, ugly, or stringy eggs.

5. Wash a few times with MMR to take away skin and other detritus and remove bad eggs. Pour off MMR and add 250 ml dejellying solution. When laid, eggs are enveloped in a transparent jelly coat and do not pack closely together. Swirl the beaker frequently and change the dejellying solution. After removal of the jelly coat, eggs pack together. This takes about 5 min. Eggs left for too long in cysteine will lyse.
6. Pour off the dejellying solution and add 500 ml MMR. Repeat the rinse one more time. After removal of the jelly coat, the eggs become fragile. They lyse easily and can activate if in contact with air. They must always remain immersed in buffer. Remove all bad-looking eggs: white and puffy, flattened, or activated ones (darker pole retracted), and those with mottled pigmentation.
7. Wash three times with 100–200 ml XB.
8. Remove as much buffer as possible, keeping all eggs immersed. Wash three times with 100–200 ml CSF-XB and finally keep eggs in CSF-XB with PI.
9. Transfer eggs using the cut glass Pasteur pipette to SW50 tubes containing 1 ml CSF-XB with PI plus 3 μl cytochalasin D (30 μg/ml). Always immerse the pipette tip in solution before expelling eggs to prevent contact with air. Transfer the SW50 tubes to 13-ml Sarstedt adaptor tubes, which contain 0.5 ml of water to prevent the tubes from collapsing.
10. Centrifuge in a clinical centrifuge at 16°C at 600 *g* for 1 minute and then immediately increase the speed to 1200 *g* for an additional 30s. After centrifugation, remove all excess buffer from the top of the packed eggs. Removal of buffer is critical to obtain a concentrated cytoplasm.
11. Place the tubes in an HB-4 or HB-6 rotor containing rubber adaptors. Centrifuge at 10,500 rpm for 15 min at 16°C to crush the eggs.
12. Place the tubes on ice. A yellow lipid layer is at the top of the tube. Underneath is the cytoplasmic layer, then heavy membranes, and yolk particles at the bottom of the tube. Wipe the sides of the tubes with a tissue before piercing with a 18-gauge needle at the bottom of the cytoplasmic layer. Slowly and carefully aspirate the cytoplasm using a 1-ml syringe with the needle opening facing upward. Remove needle from syringe and carefully expel the cytoplasm into a 1.5-ml tube. Place on ice.
13. Add LP and aprotinin to 10 μg/ml (1:1000 dilution of stocks) and cytochalasin D to 20 μg/ml (1:500 dilution of stock). Add energy mix (1:20 dilution of stock) and mix gently.

Comments on Sperm Nucleus Preparation

Based on Gurdon (1976) and modified by Murray (1991), our only modification is to mash fragments of testes between two frosted (rough) slides before filtering through a cheesecloth mesh.

B. Preparation of DNA Beads

For preparation of DNA beads, linearize plasmid DNA and fill in with nucleotides so that one end contains biotinylated bases and the other thio-nucleotides to inhibit exonuclease activity and the formation of large aggregates of beads. Couple the DNA to 2.8 μm magnetic beads. Incubation in the extract will allow the association of chromatin proteins and the consequent formation of chromatin (Heald *et al.*, 1996).

Solutions

1. *TE*: 10 m*M* Tris, 1 m*M* EDTA, pH 8. Adjust pH with HCl, autoclave, and store at room temperature.

2. *Washing and binding solutions*: Included with Kilobase BINDER kit.
3. *Bead buffer*: 2*M* NaCl, 10 m*M* Tris, 1 m*M* EDTA, pH 7.6. Adjust pH with HCl and store at room temperature.
4. *Hoechst dye solution*: Dissolve bisbenzimide in water to a final concentration of 10 mg/ml. Store in the dark at 4°C.

Steps

1. Prepare plasmid DNA by Qiagen column purification. While the sequence of the DNA is not important, the plasmid should be more than 5 kb to effectively induce chromatin assembly. Cut 50 μg of the DNA with two restriction enzymes that have unique sites in the polylinker to produce one short and one long DNA fragment. One end of the long fragment should terminate in an overhang containing Gs and Cs and the other should contain only As and Ts (e.g., *Not*1, *Bam*H1). See Fig. 19.1.
2. Ethanol precipitate the DNA and resuspend in 25 μl TE. Quantify recovery by OD_{260} measurement.
3. Prepare fill-in reaction in 70 μl containing 1X Klenow buffer, 30 μg DNA, 50 μ*M* nucleotides (biotin-dATP, biotin-dUTP, thio-dCTP, and thio-dGTP) and 20 units Klenow. Incubate for 2 h at 37°C.
4. Remove unincorporated nucleotides, following instructions supplied with the Sephadex G-50 gel filtration column (NICK column). The DNA is eluted in a large volume (400 μl), but the recovery is better than with spin columns. Quantify recovery by OD_{260} measurement.
5. Prepare coupling mix by combining 400 μl biotinylated DNA and 400 μl binding solution. Set aside 25 μl of the coupling mix for later evaluation of coupling efficiency.
6. Prepare 3 μl of streptavidin-conjugated dynabeads for each microgram of DNA recovered, so 120 μl for 30 μg. Retrieve beads using the MPC (magnet) and wash once with 5 volumes of binding solution (600 μl for 120-μl beads). Retrieve the beads and resuspend them in coupling mix containing DNA.
7. Incubate bead/coupling mixture for several hours (or overnight) on a rotating wheel at 4°C.
8. Retrieve the beads and save the supernatant. Compare the OD_{260} of the supernatant to the sample taken before coupling to determine the amount of DNA immobilized. Typically two-thirds of the DNA is coupled. Stain 1 μl of beads with 5 μg/ml Hoechst dye and observe them by fluorescence microcopy. The beads should appear as very bright dots with no dark patches. If the amount of DNA coupled in the first round does not seem sufficient, incubate the beads a second time with biotinylated DNA.

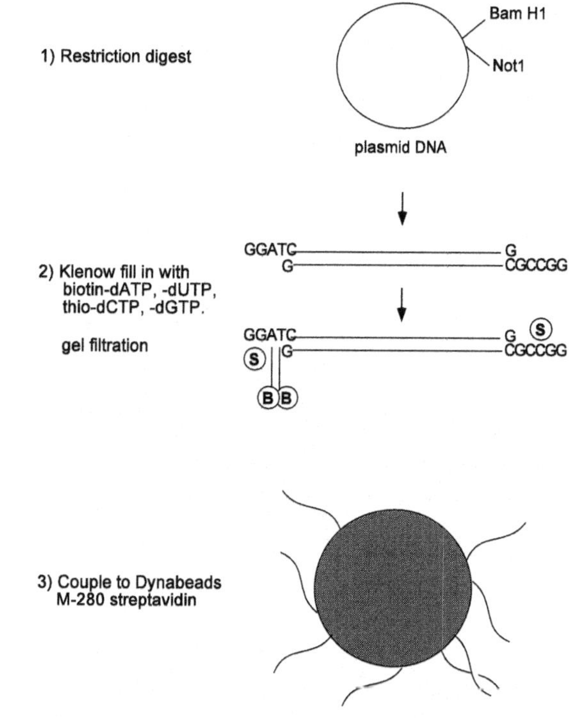

FIGURE 19.1 Steps in preparation of DNA beads.

9. Wash beads twice with washing solution and then twice with bead buffer. After the last wash, resuspend the beads in bead buffer so that the final concentration of immobilized DNA is 1 μg/5 μl of beads. Store at 4°C.

C. Spindle and Aster Assembly *in vitro*

Chromatin triggers spindle assembly in CSF Xenopus egg extract. Experimentally there are two ways to induce formation of the spindle: by the addition of sperm nuclei (Desai *et al.*, 1999) or DNA beads to the extract (Heald *et al.*, 1996). To decide on which method to use it is important to understand the characteristics of the two systems. Each sperm nucleus is tightly associated to a centriole. In a very simple process, half spindles can assemble around sperm nuclei just by incubation of the sperm nuclei in extract. Over time these half spindles fuse to form bipolar spindles. If the extract is cycled through interphase the DNA replicates and each centrosome duplicates. The addition of fresh CSF extract sends the extract back into mitosis, chromosomes with their paired kinetochores condense, the duplicated centrosomes move apart, and a bipolar spindle assembles. Anaphase can be induced at this point by the addition of calcium to the system.

DNA beads are made from any kind of plasmid DNA that is long enough to induce chromatin assembly. They are unlikely to assemble kinetochores, as centromere sequences are not present. Spindle assembly around DNA beads is achieved by first assembling chromatin on the beads in an extract that is cycled through interphase. The beads are then retrieved and resuspended in fresh CSF extract. Using this method, spindles assemble in the absence of kinetochores and centrosomes. It is then ideal to study certain processes of spindle assembly that are independent of these structures.

A simplified method to study microtubule dynamics and focusing of microtubule minus ends is the assembly of asters in CSF extract. Asters can be assembled by the addition of human centrosomes purified from KE37 lymphoid cells (Bornens *et al.*, 1987), DMSO, or taxol to the extract (Wittmann *et al.*, 1998).

Solutions

1. *10X calcium solution*: 4 m*M* $CaCl_2$, 100 m*M* KCl, 1 m*M* $MgCl_2$. Store in aliquots at −20°C.
2. *4.9M $MgCl_2$ stock solution (Sigma)*
3. *10X MMR*: See Section III,A.
4. *Hoechst dye solution*: See Section III,B.
5. *Spindle fix*: 48% glycerol, 11.1% formaldehyde, 5 μg/ml Hoechst dye in 1X MMR. Always prepare fresh on day of use.
6. *BRB80*: 80 m*M* PIPES, 1 m*M* $MgCl_2$, 1 m*M* EGTA, pH 6.8. Prepare a 5X BRB80 solution. Dissolve components in sterile water and, while stirring, add KOH pellets until the PIPES dissolves. Adjust to pH 6.8 with KOH solution. Sterilize by filtration and store at 4°C.
7. *Dilution buffer*: 1X BRB80 containing 30% glycerol and 1% Triton X-100. Store at room temperature. Better preservation of the spindles can be achieved by adding 0.25% glutaraldehyde to the dilution buffer. Use a full vial of glutaraldehyde and store in aliquots at −20°C.
8. *Aster dilution buffer*: BRB80 containing 10% glycerol, 0.25% glutaraldehyde, 1 m*M* GTP, 0.1% Triton X-100. Use a fresh vial of glutaraldehyde to prepare buffer and store in aliquots at −20°C.
9. *Cushion*: 40% glycerol, BRB80. Store at 4°C.
10. *Aster cushion*: 25% glycerol, BRB80. Store at 4°C.
11. *PBS*: 137 m*M* NaCl, 2.7 m*M* KCl, 4.3 m*M* Na_2HPO_4, 1.4 KH_2PO_4 m*M*, pH 7.4. Sterilize by autoclaving and store at room temperature.
12. *Immunofluorescence buffer (IF buffer)*: PBS containing 2% BSA and 0.1% Triton X-100. Add azide and store at 4°C.

13. *Mowiol*: 10% Mowiol 4–88, 25% glycerol, and 0.1*M* Tris, pH 8.5. To prepare, mix 6 g Mowiol with 6 g glycerol. Add 6 ml sterile water and stir for several hours at room temperature. Add 12 ml 0.2*M* Tris, pH 8.5, and place the mixture in a heating plate at 50°C for 10 min with continues stirring. After the Mowiol dissolves, centrifuge at 5000 *g* for 15 min. Store in aliquots at −20°C.

Steps for Spindle Assembly around Sperm Nuclei (see Fig. 19.2)

1. Before using an extract to assemble spindles, its quality should be tested by setting up a "half spindle" reaction. Combine 20 μl CSF extract, 0.2 μl rhodamine-labeled tubulin and 0.8 μl sperm nuclei (about 75 nuclei/μl extract) in a 1.5-ml tube and mix. The extract can be mixed by gently moving the pipette in circles or by pipetting up and down, always avoiding making bubbles. Incubate the reaction mixture at 20°C and take a squash at 30–40 min. To make a squash, using a cut-off tip, transfer 1 μl of reaction mixture to a microscope slide, carefully place 5 μl of spindle fix solution on top, and gently place a 18 × 18-mm coverslip on top. Analysis of the squash by fluorescence microscopy should reveal half spindles with condensed chromosomes and occasionally also spindles. There should be no free microtubule nucleation.
2. If the extract is good, proceed with the spindle assembly protocol. For each reaction add on ice 0.2 μl rhodamine-labeled tubulin and 0.8 μl sperm nuclei to 20 μl CSF extract in a 1.5-ml tube and mix. Incubate for 10 min at 20°C and then release extract into interphase by the addition of 2 μl 10X calcium solution. Mix gently.
3. Incubate for 80 min at 20°C. Check that the extract is in interphase by taking a squash as described in step 1. If the sample is to be saved, seal the coverslip to the slide with nail polish. At this stage, nuclei should appear large, round, and uniform, and microtubules should be long and abundant.
4. At 90 min postcalcium addition, add 20 μl of fresh CSF extract containing 0.2 μl rhodamine-labeled tubulin to the reaction.
5. Incubate further at 20°C. Take squashes (step 1) at different time points to assess the stage of spindle formation. During the incubation, prepare 15 ml Corex tubes with plastic adapters, a round 12-mm coverslip, and 5 ml of cushion.
6. Forty-five to 60 min after mitosis reentry (step 4), quickly add 1 ml dilution buffer and mix by gently inverting the tube a couple of times. Carefully layer the mixture over the cushion using a cut pipette tip. Centrifuge at 16°C for 12 min at 12,000 rpm in an HB-4 or HB-6 rotor. Aspirate supernatant and cushion before removing the coverslip. Fix coverslips in −20°C methanol for 5 minutes. Rehydrate samples by placing the coverslips in PBS. If dilution buffer contains glutaraldehyde, incubate the coverslips twice for 10 min in 0.1% $NaHB_4$ in PBS and wash with PBS.
7. To perform immunofluorescence, place the coverslip face up on a surface covered with parafilm. Incubate the coverslips with primary antibodies diluted in IF buffer for 20–30 min. Wash the coverslips two times for 5 min with PBS and incubate with IF buffer containing the secondary antibody and 5 μg/ml Hoechst for 20–30 min. Wash three times for 5 min with PBS and carefully place the coverslips upside down on a 3- to 4-μl drop of Mowiol on a microscope slide, avoiding air bubbles. Prior to observation, allow the Mowiol to set for 15 min at 37°C or at room temperature for several hours.
8. To visualize the early steps of anaphase, add 4 μl 10X calcium solution to the metaphase spindles before step 6. Take samples before calcium addition and every 5 min after calcium addition for 20–30 min and process

them as described earlier. It is not always possible to visualize separation of the sister chromatids and only high-quality extracts can be used successfully to study anaphase.

Steps for Spindle Assembly around Chromatin Beads (see Fig. 19.2)

1. Transfer 3 μl of DNA beads (about 0.5 μg DNA) to a 0.5-ml Eppendorf tube and place on a magnet. Remove supernatant, and wash beads by resuspending them in 20 μl of extract.
2. Retrieve beads and resuspend in 100 μl CSF extract. Transfer to a 1.5-ml Eppendorf tube and incubate at 20°C.
3. After 10 min, release the CSF extract into interphase by adding 10 μl of calcium solution. Incubate for 2 h at 20°C.
4. Return the extract containing the beads to mitosis by adding 50 μl of fresh CSF extract. Incubate for 30 more min at 20°C. The procedure can be continued or stopped at this point by freezing the beads in the extract using liquid ethane and storing them in liquid nitrogen.

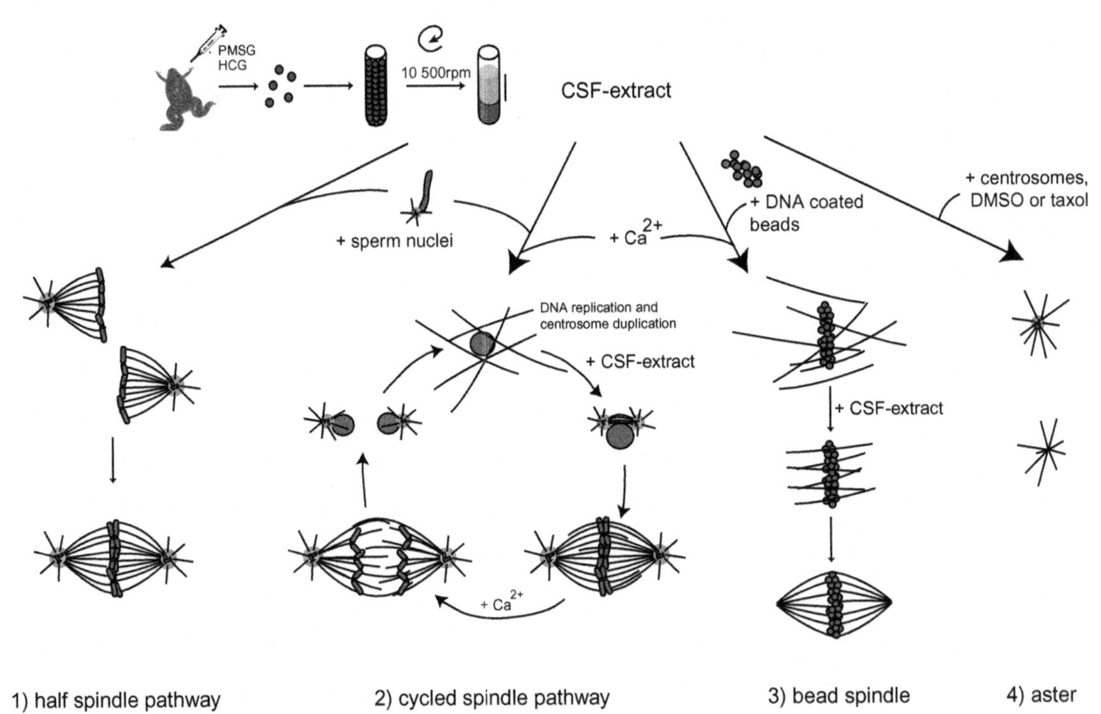

FIGURE 19.2 Preparation of CSF Xenopus egg extract and structures that assemble after addition of different components. (1) Addition of sperm nuclei to the CSF extract leads to the formation of half spindles that can fuse to form bipolar spindles. (2) Addition of calcium together with sperm nuclei sends the system into interphase, allowing DNA replication and centrosome duplication to occur. Addition of fresh CSF extract sends the system back to mitosis and bipolar spindles form. Calcium addition at this point induces chromosome segregation and entry into interphase. (3) Bead spindle: addition of DNA-coated beads and calcium to an extract leads to the formation of chromatin and nuclei around beads. Addition of CSF extract followed by incubation in fresh CSF extract allows the formation of bipolar spindle. (4) Asters: Addition of purified centrosomes or DMSO or taxol leads to the formation of microtubule asters with microtubule minus ends focused at the centrosome or at the center of the aster.

5. Incubate the bead mixture on ice for several minutes. Retrieve the beads on ice over 10–15 minutes. Due to the viscosity of the extract, bead retrieval is slow. Pipette the mixture every several minutes, keeping the tube on the magnet.
6. Remove the supernatant and resuspend the beads in 150 μl of fresh CSF extract containing 1.5 μl rhodamine-labeled tubulin.
7. Incubate at 20°C. Monitor the spindle assembly by taking 1-μl samples and squashing with fixative as described earlier. Spindle assembly requires between 30 and 90 min, depending on the extract. For immunofluorescence studies, spin the bead spindles through a cushion onto a coverslip and perform immunofluorescence as described previously for spindle assembled around sperm nuclei.

Steps for Aster Formation (see Fig. 19.2)

1. For each reaction, add on ice 0.2 μl rhodamine-labeled tubulin to 20 μl CSF extract in a 1.5-ml tube. To assemble asters, add either purified human centrosomes (Bornens *et al.*, 1987) or 5% DMSO or 1 μ*M* taxol.
2. Incubate the reaction for 30–60 min in a 20°C water bath. During this incubation, prepare as before 15-ml Corex tubes with plastic adapters, round 12-mm coverslip, and 5 ml aster cushion.
3. At the end of the incubation time, dilute the reactions with aster dilution buffer and carefully layer them on top of the cushion using a pipette with a cut tip and subsequently centrifuge onto the coverslips in a HB4 or HB6 rotor at 12,000 rpm for 12 min at 16°C.
4. As before, remove the cushion with a vacuum pump and postfix the coverslips in −20°C methanol for 5 min. To quench the glutaraldehyde, incubate twice for 10 min in 0.1% $NaHB_4$ in PBS and process for immunofluorescence.

D. Functional Studies of Proteins Involved in Spindle Assembly

The Xenopus egg extract system is a powerful tool to assess the involvement of individual proteins in the process of spindle assembly. Different methods have been developed to address specific questions.

Localization studies can be performed by classical immunofluorescence methods or by direct visualization of GFP-tagged proteins added to the reaction mixture or by a combination of both by adding GST-tagged proteins followed by immunofluorescence with anti-GST antibodies.

Functional studies can be performed by adding dominant-negative constructs to the reaction mixture or by depleting the protein or antibodies under study from the extracts before use (Antonio *et al.*, 2000; Boleti *et al.*, 1996). Depletion experiments should be complemented by "add back" or rescue experiments in which a recombinant protein is added to the depleted extracts (Fig. 19.3). In all three cases, a careful quantification of the structures found in control or treated samples has to be performed.

Solutions

1. *CSF-XB with PI*: See Section III,A.
2. *PBS-TX*: PBS containing 0.1% Triton X-100. Store at 4°C.

Steps for Immunodepletion

1. Usually around 10 μg of specific antibody is needed to deplete 150 μl of extract. The amount of antibody has to be adjusted for each case depending on the antibody itself and on the abundance of the protein in the extract. Prepare a control mock-depleted extract using the same amount of unspecific purified IgG.
2. Transfer 40 μl of Dynal beads coupled to protein A to a 1.5-ml tube and retrieve them using a magnet. Wash them twice by

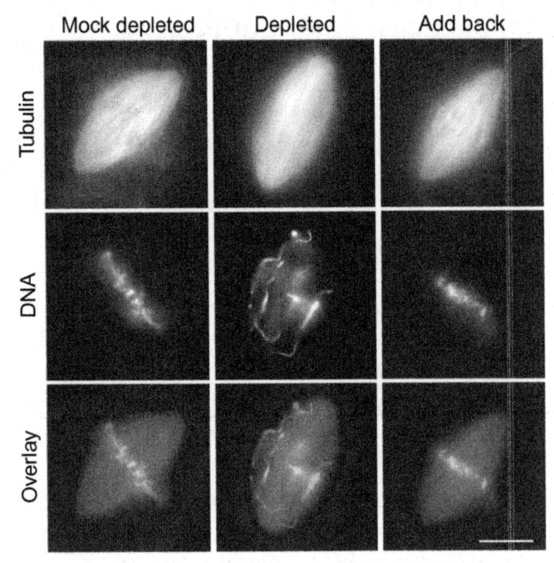

FIGURE 19.3 Spindles assembled in CSF Xenopus egg extract following the cycled spindle pathway. The extract was mock depleted, Xkid depleted, or Xkid depleted supplemented with recombinant Xkid protein (add back). When Xkid was depleted, mitotic chromosomes failed to align at the metaphase plate. Alignment of chromosomes was rescued by the addition of recombinant protein to the extract. Scale bar: 10 μm.

resuspending the beads in 500 μl PBS-TX and retrieving them with the magnet.

3. Incubate the beads with 10 μg of antibody diluted in 200 μl of PBS-TX for 1 h on a rotating wheel at 4°C
4. Wash twice as just described with PBS-TX and once with CSF-XB containing PI. Remove as much wash buffer as possible and carefully resuspend the beads in 150 μl of extract by pipetting up and down. Keep the extract (containing the beads) on ice for 90 min with occasional mixing by pipetting up and down.
5. Retrieve the beads by placing the tube on the magnet for 5 min on ice. Collect the extract into a new tube and keep it on ice. To remove the beads more efficiently, repeat this procedure once more. Check the efficiency of depletion by Western blot analysis of mock and protein-depleted extracts. To visualize the immunoprecipitated protein, wash the beads once with CSF-XB containing PI and three times with PBS-TX. Resuspend the beads in Laemmli sample buffer and incubate at room temperature for 5 min. Retrieve the beads on a magnet and boil the supernatant before loading on SDS–PAGE.
6. An important control to include in depletion experiments is the add back of the purified recombinant protein to the depleted extract in a range of concentrations close to that of the endogenous protein. In case the protein is degraded during interphase, the recombinant protein should be added only at mitosis reentry.
7. The depleted, the control mock depleted, and add-back extracts can now be used in parallel to assemble asters and spindles following the methods described in Section III,C.
6. Compare the spindles or asters formed in each reaction either by taking samples and squashing them in a slide with fixative solution or by centrifuging them onto coverslips and performing immunofluorescence.

Steps for Specific Antibodies or Dominant-Negative Protein Addition

1. In the case of addition of specific antibodies, prepare a concentrated solution (at least 2 mg/ml) of an affinity-purified polyclonal or monoclonal antibody in PBS or CSF-XB. It is a good idea to try different concentrations of antibody (ranging from 200 to 400 μg/ml). For dominant-negative addition, prepare constructs expressing different fragments of the protein under study fused to GFP or GST. Purify the fusion proteins and dialyze them extensively against CSF-XB.
2. Add the antibodies or dominant-negative protein to the extract in a volume corresponding to one-tenth of the total

volume and follow the protocol for spindle assembly or aster formation as described in Section III,C. The antibodies or dominant negative can be added at the beginning of the spindle assembly procedure or at mitosis reentry.

3. When adding antibodies, two control samples should be run in parallel: one containing a similar amount of a control antibody and another containing the same volume of buffer. In the case of dominant-negative addition, include also the same volume of buffer as control and a sample containing GST or GFP protein.
4. Samples taken at different time points are fixed and analyzed as before. To examine the localization of the antibody, process the samples by immunofluorescence using a fluorochrome-conjugated secondary antibody. Localization of the dominant negative can be observed directly if the tag is GFP or by using an anti-GST antibody.

IV. PITFALLS

The biggest pitfall in studying spindle assembly with this system is the problem of reproducibility in Xenopus egg extracts. To obtain good eggs, the frog colony must be healthy and well cared for. This requires a substantial commitment on the part of the laboratory. Even with healthy frogs, there is seasonal variation in the quality of eggs, with summer being the off season. Furthermore, even experienced extract makers do not manage to prepare functional extracts every time, which can be frustrating. Therefore, experiments must be repeated several times to ensure a valid interpretation.

References

Antonio, C., Ferby, I., Wilhelm, H., Jones, M., Karsenti, E., Nebreda, A. R., and Vernos, I. (2000). Xkid, a chromokinesin required for chromosome alignment on the metaphase plate. *Cell* **102**(4), 425–435.

Boleti, H., Karsenti, E., and Vernos, I. (1996). Xklp2, a novel Xenopus centrosomal kinesin-like protein required for centrosome separation during mitosis. *Cell* **84**, 49–59.

Bornens, M., Paintrand, M., Berges, J., Marty, M. C., and Karsenti, E. (1987). Structural and chemical characterization of isolated centrosomes. *Cell Motil. Cytoskel.* **8**(3), 238–249.

Desai, A., Maddox, P. S., Mitchison, T. J., and Salmon, E. D. (1998). Anaphase A chromosome movement and poleward spindle microtubule flux occur at similar rates in Xenopus extract spindles. *J. Cell Biol.* **141**(3), 703–713.

Desai, A., Murray, A., Mitchison, T. J., and Walczak, C. E. (1999). The use of Xenopus egg extracts to study mitotic spindle assembly and function *in vitro*. *Methods Cell Biol.* **61**, 385–412.

Evans, L., Mitchison, T. J., and Kirschner, M. W. (1985). Influence of the centrosome on the structure of nucleated microtubules. *J. Cell Biol.* **100**, 1185–1191.

Gurdon, J. B. (1976). Injected nuclei in frog oocytes: Fate, enlargement, and chromatin dispersal. *J. Embryol. Exp. Morphol.* **36**, 523–540.

Heald, R., Tournebize, R., Blank, T., Sandaltzopoulos, R., Becker, P., Hyman, A., and Karsenti, E. (1996). Self organization of microtubules into bipolar spindles around artificial chromosomes in Xenopus egg extracts. *Nature* **382**, 420–425.

Hyman, A., Drechsel, D., Kellogg, D., Salser, S., Sawin, K., Steffen, P., Wordeman, L., and Mitchison, T. (1991). Preparation of modified tubulins. *Methods Enzymol.* **196**, 478–485.

Kalab, P., Weis, K., and Heald, R. (2002). Visualization of a Ran-GTP gradient in interphase and mitotic Xenopus egg extracts. *Science* **295**(5564), 2452–2456.

Lohka, M., and Maller, J. (1985). Induction of nuclear envelope breakdown, chromosome condensation, and spindle formation in cell-free extracts. *J. Cell Biol.* **101**, 518–523.

Murray, A. (1991). Cell cycle extracts. *In "Methods in Cell Biology"* (B. K. Kay, H. B. Peng, eds.), vol. 36, pp. 581–605. Academic Press, San Diego.

Murray, A. W., and Kirschner, M. W. (1989). Cyclin synthesis drives the early embryonic cell cycle. *Nature* **339**, 275–280.

Sawin, K. E., and Mitchison, T. J. (1991). Mitotic spindle assembly by two different pathways *in vitro*. *J. Cell Biol.* **112**, 925–940.

Shamu, C. E., and Murray, A. W. (1992). Sister chromatid separation in frog egg extracts requires DNA topoisomerase II activity during anaphase. *J. Cell Biol.* **117**, 921–934.

Wittmann, T., Boleti, H., Antony, C., Karsenti, E., and Vernos, I. (1998). Localization of the kinesin-like protein Xklp2 to spindle poles requires a leucine zipper, a microtubule-associated protein, and dynein. *J. Cell Biol.* **143**(3), 673–685.

CHAPTER

20

In Vitro Motility Assays with Actin

James R. Sellers

OUTLINE

I. INTRODUCTION

The interaction between actin and myosin has been studied for years using a variety of techniques, including ultracentrifugation, light scattering, chemical cross-linking, fluorescence, and measurement of the effect of actin on the MgATPase activity of myosin. The sliding actin *in vitro* motility assay constitutes a relatively recent technique for studying actin–myosin interaction. This assay, developed by Kron and Spudich (1986), takes advantage of the ability to image rhodamine–phalloidin-labeled actin filaments by fluorescence microscopy as they interact with and are translocated by myosin bound to a coverslip surface. The sliding actin *in vitro* motility assay is among the most elegant biochemical assays, reproducing the most fundamental property of a muscle, the ability of myosin to translocate actin using only the two highly purified proteins. It is a close *in vitro* correlate of the maximum unloaded shortening velocity of muscle fibers (Homsher *et al.*, 1992). As shown here, it is simple to set up, reproducible, quantitative, and utilizes as little as 1 μg of myosin per assay. The assay is now used routinely in a large number of laboratories studying myosin and actin biochemistry. Although originally developed for studying the conventional class II myosin, it can be adapted to study unconventional myosins also.

This article discusses the design of the assay, describes the equipment required for its setup, and deals with methods for quantification and presentation of the results. Because different myosins exhibit a range of actin translocation speeds from 0.02 to 60 μm/s (Sellers, 1999), it will be necessary to discuss modification of the experimental setup for fast and slow myosins. Also, differences between low- and high-duty cycle myosins are discussed. We will describe the instrumentation that we use in our system and elaborate on other options where applicable.

II. MATERIALS AND INSTRUMENTATION

The following reagents are from Sigma Chemical Company (www.sigmaaldrich.com): MOPS (Cat. No. M-5162), EGTA (Cat. No. E-4378), ATP (Cat. No. A-5394), glucose (Cat. No. G-7528), glucose oxidase (Cat. No. G-6891), catalase (Cat. No. C-3155), and methylcellulose (Cat. No. M-0512). Rhodamine–phalloidin (Cat. No. R415) is from Molecular Probes, Inc. (www.probes.com). Nitrocellulose (superclean grade, Cat. No. 11180) is from Ernest F. Fullam, Inc. (www.fullam.com). Bovine serum albumin (BSA, Cat. No. 160069) and dithiothreitol (DTT, Cat. No. 856126) are from ICN (www.icnbiomed.com). The following items are from Thomas Scientific (www.thomassci.com): microscope slides (Cat. No. 6684-H30) and microscope 18-mm^2 No. 1 thickness coverslips (Cat. No. 6667-F24). Double sticky cellophane tape (Scotch Brand) and Sony sVHS videotapes (ST120) are from local suppliers.

The following instrumentation is used in our laboratory for the following procedures. Zeiss Axioplan microscope and objectives (www.zeiss.com), Air Therm Heater (www.wpiinc.com), intensified CCD camera from Videoscope, International (www.videoscopeintl.com), TR black-and-white videomonitor and sVHS videotape recorder (Model AG7350) from Panasonic (www.panasonic.com), Argus 10 image processor from Hamamatsu Photonics (www.hamamatsu.com), and VP110 digitizer from Motion Analysis (www.motionanalysis.com).

III. PROCEDURES

A. Construction of Flow Cells

Steps

1. Prepare nitrocellulose-coated coverslips by first placing 3 μl of a 1% solution of nitrocellulose in isoamylacetate directly on a No. 1 thickness 18-mm^2 coverslip and spreading with the broad side of the micropipette tip. Dry the coverslip to create the film and use within 1 day. Some investigators use silicon coating of the coverslips (Fraser and Marston, 1995)
2. Place two 5 × 25-mm strips of double sticky Scotch cellophane tape about 10 mm apart on a 25 × 75-mm glass microscope slide. Place a nitrocellulose-coated coverslip with the coated side down onto the tracks. Press gently to create a tight seal.

B. Preparation of Rhodamine–Phalloidin-Labeled Actin

1. Place 60 μl of 3.3 μ*M* rhodamine–phalloidin (in methanol) into an Eppendorf tube and dry using a Speed-Vac concentrator.
2. Redissolve the rhodamine–phalloidin powder in 3–5 μl of methanol.
3. Add 85 μl of 20 m*M* KCl, 20 m*M* MOPS (pH 7.4), 5 m*M* $MgCl_2$, 0.1 m*M* EGTA, and 10 m*M* DTT (buffer A). To make 100 ml of buffer A, add 1 ml of 2*M* KCl, 1 ml of 2*M* MOPS (pH 7.4), 0.5 ml of 1*M* $MgCl_2$, 0.02 ml of 0.5*M*

EGTA, and 154 mg DTT. Bring to 100 ml with H_2O and adjust pH to 7.4.

4. Add 10 μl of a freshly diluted 20 μM F-actin solution (in buffer A) and incubate for an hour on ice. The rhodamine–phalloidin actin can routinely be used as is at this stage.
5. If a lower background fluorescence is desired, centrifuge for 15 min at 435,000 *g* in a TL-100 ultracentrifuge (Beckman Instruments), remove the supernatant, and gently resuspend the pink rhodamine–phalloidin-labeled actin pellet in buffer A using a pipettor tip that has been cut to widen the bore size.
6. The rhodamine–phalloidin-labeled actin solution is stable for several weeks.

C. Preparation of Sample for Motility Assay

Solutions

1. *Wash solution:* 50 m*M* KCl, 20 m*M* MOPS (pH 7.4), 5 m*M* $MgCl_2$, 0.1 m*M* EGTA, and 5 m*M* DTT; to make 100 ml, add 2.5 ml of 2*M* KCl, 1 ml of 2*M* MOPS, (pH 7.4), 0.5 ml of 1*M* $MgCl_2$, 0.02 ml of 0.5*M* EGTA, and 77 mg of DTT. Bring volume to 100 ml with H_2O and pH to 7.4.
2. *Blocking solution:* 1 mg/ml BSA in 0.5*M* NaCl, 20 m*M* MOPS (pH 7.0), 0.1 m*M* EGTA, and 1 m*M* DTT solution. To make 100 ml of blocking solution, add 10 ml of 5*M* NaCl, 1 ml of 2*M* MOPS (pH 7.4), 0.02 ml of 0.5*M* EGTA, 15.4 mg of DTT, and 100 mg of BSA. Bring to 100 ml with H_2O and pH to 7.0.
3. *Rhodamine–phalloidin-labeled actin solution:* 20 n*M* rhodamine–phalloidin-labeled actin. To make 1 ml, take 10 μl of 2 μ*M* rhodamine–phalloidin-labeled actin, 10 μl of 500 m*M* DTT, and 980 μl of wash solution.
4. *ATP–actin wash:* 1 m*M* ATP and 5 μ*M* F-actin (unlabeled) in wash solution. To make 1 ml, add 10 μl of 0.1*M* ATP and 50 μl of 100 μ*M* F-actin to 940 μl of wash solution.
5. *4X stock solution:* 80 m*M* MOPS (pH 7.4), 20 m*M* $MgCl_2$, and 0.4 m*M* EGTA. To make 100 ml, add 4 ml of 2*M* MOPS (pH 7.4), 2 ml of 1*M* $MgCl_2$, and 0.08 ml of 0.5*M* EGTA. Bring volume to 100 ml with H_2O and pH to 7.4.
6. *1.4% methycellulose solution:* Dissolve 1.4 g of methylcellulose in a final volume of 100 ml of H_2O by stirring overnight. Occasionally it is necessary to homogenize the solution with a glass–Teflon homogenizer to aid in solubilization. Dialyze the dissoved methylcellulose against 4 liters of H_2O overnight. Divide into 10-ml aliquots and freeze at −20°C.
7. *Motility buffer:* 50 m*M* KCl, 20 m*M* MOPS (pH 7.4), 5 m*M* MgCl2, 0.1 m*M* EGTA, 1 m*M* ATP, 50 m*M* DTT, 2.5 mg/ml glucose, 0.1 mg/ml glucose oxidase, and 0.02 mg/ml catalase. To make 1 ml, add 250 μl of 4X stock solution, 10 μl of 0.1*M* ATP, 25 μl of 2*M* KCl, 100 μl of 0.5*M* DTT (prepare fresh each day by adding 77 mg DTT to 1 ml of H_2O), 20 μl of 125 mg/ml glucose, 20 μl of 5 mg/mol glucose oxidase, 1 μl of 20 mg/ml catalase, and 573 μl of H_2O. If methylcellulose is to be used in the motility buffer (see later), then add 500 μl of 1.4% methylcellulose and 73 μl of H_2O.

Steps

1. Apply 0.2 mg/ml myosin in 0.5*M* NaCl, 20 m*M* MOPS (pH 7.0), 0.1 m*M* EGTA, and 1 m*M* DTT to fill the flow chamber. Wait 1 min.
2. Wash with 75 μl of blocking buffer. Wait 1 min.
3. Wash with 75 μl of wash solution, followed by 75 μl of ATP–actin wash solution. Wait 1 min. This step is optional.
4. Wash with 75 μl of wash solution, followed by 75 μl of rhodamine–phalloidin-labeled actin solution. Wait 1 min.
5. Initiate reaction by the addition of 75 μl of motility buffer.
6. Place slide on microscope stage and image.

Comments

In the protocol just described, myosin is bound to the surface as monomers. If myosin is to be bound as filaments, it is necessary to block with BSA that is in a low ionic strength solution, such as the wash solution. Alternatively, heavy meromyosin or a soluble unconventional myosin can be applied to the flow chamber at either low or high ionic strength. In some cases, myosin or HMM can be attached to the surface via specific antibodies against their carboxyl-terminal sequence (Winkelmann *et al.*, 1995; Cuda *et al.*, 1993; Reck-Peterson *et al.*, 2001), which may also serve the purpose of further purifying the desired isoform of myosin. If actin filaments are binding poorly to the surface, the motility buffer can be augmented with 0.7% methlycellulose (modify motility buffer preparation to add 0.5 ml of 1.4% methylcellulose and add less water to bring the final volume to 1 ml. Note that this solution is very viscous and must be mixed well).

Although each myosin has its own characteristic velocity, the velocity of a given myosin can vary with ionic and assay conditions. In general the velocity tends to increase as the ionic strength is raised from 20 to 100 m*M* and increases with temperature. At higher ionic strengths the actin filaments typically begin to become weakly associated with the myosin-coated surface and move erratically. The velocity of actin filament translocation by some myosins, such as vertebrate smooth muscle myosin and *Limulus*-striated muscle myosin, is increased markedly (two to four times) by the inclusion of 200 n*M* tropomyosin in the motility buffer (Wang *et al.*, 1993; Umemoto and Sellers, 1990). However, tropomyosin inhibits the movement of brush border myosin I (Collins *et al.*, 1990).

Step 3 is optional. It is often included to improve the "quality" of movement by binding unlabeled actin to noncycling myosin heads that would otherwise bind and tether the labeled actin added subsequently. This step can be omitted if the myosin is capable of moving actin filaments smoothly. For myosin V, better quality movement can be obtained by first blocking the surface with 0.1–1.0 mg/ml of BSA before adding the myosin.

D. Recording and Quantifying Data Steps and Equipment

A schematic diagram of the equipment setup is shown in Fig. 20.1. The following describes the equipment used in our laboratory. There is a wide selection of video microscopy equipment represented by many manufacturers.

1. Place the microscope slide under a 100X, 1.4 NA Plan-Neofluor objective in an Axioplan microscope (both from Carl Zeiss, Germany) equipped for epifluorescence. Other microscopes, including ones with an inverted format, are also suitable and a variety of objectives can be used, but note that high numerical apertures are required for maximal brightness. Illumination is via a 100-W mercury lamp. An IR filter should be placed between the source and the sample to attenuate heat; neutral density filters are useful to attenuate light intensity if needed. A filter set designed for rhodamine fluorescence

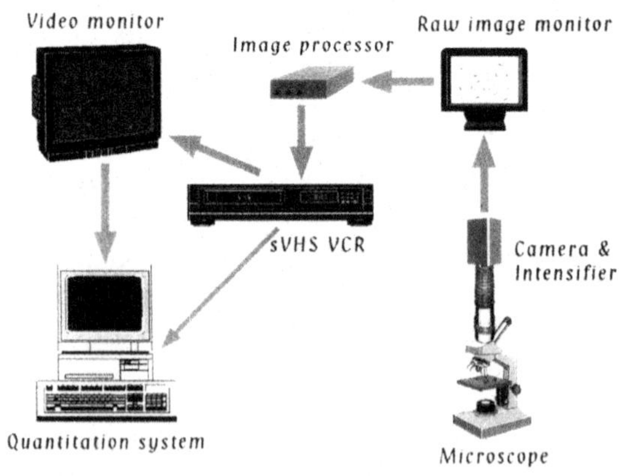

FIGURE 20.1 Schematic diagram of equipment setup.

measurements should be utilized in the filter cube. For quantitative work it is necessary to control the temperature of the assay. This can be accomplished in several ways. The most inexpensive way is to create an air curtain using a hair dryer. Other methods include fabricating a water jacket for the objective and to use a circulating water bath to regulate temperature or creating an environmental box heated with an Airtherm heater (World Precision Instruments). In our experience, commercial stage heaters are not sufficient as oil immersion objectives act as large heat sinks.

2. Image actin filaments using an ICCD 350F intensified CDD camera (Videoscope International). Other low-light systems are possible, such as an SIT camera or intensified SIT camera. Several manufacturers sell this equipment. These camera systems produce an analog output that can be viewed on a standard black-and-white video monitor (TR Panasonic) and recorded on an AG7350 sVHS recorder (Panasonic). It is also possible to collect data digitally using a cooled CCD camera linked to a computer.
3. It is useful to process the raw image using an image processor such as the Argus 10 or Argus 20 (Hamamatsu Photonics) to perform frame averaging and/or background subtraction. There are several commercial software packages that can do this also. Display the processed image on another video monitor.
4. Determine the movement of individual actin filaments using an automated tracking system equipped with a VP110 digitizer from Motion Analysis. Many investigators have created their own software routines to track actin filaments in a semiautomated manner.

Comments

Given the range of motility rates of different myosins, there is no standard number of frames to average in order to get a good image. If the myosin is moving at 5 μm/s, a 2 or 4 frame average is used along with high illumination levels that can be tolerated because of the short exposure time needed to define a filament path. With slow myosins that may move at rates of less than 0.1 μm/s, it is possible to average 64 frames and to reduce the light intensity so that longer recording periods are possible.

The quantification of the rate of actin filament sliding is perhaps the most difficult part of the motility assay. The method just described requires a fairly expensive apparatus that is accurate, very fast, and can give unbiased results (for an extensive discussion of quantification of data, see Homsher *et al.*, 1992). The user inputs the desired sampling rate and sampling time to collect data from either the live image or a prerecorded image. The computer determines the centroid position of each actin filament in each frame, connects the centroids to form paths, calculates the incremental velocity between each successive data point in a path, and, finally, calculates a mean ± SD for each filament path. This process takes only seconds for a field of 25–30 actin filaments. Several investigators use commercial frame grabbers and write their own software for semiautomated tracking of actin filaments (Work and Warshaw, 1992; Marston *et al.*, 1996). There is at least one free downloadable source of semiautomatic tracking software available (http://mc11.mcri.ac.uk/retrac/).

E. Presentation of Data

Figure 20.2 shows three frames taken at 1-s intervals of actin filaments moving over a myosin-coated surface as it would appear on the video monitor. Data from such an experiment are presented most commonly as the mean 6 SD of the velocity of the population of actin filaments. In general, the SD is typically 10–20% of the mean. There are two cases where merely reporting this number does not always accurately

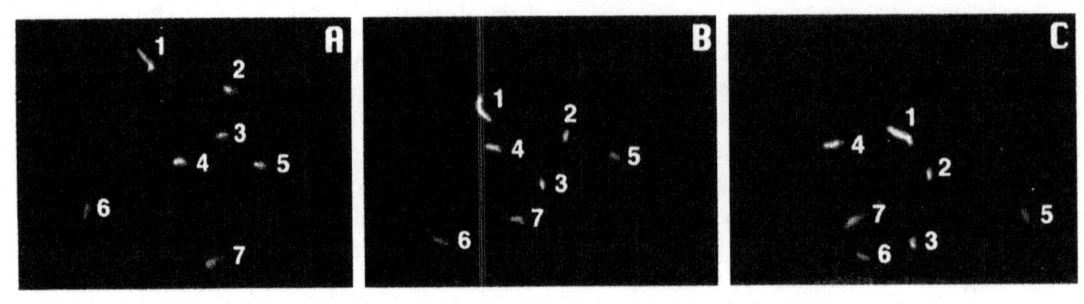

FIGURE 20.2 One-second intervals of actin filaments moving over a myosin-coated surface.

describe what is occurring in the assay. One such case is when something (perhaps a regulatory protein) is affecting the number of filaments that are moving. If, in the absence of the regulatory protein, 95% of the actin filaments are moving at 1 μm/s whereas only 5% of the filaments move at any velocity in the presence of the regulatory protein, reporting only the mean value for the velocity in each case does not reflect the difference that is observed in the assay between the two conditions. A better method for data display for this example is to display all data in the form of a histogram so that one can see that most of the actin filaments are not moving in the presence of the regulatory protein. This display also allows the reader to see whether the regulatory protein affects the speed of movement of the few actin filaments that remain moving. The other case where more complex data display is necessary is if the filaments are moving erratically. Here the mean velocity will underestimate the "instantaneous" velocity and will have a considerably larger standard deviation than that of smoothly moving filaments. One way to display these data graphically is to show a path plot in which the centroid position of the moving actin filaments is plotted in two-dimensional space as a function of time (Fig. 20.3).

IV. PITFALLS

1. Actin filaments are moving erratically or only a fraction of the actin filaments are moving. The cause of this phenomena is usually noncycling heads in the preparation. Using the actin–ATP wash solution described in Section III,C,3 usually helps or eliminates the problem, but if erratic motility persists, do the following. Bring the myosin solution to 0.5*M* in NaCl and add actin to a final concentration of 10 μM, ATP to 2 m*M*, and $MgCl_2$ to 5 m*M*. Immediately sediment at 435,000 *g* for 15 mn in a Beckman TL-100 ultracentrifuge. Remove the supernatant and use it for the motility assay.
2. Actin filaments shear quickly into small dots. Several things can contribute to this phenomenon. Poor-quality myosin containing a significant number of noncycling heads might be a problem. See Pitfall 1 for advice on how to remove these. Decreasing the density of myosin heads on the surface and/or increasing the ionic strength of the assay solution also sometimes helps, as does decreasing the light intensity. In general, myosin bound to the surface as monomers tend to shear less than myosin bound as filaments.
3. Actin filaments appear wobbly when they move or are moving in a back and forward type manner. If the assay does not contain methylcellulose, any portion of the actin filament that is not bound along its length by myosin will experience Brownian motion and appear very wobbly. Even though these filaments may be moving, their movement will be erratic and difficult to quantify.

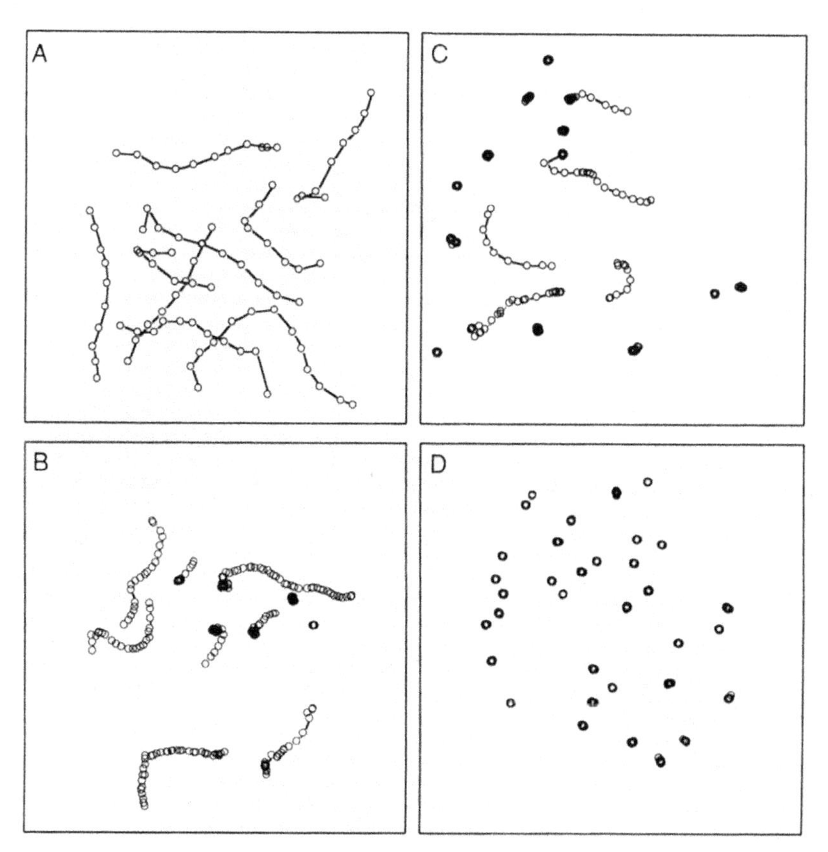

FIGURE 20.3 Actin filaments plotted in two-dimensional space as a function of time. A) smoothly moving actin filaments; B) actin filaments moving more erratically; C) a mixture of moving and nonmoving actin filaments; D) nonmoving actin filaments.

Increasing the density of myosin on the surface, decreasing the ionic strength of the assay, or using methylcellulose in the assay buffer usually helps. The back and forward motion of the actin filaments seen in the presence of methylcellulose is merely Brownian motion in the presence of the viscous solution where the actin filament is restricted to move mostly along its long axis. If the actin filament is not bound it will move back and forward. Increasing the density of myosin or decreasing the ionic strength of the solution should help.

4. Actin filaments photobleach rapidly. Decrease the light intensity if possible and use image processing to do frame averaging to improve the signal-to-noise ratio. Degas the solutions. Make sure the glucose, glucose oxidase, and catalase components of the motility buffer are good. The presence of 50 m*M* DTT also aids in preventing photobleaching.
5. Actin filaments leave comet tail-like images as they move. If you are frame averaging, merely decrease the number of frames averaged. If not, the problem is likely to be encountered when the actin filaments are moving fast and a non-CCD type camera is used. The streaking or persistence in this case is related to the fact that the tube cameras effectively average about four frames in producing their image. The persistence can be attenuated by increasing the light level or by switching to a lower magnification objective.

Acknowledgments

I thank Qian Xu and Takeshi Sakamoto for critical reading of the manuscript.

References

Collins, K., Sellers, J. R., and Matsudaira, P. (1990). Calmodulin dissociation regulates brush border myosin I (110-kD-calmodulin) mechanochemical activity *in vitro*. *J. Cell Biol.* **110**, 1137–1147.

Cuda, G., Fananapazir, L., Zhu, W.-S., Sellers, J. R., and Epstein, N. D. (1993). Skeletal muscle expression and abnormal function of β-myosin in hypertrophic cardiomyopathy. *J. Clin. Invest.* **91**, 2861–2865.

Fraser, I. D. C., and Marston, S. B. (1995). *In vitro* motility analysis of smooth muscle caldesmon control of actin-tropomyosin filament movement. *J. Biol. Chem.* **270**, 19688–19693.

Homsher, E., Wang, F., and Sellers, J. R. (1992). Factors affecting movement of F-actin filaments propelled by skeletal muscle heavy meromyosin. *Am. J. Physiol. Cell Physiol.* **262**, C714–C723.

Kron, S. J., and Spudich, J. A. (1986). Fluorescent actin filaments move on myosin fixed to a glass surface. *Proc. Natl. Acad. Sci. USA* **83**, 6272–6276.

Marston, S. B., Fraser, I. D. C., Bing, W., and Roper, G. (1996). A simple method for automatic tracking of actin filaments in the motility assay. *J. Musc. Res. Cell Motil.* **17**, 497–506.

Reck-Peterson, S. L., Tyska, M. J., Novick, P. J., and Mooseker, M. S. (2001). The yeast class V myosins, Myo2p and Myo4p, are nonprocessive actin- based motors. *J. Cell Biol.* **153**, 1121–1126.

Sellers, J. R. (1999). "Myosins." Oxford Univ. Press, Oxford.

Umemoto, S., and Sellers, J. R. (1990). Characterization of *in vitro* motility assays using smooth muscle and cytoplasmic myosins. *J. Biol. Chem.* **265**, 14864–14869.

Wang, F., Martin, B. M., and Sellers, J. R. (1993). Regulation of actomyosin interactions in *Limulus* muscle proteins. *J. Biol. Chem.* **268**, 3776–3780.

Winkelmann, D. A., Bourdieu, L., Kinose, F., and Libchaber, A. (1995). Motility assays using myosin attached to surfaces through specific binding to monoclonal antibodies. *Biophys. J.* **68**(*Suppl.*), 72S.

Work, S. S., and Warshaw, D. M. (1992). Computer-assisted tracking of actin filament motility. *Anal. Biochem.* **202**, 275–285.

CHAPTER

21

Use of Brain Cytosolic Extracts for Studying Actin-Based Motility of *Listeria monocytogenes*

Antonio S. Sechi

OUTLINE

I. INTRODUCTION

Cell motility is essential for numerous biological events. Unicellular organisms, for instance, use directed movement to find and ingest food. In multicellular organisms, cell motility is required for the morphogenetic movements that accompany embryogenesis, fibroblast migration during wound healing, and the chemotactic movement of immune cells during an immune response. Cell motility is characterised by the formation of cellular extensions that, depending on their morphology and cellular context, are called lamellipodia, ruffles, or filopodia. In recent decades many cell biological, biochemical, and biophysical studies have established that the formation of these structures depends on the activity of the actin cytoskeleton and its associated proteins. More specifically, the assembly of a network of actin filaments at the leading edge of motile cells provides the propulsive force for the extension of these structures (see Small *et al.*, 2002). Despite much effort, however, the complexity of cell motility has precluded the detailed analysis of the molecular mechanisms and components that govern this process.

Since the mid-1980s, much work focused on the intracellular actin-based motility of the Gram-positive bacterium *Listeria monocytogenes*. *Listeria* can induce its own uptake by phagocytic and nonphagocytic cells and, once free in the cytoplasm, recruits host cell cytoskeletal components, which are then rearranged into phase-dense actin tails. The assembly of actin monomers at the actin filament (+) ends abutting the bacterial surface provides the propulsive force that allows *Listeria* to move within the infected cells and spread to adjacent cells while avoiding exposure to the host's humoral immune system. As these bacteria imitate the protrusive behavior of lamellipodial edges, *Listeria* motility is considered a simplified model system for actin filament dynamics during cell motility (Cossart and Bierne, 2001; Frischknecht and Way, 2001). As one approach towards defining the molecular basis of bacterial motility, we and others have developed simple *in vitro* systems that support actin-based *Listeria* motility based on *Xenopus*, platelets, and mouse brain extracts (Theriot *et al.*, 1994; Marchand *et al.*, 1995; Laurent and Carlier, 1998; Laurent *et al.*, 1999; May *et al.*, 1999). These cell-free systems in combination with bacterial genetics and cell biological studies have been essential for the characterisation of two key regulators of actin cytoskeleton dynamics: Ena/VASP proteins and the Arp2/3 complex (Pistor *et al.*, 1995, 2000; Smith *et al.*, 1996; Niebuhr *et al.*, 1997; May *et al.*, 1999; Skoble *et al.*, 2000, 2001; Geese *et al.*, 2002). They also provided the basis for further development of *in vitro* motility systems that culminated in the reconstruction of bacterial motility using a limited set of purified proteins (Loisel *et al.*, 1999). This article describes procedures for the preparation and use of mouse brain extracts for studying *Listeria* motility.

II. MATERIALS AND INSTRUMENTATION

Calcium chloride (Cat. No. 102378), magnesium chloride hexahydrate (Cat. No. 105832), potassium chloride (Cat. No. 4936), HEPES (Cat. No. 10110), sucrose (Cat. No. 1.07654), sodium chloride (Cat. No. 1.06404), disodium hydrogen phosphate dihydrate (Cat. No. 1.06580), sodium dihydrogen phosphate hydrate (Cat. No. 6346), and paraffin (Cat. No. 1.07160) are from Merck. EGTA (Cat. No. E-3889), methylcellulose (Cat. No. M-0555), ATP (Cat. No. A-2383), erythromycin (Cat. No. E-6376), and lanoline (Cat. No. L-7387) are from Sigma. Chymostatin (Cat. No. 17158), leupeptin (Cat. No. 51867), pepstatin (Cat. No. 52682), PEFABLOCK (Cat. No. 31682), aprotinin (Cat. No. 13178), and creatine kinase (Cat. No. 127566) are from Boehringer-Ingelheim. Dithiothreitol (DTT, Cat. No. 43815), Tris–HCl (Cat. No. 93363), and glycerol (Cat. No. 49770) are from Fluka. Cell-Tak (Cat. No. 354241) is from BD Biosciences. Creatine phosphate (Cat. No. 621714) is from Roche. Vaseline (Cat. No. 16415) is from Riedel-de Haen. Brain heart infusion (BHI) culture medium and Bacto agar are from Difco Laboratories. Rhodamine-labeled actin (Cat. No. AR05) is from Cytoskeleton. A glass homogeniser equipped with a Teflon piston, glass slides (76 × 26 mm), glass coverslips (22 × 22 mm), forceps, scissors, and razor blades are from local suppliers. Bacterial motility is observed using an Axiovert 135TV microscope (Zeiss) equipped with a Plan-Apochromat 100×/1.4NA oil immersion objective. Images can be acquired with a cooled, back-illuminated CCD camera (TE/CCD-1000 TKB; Roper Scientific) driven by IPLab Spectrum software (Scanalitics).

III. PROCEDURES

A. Explantation of Mouse Brains

Solution

Phosphate-buffered saline (PBS): 130 m*M* NaCl, 1.5 m*M* $NaH_2PO_4 \cdot H_2O$, and 4 m*M* $Na_2HPO_4 \cdot 2H_2O$, pH 7.4. For 1 litre, weigh out 7.65 g NaCl, 0.21 g $NaH_2PO_4 \cdot H_2O$, and 0.72 g $Na_2HPO_4 \cdot 2H_2O$. Dissolve in 900 ml H_2O, adjust pH to 7.4 with 1*N* NaOH, and bring volume to 1 litre. Store at 4°C.

Steps

1. Kill mice using CO_2 or by cervical dislocation.
2. Remove skin and fur from the head using thin-tipped scissors and discard.
3. Cut the skull bone by sliding the scissors along the sagittal suture (Fig. 21.1a, dashed red line).
4. Cut the frontal, parietal, and interparietal bones along their sutures and remove them (Fig. 21.1a, dashed line).
5. Gently pinch the surface of the brain to lift the meninges up and gently ease the brain out of the skull.
6. Put explanted brains in ice-cold PBS.
7. Wash brains three times in ice-cold PBS to remove tissue debris and blood residues.
8. Proceed to Section III,B or flash freeze the brains in liquid nitrogen. Store frozen brains at −80°C.

B. Preparation of Cytosolic Extracts

Solutions

1. *Homogenisation buffer (HB)*: 20 m*M* HEPES, pH 7.5, 100 m*M* KCl, 1 m*M* $MgCl_2 \cdot H_2O$, 1 m*M* EGTA, and 0.2 m*M* $CaCl_2$. For 1 litre, weigh out 4.76 g HEPES, 7.45 g KCl, 0.2 g $MgCl_2 \cdot H_2O$, 0.38 g EGTA, and 0.02 g $MgCl_2$. Dissolve in 900 ml H_2O, adjust pH to 7.5 with 1*N* NaOH, and bring volume to 1 litre. Store at 4°C.
2. *Protease inhibitors*: 20 mg/ml chymostatin, 1 mg/ml leupeptin, 1 mg/ml pepstatin, 167 m*M* Pefabloc, and 10 mg/ml aprotinin.

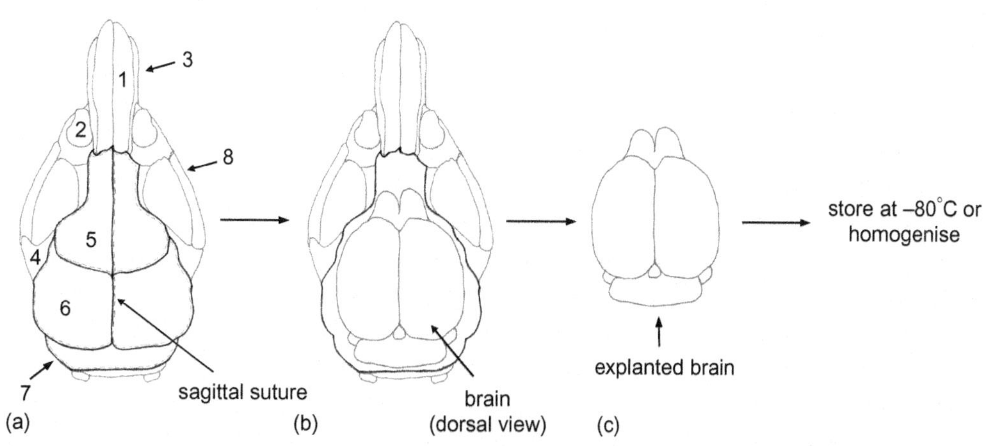

FIGURE 21.1 Diagram of the explantation of mouse brains. (a) Dorsal view of mouse skull showing the nasal bone (1), eye socket (2), nasal process of incisive bone (3), zygomatic process (4), frontal bone (5), parietal bone (6), interparietal bone (7), and zygomatic bone (8). (b) Dorsal view of skull after cutting (along the dashed green and red lines) and removing the frontal, parietal, and interparietal bones. (c) Dorsal view of explanted brain.

To prepare stock solutions, dissolve 1 mg chymostatin in 50 μl dimethyl sulfoxide, 0.5 mg leupeptin in 500 μl H_2O, 0.5 mg pepstatin in 500 μl methanol, 20 mg Pefabloc in 500 μl H_2O, and 0.5 mg aprotinin in 50 μl H_2O. Aliquot and store at −20°C.

3. *0.1 M ATP*: For 50 ml, weigh out 2.75 g ATP. Dissolve in H_2O, aliquot, and store at −20°C.
4. *0.1 M DTT*: For 50 ml, weigh out 0.77 g DTT. Dissolve in H_2O, aliquot, and store at −20°C.
5. *2 M sucrose*: For 20 ml, weigh out 13.69 g sucrose. Dissolve in H_2O, aliquot, and store at −20°C.
6. *Homogenisation buffer supplemented with protease inhibitors (HBI)*: HB containing 60 μg/ml chymostatin, 5 μg/ml leupeptin, 10 μg/ml pepstatin, 4 m*M* Pefabloc, 2 μg/ml aprotinin, 0.5 m*M* ATP, and 1 m*M* DTT. Shortly before use add to 20 ml of HB 60 μl of 20 mg/ml chymostatin, 100 μl of 1 mg/ml leupeptin, 200 μl of 1 mg/ml pepstatin, 476 μl of 167 m*M* Pefabloc, 4 μl of 10 mg/ml aprotinin, 200 μl of 0.1 *M* DTT, and 50 μl of 0.1 *M* ATP.

Steps

1. After the last wash in PBS (see Section III,A, step 7) remove PBS and weigh the brains.
2. Cut the brains into small pieces using a razor blade (keep brains on ice).
3. Add 0.75 ml of HBI per gram of wet tissue (keep brain suspension on ice).
4. Transfer brain suspension into a glass homogeniser on ice.
5. Grind brain tissue for 20 passages of the pestle on ice.
6. Centrifuge crude extract at 15,000 *g* for 1 h at 4°C.
7. Recover clarified supernatant (cytosolic brain extract) and supplement it with 150 m*M* sucrose, 50 mg/ml creatine kinase, 30 m*M* creatine phosphate, and 0.5 m*M* ATP.
8. Aliquot and flash freeze in liquid nitrogen. Store frozen aliquots at −80°C.

C. Preparation of Bacteria

Solutions

1. *Brain heart infusion (BHI) broth*: Prepare liquid medium according to the manufacturer's instruction, autoclave, filter, and store at 4°C. For agar plates, add Bacto-agar (15 g/liter of BHI broth), autoclave, and pour 30 ml in a 10-cm petri dish. Store plates at 4°C.
2. *Erythromycin stock solution*: Dissolve 50 mg of erythromycin in 10 ml of pure ethanol. Store at 4°C.

Bacterial Culture

1. Streak the bacteria onto BHI agar plates. Incubate at 37°C for 24 h.
2. Put 5 ml of BHI (supplemented with 50 μg/ml erythromycin) in a 15-ml sterile Falcon tube. Scrape a few colonies off the BHI plate using a sterile pipette tip or a flamed bacteriological loop. Inoculate the broth and grow bacteria overnight at 37°C with vigorous shaking.
3. Transfer bacterial culture to a centrifuge tube and pellet the bacteria at 10,000 *g* for 3 min.
4. Wash bacterial pellet three times in homogenisation buffer. After the final washing step, resuspend pellet in a final volume of homogenisation buffer corresponding to the initial volume of bacterial culture.
5. Alternatively, supplement the overnight culture with 20% glycerol, aliquot, and store at −80°C.

D. Listeria Motility Assay

Solutions

1. *2% methycellulose in homogenisation buffer (stock solution)*: Heat 100 ml of HB to 60°C and then add 2 g of methylcellulose. Stir vigorously until the methycellulose dissolves. Cool down and store at room temperature.

2. *0.5% methycellulose in homogenisation buffer (working solution)*: To make 10 ml, mix 7.5 ml of HB with 2.5 ml of 2% methycellulose. Store at 4°C.
3. *VALAP*: Mix vaseline, lanoline, and paraffin in a 1:1:1 ratio (w/w/w) and homogenise at 75°C. Store at room temperature.
4. *G buffer*: 5 m*M* Tris–HCl, pH 7.6, 0.5 m*M* ATP, 0.1 m*M* $CaCl_2$, and 0.5 m*M* DTT. For 1 litre, weigh out 0.8 g Tris–HCl, 0.01 g $CaCl_2$, and then add 5 μl of 0.1 *M* ATP and 5 μl of 0.1 *M* DTT. Dissolve in 900 ml H_2O, adjust pH to 7.6 with 1*N* NaOH, and bring volume to 1 litre. Store at 4°C.
5. *Rhodamine-labeled actin*: Add 6 μl of G buffer to one aliquot of rhodamine-labeled actin. Mix gently and store on ice.

Steps

1. Wash bacteria three times in homogenisation buffer. Resuspend pellet in 20 μl homogenisation buffer.
2. In a small Eppendorf tube, mix 4 μl brain extract, 4 μl of 0.5% methycellulose, 0.5 μl bacteria suspension, and 0.2 μl rhodamine-labeled actin. Mix by pipetting up and down gently. Do not vortex. Incubate mixture at room temperature for 10 min.
3. Remove 1.7 μl of motility mixture and spot it onto a glass slide. Gently place a 22 × 22-mm glass coverslip over the drop and press down until the drop spreads to the edges. Seal the coverslip edges with VALAP.
4. Observe slide with an upright or inverted microscope. Actin comet tails can be observed easily by phase contrast or epifluorescence using a Plan-Apochromat 100×/1.4 NA oil immersion objective (Figs 21.2 and 21.3).

IV. COMMENTS AND PITFALLS

Although cell-free systems based on egg extracts of *Xenopus laevis* or human platelet extracts provide excellent *in vitro* systems for supporting actin-based bacterial motility, their ability to do so can be affected negatively by various factors, such as health of eggs or quality and age of platelet preparations. In this context, mouse brain extracts offer a more "robust" *in vitro* system that does not seem to be influenced by external factors such as mouse strain and age.

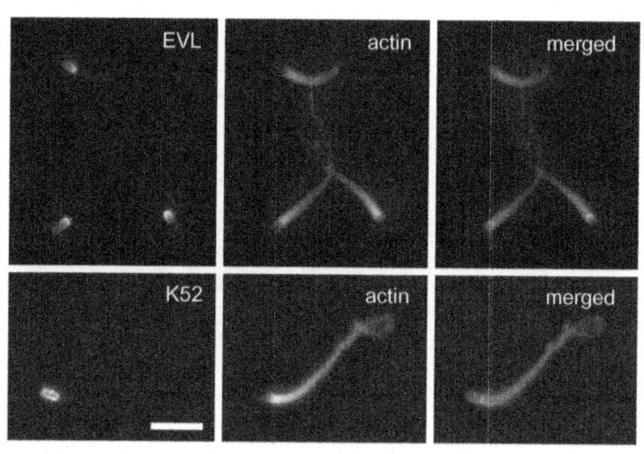

FIGURE 21.2 Immunofluorescence microscopy showing *Listeria* actin tails formed in mouse brain extracts. Bacteria were incubated in mouse brain extract for 30 min at room temperature and then 5 μl of motility mixture was applied onto Cell Tak-coated cover slips (the coating procedure was done according to the manufacturer's instructions) and incubated for 5 min on ice. Afterwards, bacteria were fixed with 4% PFA for 20 min at room temperature. Immunolabeling was done according to Geese *et al.* (2002) using the affinity-purified polyclonal antibody K52 to label bacterial surface, the monoclonal antibody 84H1 to label EVL, and Texas red-labeled phalloidin to label the actin tails. Primary antibodies were detected using Alexa 488-conjugated secondary antibodies. Scale bar: 5 μm.

The reconstitution of *Listeria* motility using mouse brain extracts can have a variety of uses. For instance, it can be used to study the role of actin cytoskeletal components involved in *Listeria* motility by interfering with their function using specific inhibitors or antibodies, as described in May *et al.* (1999). Moreover, the procedure described here may be further developed and adapted to obtain cell-free extracts from normal cultured cells or cells that lack or express mutated versions of the actin cytoskeletal proteins

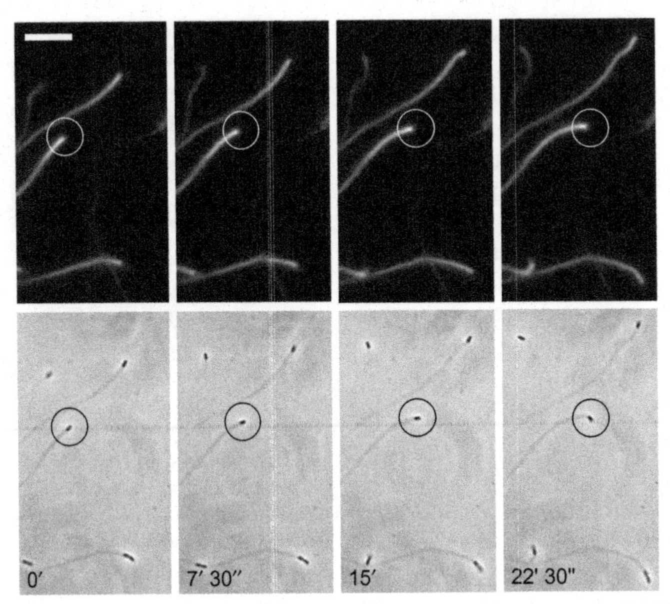

FIGURE 21.3 Dynamics of *Listeria* motility in mouse brain extracts. Bacteria were mixed with 4 μl brain extract, 4 μl of 0.5% methycellulose, and 0.2 μl rhodamine-labeled actin and were incubated at room temperature for 10 min. Thereafter, 1.7 μl of motility mixture was spotted onto a glass slide and a 22 × 22-mm glass coverslip overlaid on it. Bacterial motility was observed using a Axiovert 135 TV inverted microscopy equipped with phase-contrast and epifluorescence optics using a Plan-Apochromat 100×/1.4NA oil immersion objective. Images were acquired using a cooled, back-illuminated CCD camera (TE/CCD-1000 TKB; Roper Scientific) driven by IPLab Spectrum software (Scanalitics). (Top) Rhodamine-labeled actin and (bottom) the corresponding phase-contrast images. Scale bar: 10 μm.

of interest, thus widening the spectrum of *in vitro* systems available for studying bacterial motility.

Three main parameters have to be considered to achieve optimal *Listeria* motility with mouse brain extracts. First, the protein ActA must be expressed at high levels on the bacterial surface. As most wild-type strains express low levels of this protein under standard culture conditions, a *Listeria* strain (see Lingnau *et al.*, 1996) that constitutively expresses high levels of this protein must be used in this assay. The second critical parameter is the total protein concentration of the brain extract, which must be at least 10 mg/ml. The total protein concentration can be increased by reducing the amount of homogenisation buffer per gram of wet tissue. Moreover, mouse brain extracts should not be diluted more than fourfold as further dilution leads to loss of activity in the motility assay. Finally, the high amount of actin present in these extracts induces its spontaneous polymerisation, characterised by the formation of short actin bundles. As this actin network may affect bacterial motility, mouse brain extracts must be kept on ice to reduce the tendency of actin to polymerise.

Acknowledgments

I thank David A. Monner and Jürgen Wehland (GBF, Department of Cell Biology) for helpful discussions and valuable support.

References

Cossart, P., and Bierne, H. (2001). The use of host cell machinery in the pathogenesis of *Listeria monocytogenes*. *Curr. Opin. Immunol.* **13**, 96–103.

Frischknecht, F., and Way, M. (2001). Surfing pathogens and the lessons learned for actin polymerization. *Trends Cell Biol.* **11**, 30–38.

Geese, M., Loureiro, J. J., Bear, J. E., Wehland, J., Gertler, F. B., and Sechi, A. S. (2002). Contribution of Ena/VASP proteins to intracellular motility of *Listeria* requires phosphorylation and proline-rich core but not F-actin binding or multimerization. *Mol. Biol. Cell* **13**, 2383–2396.

Laurent, V., and Carlier, M. F. (1998). Use of platelet extracts for actin-based motility of *Listeria monocytogenes*. *In* "Cell Biology: A Laboratory Handbook" (J. Celis, ed.), pp. 359–365. Academic Press, New York.

Laurent, V., Loisel, T. P., Harbeck, B., Wehman, A., Grobe, L., Jockusch, B. M., Wehland, J., Gertler, F. B., and Carlier, M. F. (1999). Role of proteins of the Ena/VASP family in actin-based motility of *Listeria monocytogenes*. *J. Cell Biol.* **144**, 1245–1258.

Lingnau, A., Chakraborty, T., Niebuhr, K., Domann, E., and Wehland, J. (1996). Identification and purification of novel internalin-related proteins in *Listeria monocytogenes* and *Listeria ivanovii*. *Infect. Immun.* **64**, 1002–1006.

Loisel, T. P., Boujemaa, R., Pantaloni, D., and Carlier, M. F. (1999). Reconstitution of actin-based motility of *Listeria* and *Shigella* using pure proteins. *Nature* **401**, 613–616.

Marchand, J. B., Moreau, P., Paoletti, A., Cossart, P., Carlier, M. F., and Pantaloni, D. (1995). Actin-based movement of *Listeria monocytogenes:* Actin assembly results from the local maintenance of uncapped filament barbed ends at the bacterium surface. *J. Cell Biol.* **130**, 331–343.

Niebuhr, K., Ebel, F., Frank, R., Reinhard, M., Domann, E., Carl, U. D., Walter, U., Gertler, F. B., Wehland, J., and Chakraborty, T. (1997). A novel proline-rich motif present in ActA of *Listeria monocytogenes* and cytoskeletal proteins is the ligand for the EVH1 domain, a protein module present in the Ena/VASP family. *EMBO J.* **16**, 5433–5444.

Pistor, S., Chakraborty, T., Walter, U., and Wehland, J. (1995). The bacterial actin nucleator protein ActA of Listeria monocytogenes contains multiple binding sites for host microfilament proteins. *Curr. Biol.* **5**, 517–525.

Pistor, S., Grobe, L., Sechi, A. S., Domann, E., Gerstel, B., Machesky, L.M., Chakraborty, T., and Wehland, J. (2000). Mutations of arginine residues within the 146-KKRRK-150 motif of the ActA protein of *Listeria monocytogenes* abolish intracellular motility by interfering with the recruitment of the Arp2/3 complex. *J. Cell Sci.* **113**, 3277–3287.

Skoble, J., Auerbuch, V., Goley, E. D., Welch, M. D., and Portnoy, D. A. (2001). Pivotal role of VASP in Arp2/3 complex-mediated actin nucleation, actin branch-formation, and Listeria monocytogenes motility. *J. Cell Biol.* **155**, 89–100.

Skoble, J., Portnoy, D. A., and Welch, M. D. (2000). Three regions within ActA promote Arp2/3 complex-mediated actin nucleation and *Listeria monocytogenes* motility. *J. Cell Biol.* **150**, 527–538.

Small, J. V., Stradal, T., Vignal, E., and Rottner, K. (2002). The lamellipodium: Where motility begins. *Trends Cell Biol.* **12**, 112–120.

Smith, G. A., Theriot, J. A., and Portnoy, D. A. (1996). The tandem repeat domain in the *Listeria monocytogenes* ActA protein controls the rate of actin-based motility, the percentage of moving bacteria, and the localization of vasodilator-stimulated phosphoprotein and profilin. *J. Cell Biol.* **135**, 647–660.

Theriot, J. A., Rosenblatt, J., Portnoy, D. A., Goldschmidt-Clermont, P. J., and Mitchison, T. J. (1994). Involvement of profilin in the actin-based motility of *L. monocytogenes* in cells and in cell-free extracts. *Cell* **76**, 505–517.

CHAPTER

22

Pedestal Formation by Pathogenic *Escherichia coli*: A Model System for Studying Signal Transduction to the Actin Cytoskeleton

Silvia Lommel, Stefanie Benesch, Manfred Rohde, and Jürgen Wehland

OUTLINE

I. INTRODUCTION

Dynamic rearrangement of the actin cytoskeleton in response to signal transduction plays a fundamental role in the regulation of cellular functions. Understanding actin dynamics therefore represents one of the challenges of modern cell biology. Several pathogens have evolved diverse strategies to trigger rearrangement of the host actin cytoskeleton to facilitate and enhance their infection processes. By manipulating the actin assembly machinery or signaling routes leading to its activation, pathogens can either block or induce phagocytosis, drive intracellular motility, and exploit their host cells in other ways. Analyses of host pathogen interactions have not only broadened our knowledge of how these pathogens cause disease, but have also emerged as model systems in the study of cellular actin dynamics (for examples, see Frischknecht and Way, 2001). The focus of this article is on the formation of cytoskeletal rearrangements called actin pedestals induced by the diarrhoeagenic extracellular bacterial pathogens enteropathogenic *Escherichia coli* (EPEC) and enterohaemorrhagic *E. coli* (EHEC) as systems to study signaling and actin assembly at the plasma membrane.

During infection of the intestinal mucosa, EPEC and EHEC induce specific, so-called attaching and effacing (A/E) lesions on intestinal epithelial cells (for review, see Nataro and Kaper, 1998). A/E lesions are characterized by a localized loss of microvilli and intimate adherence of bacteria to the cell surface followed by recruitment of the cellular actin assembly machinery to sites of bacterial attachment, resulting in the formation of actin-rich pseudopod-like structures termed pedestals to which the bacteria intimately adhere.

Importantly, the histopathological changes associated with the A/E phenotype *in vivo* can be mimicked in cell culture [(Knutton *et al.*, 1987), see Fig. 22.1], which allows to define the molecular mechanisms employed by EPEC and EHEC to induce cytoskeletal rearrangements. The ability to form actin pedestals on cultured cells furthermore correlates with the ability of EPEC and EHEC to colonize the intestine and cause disease in human and other animal hosts (e.g., Donnenberg *et al.*, 1993).

The genes necessary for A/E lesion formation in EPEC map to an about 35-kb chromosomal pathogenicity island, designated the locus of enterocyte effacement (LEE), which is

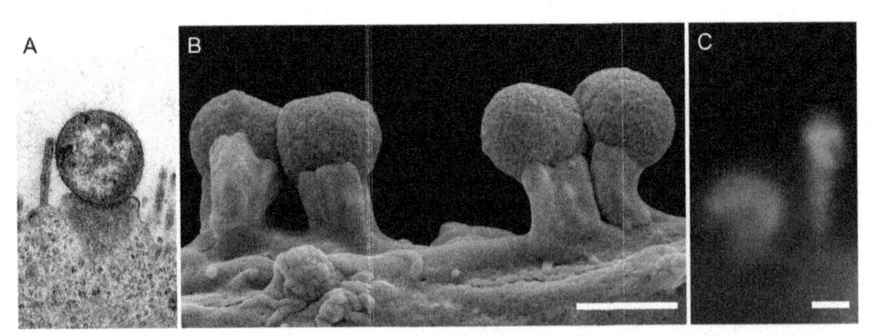

FIGURE 22.1 Actin pedestal formation induced by EHEC and EPEC *in vivo* and on the surface of cultured cells. (A) Transmission electron micrograph showing the characteristic attaching and effacing (A/E) lesion formation of the enterohaemorrhagic *Escherichia coli* (EHEC) O157:H7 strain 86/24 observed in piglet colon. Note the intimate attachment, localized loss of microvilli, and formation of a raised, pedestal-like structure beneath the bacterium that characterizes this lesion (courtesy of Florian Gunzer, Institute of Medical Microbiology, Hannover Medical School, Germany). **(B)** Scanning electron micrograph of enteropathogenic *E. coli* (EPEC) O127:H6 strain E2348/69 sitting on top of pedestals induced on the surface of cultured murine embryonic fibroblast cells upon infection that resemble A/E lesions formed by EPEC *in vivo*. **(C)** EHEC O157:H7 strain 86/24-induced actin pedestal formation as visualized by fluorescence actin staining using AlexaFluor 594 phalloidin to specifically label F-actin. EHEC bacteria were detected with monoclonal antibodies against EspE (Deibel *et al.*, 1998) and AlexaFluor 488-conjugated goat antimouse secondary antibodies. Bars: 1 μm.

highly conserved in EHEC. Although EPEC and EHEC produce highly similar lesions, EPEC in the small intestine and EHEC in the large intestine, the molecular mechanisms of pedestal formation employed by EPEC versus EHEC differ (for review, see Campellone and Leong, 2003).

Both pathogens translocate their own receptor, the translocated intimin receptor (Tir, EspE), which binds to the bacterial surface protein intimin, via a type III secretion system into the underlying host cell. EPEC Tir becomes tyrosine phosphorylated upon insertion, into the host cell membrane and binding to intimin. This phosphorylation is critical for actin pedestal formation by EPEC (Kenny, 1999), as it allows recruitment of the cellular signaling adaptor protein Nck to bacterial attachment sites. Nck, in turn, triggers by a so far unknown mechanism recruitment and activation of N-WASP. Both Nck and N-WASP are essential host proteins for pedestal formation induced by EPEC, as cells lacking either Nck or N-WASP are resistant to actin pedestal formation by EPEC (Gruenheid *et al.*, 2001; Lommel *et al.*, 2001) (for N-WASP, see Fig. 22.2). N-WASP is a member of the WASP/Scar family of cellular nucleation-promoting factors and has emerged as a central node protein that regulates actin polymerization by activating the Arp2/3 complex, a main factor for nucleation of actin filaments in response to multiple upstream signals.

In contrast, EHEC Tir is not tyrosine phosphorylated and pedestals are formed independently of Nck. Despite that, EHEC-induced pedestal formation still depends on N-WASP function (Lommel *et al.*, 2004) to promote Arp2/3 complex-mediated actin polymerization. Thus, EPEC and EHEC have evolved different strategies to trigger cellular signaling routes leading to actin assembly, which converge in recruitment and activation of N-WASP to promote Arp2/3-mediated actin polymerization.

The aim of this article is to outline a methodology for the analysis of actin pedestal formation by EPEC and EHEC using cultured cells as

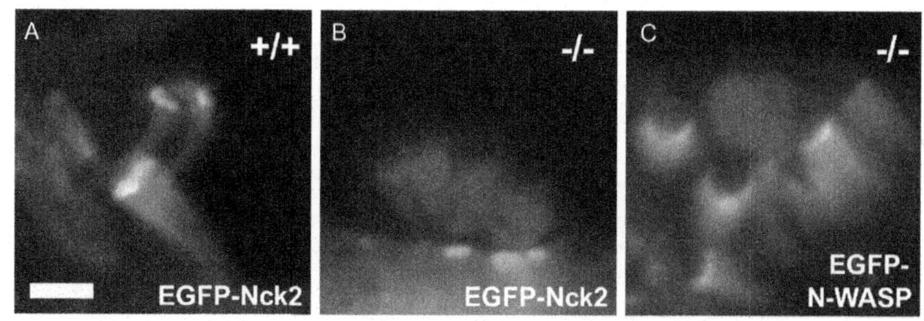

FIGURE 22.2 **EGFP-tagging and knockout cell lines as tools for analysis of the molecular mechanism of actin pedestal formation induced by EPEC.** Cultured murine embryonic fibroblasts expressing (+/+, **A**) or lacking N-WASP (−/−, **B** and **C**) were infected with EPEC and examined by immunofluorescence microscopy. F-actin is shown as detected by fluorescence actin staining with AlexaFluor 594 phalloidin, bacteria are shown as detected with anti-EspE monoclonal antibodies in combination with AlexaFluor 350-conjugated goat antimouse secondary antibodies, and the host proteins Nck2 (**A** and **B**) and N-WASP (**C**) expressed ectopically with an EGFP tag are shown. Whereas EPEC induced the formation of prominent pedestals in N-WASP-expressing cells (+/+, **A**), they were unable to trigger actin accumulation on N-WASP-defective cells (−/−, **B**). The bacterial ability to direct actin reorganization in N-WASP-defective fibroblasts was restored by providing EGFP-tagged N-WASP by transient transfection (**C**). Together, this clearly demonstrates that N-WASP is a host cell protein essential for pedestal formation induced by EPEC. Recruitment of the host cell signaling adaptor protein Nck2 is triggered by EPEC independently of actin accumulation, as was revealed by ectopic expression of EGFP-tagged Nck2 in either N-WASP-expressing (+/+, **A**) or N-WASP-defective cells (−/−, **B**). Bar: 2 μm, (valid for A–C).

model systems to study regulatory mechanisms controlling actin assembly at the plasma membrane. This article describes protocols for the coculture of bacteria with cell lines and for basic immunofluorescence microscopy techniques that are used to examine bacterial–host cell interactions in terms of bacterial attachment and effects on the actin cytoskeleton. These may be combined with ectopical expression of host cell proteins tagged with green fluorescent protein (GFP) for analysis of subcellular localization and dynamic reorganisation of a protein during the infection process using digital fluorescence microscopy (for examples, see Fig. 22.2). A further advance to unravel the molecular mechanism of pedestal formation is the use of cell lines derived from knockout mice. Cell lines derived from such mice can be reconstituted by expressing wild-type or mutated proteins tagged with GFP, which facilitates the analysis of the contribution of specific protein domains to actin pedestal formation (for examples, see Fig. 22.2).

II. MATERIALS AND INSTRUMENTATION

A Cell Lines, Bacterial Pathogen Strains, and Culture Media Reagents

1. *Cells*

Epithelial cell lines used commonly for EPEC and EHEC infection experiments are HeLa and HEp-2 cells, as well as the intestinal epithelial cell lines T84 or Caco-2 (available from American Type Culture Collection). EPEC and EHEC will also adhere and induce the formation of actin pedestals on mouse embryonic fibroblast cell lines (MEFs), which are used routinely in our laboratory. Embryonic fibroblast cell lines offer the advantage that such cell lines can be established quite easily from conditional or conventional knockout mice, thus allowing the analysis of specific host proteins in actin pedestal formation induced by EPEC and EHEC.

2. *Bacterial Pathogen Strains*

Prototype enteropathogenic and enterohaemorrhagic *E. coli* strains used in analyses of the molecular mechanism of actin pedestal formation are enteropathogenic *E. coli* strain E2348/69 (O127:H6) (Levine *et al.*, 1978), enterohaemorrhagic *E. coli* strain 86/24 (O157:H7) [isolated from an outbreak in Walla Walla, WA., U.S.A. (Griffin *et al.*, 1988)], and enterohaemorrhagic *E. coli* strain EDL933 (O157:H7) (American Type Culture Collection #700927; for genome sequence information, see Perna *et al.*, 2001).

3. *Cell and Bacterial Culture Reagents*

Dulbecco's modified eagle medium (DMEM), low glucose (Invitrogen Corp., GIBCO #31885-023), fetal bovine serum (Sigma-Aldrich #F 7524), l-glutamine (Invitrogen Corp., GIBCO #25030-024), penicillin/streptomycin (Invitrogen Corp., GIBCO #15070-063), Luria-Bertani (LB) broth agar (e.g., BD #244520), LB broth (e.g., BD #244620), and HEPES (Sigma-Aldrich #H 3375), fibronectin (pure) (Roche #1051407).

B. Constructs and Reagents for Expression of GFP-Tagged Host Proteins

1. *Transfection reagent*, e.g., FuGENE 6 (Roche #1 814 443).
2. A set of vectors for construction of EGFP fusions is available from Clontech (BD Biosciences).
3. *EGFP-N-WASP*: as described (Lommel *et al.*, 2001).
4. *EGFP-Nck2*: as described (Scaplehorn *et al.*, 2002).

C. Additionally Needed Reagents and Plasticware

1. *Basics*

General equipment and plasticware for molecular biology and cell culture techniques.

2. *Cell Culture and Immunofluorescence Microscopy*

Twenty-four-well cell culture plates (e.g., Corning Inc. #3524), 12-mm round glass coverslips (e.g., Assistent #1001, thickness 0.17mm), absolute ethanol (Sigma-Aldrich #E 7023), HCl (37%) (Sigma-Aldrich #H 7020), lint-free absorptive paper [e.g., GB002, Schleicher&Schuell #10427736 (58 × 68cm) and #10485285 (22.2 × 22.2cm)], large square plastic dish (e.g., 24.5 × 24.5-cm polystyrene dish with lid, Sigma-Aldrich #Z37,165-3), Parafilm M (e.g., Fisher Scientific #917 00 02), forceps with curved fine tips (e.g., coverslip forceps Dumont #11251-33), NaCl (Sigma-Aldrich #S 7653), KCl (Sigma-Aldrich #P 1338), Na_2HPO_4 (Sigma-Aldrich #S 7907), KH_2PO_4 (Sigma-Aldrich #P 0662), paraformaldehyde (PFA), (Sigma-Aldrich #P 6148), NaOH (Merck/VWR International #109913), Triton X-100 (Sigma-Aldrich #T 8532), normal goat serum (Invitrogen Inc.: GIBCO #16210-064), bovine serum albumin (BSA, Sigma-Aldrich #A 2153), goat antimouse Alexa Fluor 350- or Alexa Fluor 488-conjugated secondary antibodies (Molecular Probes #A-11045 and #A-11001), DAPI (e.g., Molecular Probes, #D-1306), fluorophore-coupled phalloidin [e.g., Alexa Fluor 594 phalloidin (red fluorescence, Molecular Probes #A-12381)], glycerol (87%, analytical grade) (e.g., Merck/VWR International #104094), Mowiol 4-88 (Calbiochem #475904), Tris base (e.g., Trizma base, Sigma-Aldrich #T 1503), n-propyl gallate (Sigma-Aldrich #P 3130), and SuperFrost microscope glass slides (Fisher Scientific #9161161).

D. Instrumentation and Laboratory Equipment

1. *Centrifuges*

Tabletop centrifuge (e.g., centrifuge 5414D, Eppendorf #5425 000.219), centrifuge equipped with rotor suitable for microtiter plates, and 15- and 50-ml polypropylene tubes (e.g., centrifuge 5810, Eppendorf #5810 000.017 with rotor A-4-81 and A-4-81-MTP).

2. *Clean Benches and Cell Incubators*

Clean bench and cell culture incubator suitable for work with EPEC and EHEC pathogens in accordance with respective national safety regulations.

3. *Microscope*

Inverted microscope (e.g., Axiovert 135TV, Carl Zeiss Jena GmbH) equipped for epifluorescence microscopy with 40×/1.3NA and 100×/1.3NA Plan-NEOFLUAR oil immersion objectives, 1.6 and 2.5 optovar magnification, electronic shutters (e.g., Uniblitz Electronic 35-mm shutter including driver Model VMMD-1, BFI Optilas) to allow for computer-controlled opening of the light paths, excitation and emission filters (Omega Optical Inc. or Chroma Technology Corp.) to enable three-colour epifluorescence, and mercury short arc lamp (Osram, HBO103W/2) for fluorescence light path.

4. *Data Acquisition*

Preferably a back-illuminated, cooled charge-coupled device (CCD) camera (e.g., Princeton Research Instruments TKB 1000×800, SN J019820; Controller SN J0198609) driven, for instance, by IPLab (Scanalytics Inc.) or Metamorph software (Universal Imaging Corporation).

III. PROCEDURES

Solutions

1. *Cell culture growth media*: Cell culture growth medium suitable for the propagation of the cell line chosen. We use DMEM, low glucose supplemented with 10% heat-inactivated fetal bovine serum,

2 m*M* L-glutamine, and 50 U/ml each of penicillin and streptomycin for propagation of our embryonic fibroblast cell lines.

2. *Cell culture growth media for bacterial infection experiments*: For bacterial infection experiments, omit penicillin and streptomycin from growth media starting a day prior to infection.
3. *Standard bacterial cultures*: Prepare LB broth and LB agar plates according to standard protocols (e.g., Maniatis *et al.*, 1982).
4. *Preactivating culture of EPEC*: DMEM, low glucose for culturing EPEC prior to infection of cell monolayers.
5. *Preactivating culture of EHEC*: DMEM, low glucose, supplemented with 100 m*M* HEPES, pH 7.4, for culturing EHEC prior to infection of cell monolayers.
6. *Phosphate-buffered saline (PBS)*: 140 m*M* NaCl, 2.7 m*M* KCl, 10 m*M* Na_2HPO_4, 1.8 m*M* KH_2PO_4, pH 7.4
7. *Fixative*: Prepare a 4% solution of PFA in PBS, pH 7.4. Paraformaldehyde is very toxic, work in fumehood when preparing stock, do not inhale, and wear gloves. To prepare 100 ml, add 4 g PFA to 90 ml PBS. In order for the PFA to dissolve, heat the solution to 60–65°C with continuous stirring. If necessary, adjust the pH to 7.4 by adding NaOH (be patient!). Do not heat solution above 70°C, as PFA will degrade. Let cool to room temperature, check the pH, and adjust with PBS to full volume. Filter through paper filter to remove insoluble aggregates and store in aliquots (e.g., 15 and 50 ml) at −20°C.
8. *Cell permeabilization*: Fixative supplemented with 0.1% Triton X-100 just prior to use. Make up a 10% stock solution of Triton X-100 in PBS and store at 4°C.
9. *Antibody diluent*: 1% BSA in PBS. Prepare and store in suitable aliquots (1–2 ml) at −20°C.
10. *Blocking solution*: 10% normal goat serum in PBS. Prepare from frozen aliquots of serum just prior to use.
11. *Antibody Mixtures*: Dilute antibodies just prior to use to appropriate working concentration in 1% BSA in PBS. The ideal concentration will result in a strong signal with no or little background staining and has to be established experimentally for each new antibody. To start, follow instructions given by the supplier. When using a concentrated primary antibody, a 1:100 dilution resulting in about 10 μg/ml should be a good starting point for immunofluorescence microscopy. Secondary antibodies conjugated to fluorophores are available from numerous suppliers. In our hands, secondary reagents coupled to Alexa Fluor dyes from Molecular Probes have worked very well. Note, however, that most polyclonal antisera will exhibit unspecific cross-reaction with EPEC and EHEC bacteria. A way to avoid problems associated with unspecific labeling of bacteria is to use monoclonal reagents (see general comment in Section IV). We store our antibodies in the dark at 4°C in a refrigerator. For long-term storage, antibodies may be stored at −20 or −80°C. Follow the recommendations given by the supplier.
12. *Mounting medium*: Weigh out 6 g glycerol and add 2.4 g Mowiol 4-88. Stir thoroughly. Add 6 ml aqua dest and mix for several hours at room temperature. Add 0.2 ml 0.2*M* Tris–Cl, pH 8.5, and heat to 60°C for 10 min. Remove insoluble material by centrifugation at 6000 *g* for 30 min. Store in aliquots at −20°C. In order to reduce photobleaching during fluorescence microscopy, add 2.5–5 mg/ml *n*-propyl gallate prior to use (Giloh and Sedat, 1982).

A. Basic Protocol: Infection of Cell Monolayers with EPEC or EHEC

EPEC and EHEC pose a significant threat to human health, especially EHEC with its low infectious dose of ~10–100 cfu. When working with EPEC and EHEC, always wear protective clothing and work under an appropriate clean bench in accordance with national safety regulations. Decontaminate all materials that have been in contact with the pathogen.

Steps

1. Pretreat 12-mm round glass coverslips as follows: wash coverslips in a mixture of 60% absolute ethanol and 40% concentrated HCl for 30 min and rinse extensively with aqua dest (heating in a microwave oven is helpful). To dry coverslips, spread on lint-free absorptive paper. Sterilize for tissue culture either by autoclaving on the dry cycle at 220°C or by exposing to ultraviolet light for 45 min in a culture dish.
2. On day 1, streak EPEC and EHEC from frozen glycerol stocks onto fresh LB agar plates and incubate overnight at 37°C. Keep plates in the refrigerator until liquid overnight cultures are started.
3. On day 2, seed cells onto 12-mm pretreated coverslips in a 24-well tissue culture plate in cell culture growth media without antibiotics. Incubate overnight in a tissue culture incubator supplemented with CO_2 to allow the cells to adhere to the coverslips. For MEFs, we find microscopic analysis of actin pedestal formation easiest if cells have reached about 80% confluency at the time of analysis; the number of cells seeded should be adjusted accordingly. To reduce detachment of MEFs infected with EHEC, use fibronectin coated coverslips.
4. Start an overnight liquid culture of EPEC or EHEC in 5 ml LB medium with bacteria from the streaked plate. In our experience, small inoculation loads result in a higher infection efficiency. Grow overnight with aeration at 37°C on a rotary shaker at 180 rpm.
5. On day 3, collect bacteria from the 0.5-ml overnight culture by centrifugation for 3 min at 4500 rpm in a tabletop centrifuge. Wash bacterial pellet twice in DMEM and inoculate 25 ml DMEM (EPEC) or 25 ml DMEM supplemented with 100 m*M* HEPES, pH 7.4 (EHEC). Grow at 37°C at 180 rpm for 3 h.

Environmental cues such as temperature, pH, and osmolarity, as well as growth phase, have been described to influence the transcriptional regulation of expression of virulence factors, e.g., of the type III secretion system. Media conditions that appear to stimulate virulence are those considered to mimic the gastrointestinal tract (e.g., Beltrametti *et al.*, 1999; Kenny *et al.*, 1997). In our experience, preactivation of EPEC and EHEC by growth for 3 h in DMEM for EPEC or DMEM supplemented with 100 m*M* HEPES, pH 7.4, for EHEC is sufficient for reproducible efficient bacterial infection of cultured cell monolayers. In our hands, the morphological appearance of preactivated bacterial cultures correlates with their infectious capability: After 3 h, EPEC should be found aggregated to small clumps of bacteria but should not grow in long rows. Aggregation likely reflects the expression of type IV bundle-forming pili, rope-like appendages that represent an important additional virulence factor found to enhance initial adherence and virulence of EPEC (Bieber *et al.*, 1998). EHEC, which lack bundle-forming pili, should be present as single bacteria in preactivation cultures, but again should not grow in long rows.

6. In the meantime, exchange cell culture medium of cell monolayers with 1 ml per well of fresh cell culture medium without antibiotics.
7. Add 10 μl of the bacterial 3-h preactivation culture into each well. Initiate infection by brief centrifugation for 5 min at 650 *g* in a centrifuge prewarmed to at least room temperature, allowing bacteria and cells to make contact with each other.
8. Place in an incubator supplemented with 7.5% CO_2 at 37°C for 1 h. Incubation at lower

temperatures results in a reduced infection efficiency.

9. Remove media containing nonattached bacteria by aspiration and gently replace with 1 ml of fresh prewarmed culture medium without antibiotics per well. Be careful not to let the cells dry out when exchanging the medium.
10. Repeat medium changes every hour until a total infection time of 4.5 to 5 h has been reached.

In our hands, longer infection times generally result in pedestals of increased length, facilitating analysis of pedestal composition. In addition, after short incubation times, EPEC bacteria are found mostly adhered in aggregates, which may confound analysis. Later during infection, EPEC bacteria will distribute from aggregates, making it easier to distinguish single pedestals. However, too long incubation results in increased cell death and detachment of cells, especially when working with EHEC.

11. After the incubation period, remove the medium by aspiration and gently wash the coverslips twice with PBS prewarmed to 37°C to remove unattached bacteria.
12. For fixation, immerse each coverslip with 500 μl of fixative prewarmed to 37°C and incubate at room temperature for 20 min.
13. After fixation, gently wash twice with PBS and store in PBS at 4°C until proceeding to cell permeabilization and immunostaining.

B. Alternative Protocol: EPEC and EHEC Infections of Transiently Transfected Cell Monolayers

For experiments involving ectopic expression of host cell molecules, e.g., tagged with GFP, designed to analyse their role in actin pedestal formation induced by EPEC or EHEC (see general comments, Section IV), follow the basic protocol with the exceptions that cells are already seeded on day 1, and day 2 will be required for transfection of host cell monolayers with plasmid DNA in addition to starting bacterial overnight cultures.

Steps

1. On day 1, seed cells in regular cell culture growth media on coverslips as described in the basic protocol with the modification that the number of cells seeded should be reduced to allow for the additional day needed for transfection. Again, about 80% confluency of MEFs should be reached at the time of bacterial infection.
2. On day 2, prior to transfection, exchange the cell culture growth medium for 0.5 ml growth medium without antibiotics. Transfect cells growing on coverslips in 24-well plate 12–24 h overnight according to standard protocols, e.g., using FuGENE 6 transfection reagent. We use 0.2 μg of DNA per well of a 24-well plate with a ratio of FuGENE 6 to DNA of 3:1. In our hands, plasmid DNA prepared with MaxiPrep kits, e.g., from Qiagen or Invitrogen Corp. is sufficiently pure for cell transfection.

Good transfection efficiency is a prerequisite for analyses involving bacterial infections, as the number of cells that are both transfected and infected may be too low. You may want to consider testing different lots of fetal bovine serum, as we have detected great variability in transfection rates using different sera. Alternatively, you may have to resort to fluorescence-assisted cell sorting to enrich for the population expressing the GFP-tagged protein of interest. In this case, transfect cells in regular cell culture plates and seed onto coverslips after sorting.

C. Immunofluorescence Microscopy

See general comments about immunofluorescence microscopy analysis of EPEC- and EHEC-infected cells.

Steps

1. Permeablize cells in 0.1% Triton X-100 in PBS for 1 min.
2. Gently wash twice with PBS.
3. Prepare a humid chamber: Lay out a large square plastic dish (e.g., 24.5 × 24.5-cm polystyrene dish with lid) with a moistened piece of absorptive filter paper and coat with a layer of Parafilm M.

Perform all blocking and immunolabelling steps as follows: Using forceps, place coverslips with the cell side facing downward onto a drop of about 20 μl of blocking or antibody mixture spotted onto parafilm in the humid chamber. This helps minimize the amount of antibody needed without letting the cells dry out. In order to minimize the loss of cells when moving the coverslips after each incubation step, pipette a small amount of PBS right next to the edge of each coverslip after each incubation step. This will cause the coverslips to float on top of the PBS and will allow easy removal of the coverslips using forceps. Wash coverslips by carefully submerging them successively into three small beakers filled with PBS. In between individual washing steps, drain residual PBS by carefully streaking the edge of the coverslips across thin absorptive filter paper without allowing the cells to dry out.

4. Block with 10% normal goat serum in PBS for 20 min in humid chamber.
5. Stain with primary antibody mixture for 1 h at room temperature in humid chamber.
6. Wash three times with PBS. For polyclonal primary antibodies, include 0.1% Triton-X100 during first two washing steps.
7. Stain with secondary antibody mixture for 1 h at room temperature in humid chamber in the dark. To specifically label F-actin, use fluorescent dye-coupled phalloidin, e.g., Alexa Fluor 594 phalloidin (red fluorescence), added at a concentration of 1–2.5 U/ml. To detect Leatuis by staining bacterial DNA, blue fluorescent DAPI may be added.
8. Wash three times with PBS. For polyclonal primary antibodies, include 0.1% Triton-X100 during first two washing steps.
9. To mount coverslips onto microscope glass slides, place drops of about 5 μl mounting medium on ethanol-wiped slides. Avoid air bubbles. Using forceps, gently place coverslips on mounting medium with the cell side facing downward. Once in contact with the mounting medium, avoid moving the coverslips as this will cause cells to be ripped off. Carefully remove access mounting media or residual PBS by placing a piece of absorptive filter paper onto the glass slides without moving the coverslips.
10. Dry mount coverslips for 3–4 h at room temperature in the dark or overnight prior to microscopic observation when using oil immersion lenses.
11. Store samples in the dark at 4°C. It is advisable to complete photography of samples within 1–2 weeks, as background fluorescence increases over time.
12. Using epifluorescence and appropriate emission filters, infected cells are detected readily by fluorescence actin staining (FAS) of actin pedestals. Search for interesting cells (e.g., both transfected and infected) using low magnification (e.g., 40×). To resolve subcellular structures such as pedestals, switch to higher magnification (100×) with or without optovar magnification. Always use minimal exposure time required to get a good signal-to-noise ratio.

IV. GENERAL COMMENTS ON IMMUNOFLUORESCENT ANALYSIS OF EPEC- AND EHEC-INFECTED CELLS

It is important to consider that polyclonal antibodies are likely to exhibit cross-reactivity against *E. coli*, which will lead to unspecific labeling of EPEC or EHEC bacteria and may confound the analysis of the role of a specific protein in actin pedestal formation by immunofluorescence microscopy. Thus, it may be difficult to discern between recruitment of a protein to the very tips of pedestals or bacterial attachment sites and unspecific labeling of bacteria, especially when using low magnification. Always test secondary reagents for unspecific labeling of bacteria. Consider using affinity-purified reagents. For polyclonal primary reagents, try addition of 0.1% Triton-X100 to PBS during washing steps. Cross-reaction of polyclonal reagents is no problem when immunolabeling is only performed to detect bacteria, as is the case in the examples shown in Fig. 22.2, in which EPEC bacteria were detected with monoclonal antibodies against bacterial EspE (Deibel *et al.*, 1998) and Alexa Fluor 350 fluorescent-labeled goat antimouse secondary antibodies, whereas F-actin was stained with fluorescent Alexa Fluor 594 phalloidin, and the host cell protein Nck was visualized with the help of an EGFP tag. There are reports in the literature about preabsorption of antibodies against fixed bacteria prior to immunolabeling, which when tried in our laboratory was of little success. Another possible solution to overcome the problem may be the use of monoclonal secondary reagents (e.g., rat antimouse), which are available from various suppliers (e.g., Zymed Laboratories), although so far not as conjugates with AlexaFluor dyes.

An alternative approach to detect the subcellular localisation of a given protein during EPEC- or EHEC-induced pedestal formation is the use of the enhanced green fluorescent protein (EGFP) as a tag (see Fig. 22.2). It is important to keep in mind that tagging may significantly alter folding, activity, and proper subcellular localisation of the protein of interest. Thus, we recommend testing N- and C-terminal fusions and confirming that the localization pattern of an EGFP fusion protein equals the localization pattern of the untagged protein as detected by immunofluorescence microscopy. In addition, with the increasing availability and use of knockout cell lines, testing an EGFP-tagged protein for its ability to complement a null mutation may become a feasible possibility to ensure proper function of the EGFP-tagged protein.

Acknowledgments

We thank Florian Gunzer for the transmission electron micrograph of EHEC-infected piglet colon for Fig. 22.1, Klemens Rottner and Theresia Stradal for support and helpful discussions, and Brigitte Denker for excellent technical assistance. J.W. was supported by the DFG (SFB 621) and the Fonds der Chemischen Industrie.

References

Beltrametti, F., Kresse, A. U., and Guzman, C. A. (1999). Transcriptional regulation of the esp genes of enterohemorrhagic *Escherichia coli*. *J. Bacteriol.* **181**, 3409–3418.

Bieber, D., Ramer, S. W., Wu, C.-Y., Murray, W. J., Tobe, T., Fernandez, R., and Schoolnik, G. K. (1998). Type IV pili, transient bacterial aggregates, and virulence of enteropathogenic *Escherichia coli*. *Science* **280**, 2114–2118.

Campellone, K. G., and Leong, J. M. (2003). Tails of two Tirs: Actin pedestal formation by enteropathogenic *E. coli* and enterohemorrhagic *E. coli* O157:H7. *Curr. Opin. Microbiol.* **6**, 82–90.

Deibel, C., Kramer, S., Chakraborty, T., and Ebel, F. (1998). EspE, a novel secreted protein of attaching and effacing bacteria, is directly translocated into infected host cells, where it appears as a tyrosine-phosphorylated 90 kDa protein. *Mol. Microbiol.* **28**, 463–474.

Donnenberg, M. S., Tzipori, S., McKee, M. L., O'Brien, A. D., Alroy, J., and Kaper, J. B. (1993). The role of the eae gene of enterohemorrhagic *Escherichia coli* in intimate attachment *in vitro* and in a porcine model. *J. Clin. Invest.* **92**, 1418–1424.

Frischknecht, F., and Way, M. (2001). Surfing pathogens and the lessons learned for actin polymerization. *Trends Cell Biol.* **11**, 30–38.

Giloh, H., and Sedat, J. W. (1982). Fluorescence microscopy: Reduced photobleaching of rhodamine and fluorescein protein conjugates by n-propyl gallate. *Science* **217**, 1252–1255.

Griffin, P. M., Ostroff, S. M., Tauxe, R. V., Greene, K. D., Wells, J. G., Lewis, J. H., and Blake, P. A. (1988). Illnesses associated with *Escherichia coli* O157:H7 infections: A broad clinical spectrum. *Ann. Intern. Med.* **109**, 705–712.

Gruenheid, S., DeVinney, R., Bladt, F., Goosney, D., Gelkop, S., Gish, G. D., Pawson, T., and Finlay, B. B. (2001). Enteropathogenic *E. coli* Tir binds Nck to initiate actin pedestal formation in host cells. *Nature Cell Biol.* **3**, 856–859.

Kenny, B. (1999). Phosphorylation of tyrosine 474 of the enteropathogenic *Escherichia coli* (EPEC) Tir receptor molecule is essential for actin nucleating activity and is preceded by additional host modifications. *Mol. Microbiol.* **31**, 1229–1241.

Kenny, B., Abe, A., Stein, M., and Finlay, B. (1997). Enteropathogenic *Escherichia coli* protein secretion is induced in response to conditions similar to those in the gastrointestinal tract. *Infect. Immun.* **65**, 2606–2612.

Knutton, S., Lloyd, D. R., and McNeish, A. S. (1987). Adhesion of enteropathogenic *Escherichia coli* to human intestinal enterocytes and cultured human intestinal mucosa. *Infect. Immun.* **55**, 69–77.

Levine, M. M., Bergquist, E. J., Nalin, D. R., Waterman, D. H., Hornick, R. B., Young, C. R., and Sotman, S. (1978). *Escherichia coli* strains that cause diarrhoea but do not produce heat-labile or heat-stable enterotoxins and are non-invasive. *Lancet* **1**, 1119–1122.

Lommel, S., Benesch, S., Rottner, K., Franz, T., Wehland, J., and Kuhn, R. (2001). Actin pedestal formation by enteropathogenic *Escherichia coli* and intracellular motility of *Shigella flexneri* are abolished in N-WASP-defective cells. *EMBO Rep.* **2**, 850–857.

Lommel, S., Benesch, S., Rohde, M., Wehland, J., and Rottner, K. (2004) Enterohaemorrhagic and enteropathogenic *Escherichia coli* use different mechanisms for actin pedestal formation that converge on N-WASP, *Cell Microbiol.* **6**, 243–254.

Maniatis, T., Fritsch, E. F., and Sambrook, J. (1982). *"Molecular Cloning: A Laboratory Manual."* Cold Spring Harbor Laboratory, Cold Spring Harbor, NY.

Nataro, J. P., and Kaper, J. B. (1998). Diarrheagenic *Escherichia coli*. *Clin. Microbiol. Rev.* **11**, 142–201.

Perna, N. T., Plunkett, G., 3rd, Burland, V., Mau, B., Glasner, J. D., Rose, D. J., Mayhew, G. F., Evans, P. S., Gregor, J., Kirkpatrick, H. A., Posfai, G., Hackett, J., Klink, S., Boutin, A., Shao, Y., Miller, L., Grotbeck, E. J., Davis, N. W., Lim, A., Dimalanta, E. T., Potamousis, K. D., Apodaca, J., Anantharaman, T. S., Lin, J., Yen, G., Schwartz, D. C., Welch, R. A., and Blattner, F. R. (2001). Genome sequence of enterohaemorrhagic *Escherichia coli* O157:H7. *Nature* **409**, 529–533.

Scaplehorn, N., Holmstrom, A., Moreau, V., Frischknecht, F., Reckmann, I., and Way, M. (2002). Grb2 and Nck act cooperatively to promote actin-based motility of vaccinia virus. *Curr. Biol.* **12**, 740–745.

CHAPTER

23

Listeria monocytogenes: Techniques to Analyze Bacterial Infection *In Vitro*

Javier Pizarro-Cerdá and Pascale Cossart

OUTLINE

I. INTRODUCTION

Listeria monocytogenes is a Gram-positive food-borne pathogen responsible for listeriosis, a human infection with an overall 30% mortality rate, ranging from a clinically nonapparent fecal carriage to severe gastroenteritis, mother-to-child infections, and central nervous system infections (Dussurget *et al.*, 2004; Lecuit and Cossart, 2002; Vazquez-Boland *et al.*, 2001). At the individual level, *Listeria* has the capability to cross three internal barriers: the intestinal barrier, the blood–brain barrier, and the feto–placental barrier. At the cellular level, *Listeria* is able to enter macrophages, as well as nonprofessional phagocytes, such as epithelial cells. Within infected cells, *Listeria* disrupts its internalisation vacuole and escapes the phagocytic cascade by proliferating in the cytoplasm. In this environment, *Listeria* promotes actin polymerisation in order to move inside infected cells and induces its translocation to neighboring cells, where the intracellular infectious cycle starts again (Fig. 23.1). *Listeria* has become a paradigm

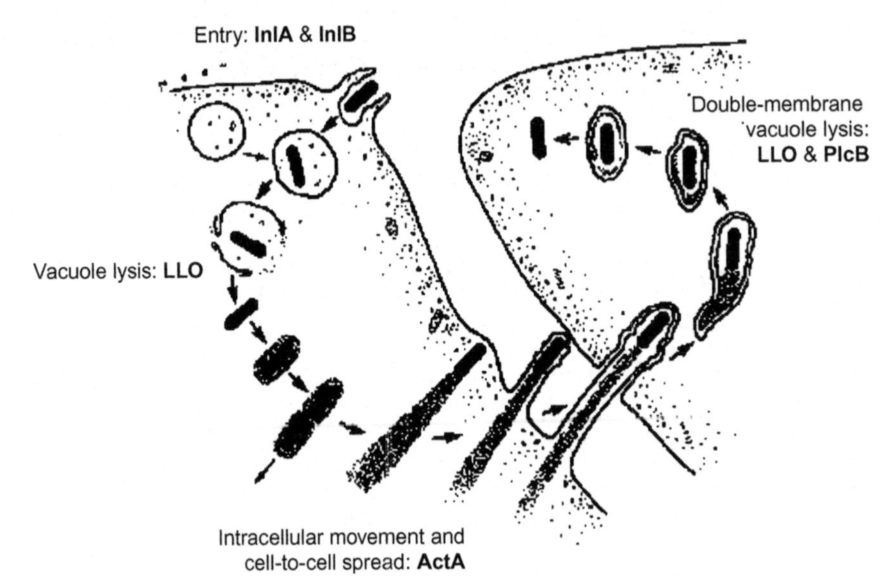

FIGURE 23.1 **Intracellular cycle of *Listeria monocytogenes*.** *Listeria* induces its own internalisation in nonphagocytic cells through interaction of the bacterial invasion proteins InlA and InlB with their cellular receptors E-cadherin and c-Met, respectively. Bacteria are initially located in a phagocytic vacuole that is disrupted by the lytic enzyme listeriolysin O (LL0), enabling *Listeria* to reach the cytoplasm and to proliferate in this environment. Intracellular actin-based movement is then induced due to the expression of the bacterial product ActA, which recruits several key players involved in actin polymerisation. Those *Listeria* that have reached the plasma membrane can form protrusions that extend to neighboring cells. Bacteria are contained in a double-membrane phagosome that is lysed through the action of listeriolysin O and the phosphatidylcholine phospholipase C PlcB, allowing the infectious cycle to start again.

of bacterial pathogenesis at both cellular and individual levels due to the extensive study of the bacterial virulence factors and of the cellular machinery necessary for the infectious process (Gaillard *et al.*, 1991; Kocks *et al.*, 1992; Dramsi *et al.*, 1995; Ireton *et al.*, 1996; Mengaud *et al.*, 1996; Lecuit *et al.*, 1999; Braun *et al.*, 2000; Shen *et al.*, 2000). This article describes some basic methods in cellular microbiology that have been used routinely to study the infection process of *Listeria in vitro* that can be adapted to study the pathogenesis of other intracellular bacteria.

II. MATERIALS AND INSTRUMENTATION

Listeria monocytogenes wild-type strain EGD (serovar 1/2a) comes from the bacterial collection in our laboratory at the Pasteur Institute. Lovo cells (Cat. No. CCL-229) are from the American Type Culture Collection. F-12K nutrient mixture (Kaighn's modification, Cat. No. 21127-022), l-glutamine (200 m*M*, 100×, Cat. No. 25030-024), fetal calf serum (Cat. No. 16000-044), trypsin/EDTA (1×, Cat. No. 25300-054), trypan blue (0.4% solution, Cat. No. 15250-061), and phosphate-buffered saline (PBS, pH 7.4, 10×, Cat. No. 70011-036) are from Gibco. Gentamicin (10 mg/ml, Cat. No. G-1272), bovine serum albumin (BSA, Cat. No. A-7030), saponin (Cat. No. S-7900), $CaCl_2$ (Cat. No. C-4901), $MgCl_2$ (Cat. No. M-8266), NaOH (Cat. No. 930-65), Tris (Cat. No. T-6066), paraformaldehyde (Cat. No. P-6148), and NH_4Cl (Cat. No. A-5666) are from Sigma. Malassez cell counting chamber (Cat. No. 99501.01), glass slides (Cat. No. 79703.01), coverslips (Cat. No. 79720.02), plastic boxes (Cat. No. 96496.02), curved forceps (Cat. No. 24434.01), Whatmann 3 mm chromatography paper (Cat. No. 10343.01), absorbant paper (Cat. No. 62908.01), Parafilm (Cat. No. 97949.01), orbital shaker (Cat. No. 08242.107), beakers (Cat. No. 11362.01), and pH meter (Cat. No. 71519.01) are from Merck-Eurolab. Twenty-four-well plates (Cat. No. 351157), 15-ml polystyrene conical tubes (Cat. No. 352099), petri dishes (Cat. No. 343633), 75-cm^2 vented cell culture flasks (Cat. No. 353110), brain–heart infusion agar (Cat. No. 241830), and brain–heart infusion broth (Cat. No. 299070) are from Becton-Dickinson Biosciences. Anti-*Listeria* rabbit polyclonal serum was developed in our laboratory at the Pasteur Institute (clone R69). The mouse monoclonal anti-E cadherin BTA-1 antibody is from R&D Systems. Rabbit polyclonal antihuman c-Met (Cat. No. C-28) is from Santa Cruz Biotechnology. Goat antirabbit antibodies coupled with Alexa Fluor 488 (Cat. No. A-11008) or Alexa Fluor 546 (Cat. No. A-11010) and goat antimouse antibodies coupled with Alexa Fluor 488 (Cat. No. A-11001) and Alexa Fluor 488 phalloidin (Cat. No. A-12379) are from Molecular Probes. Mowiol 4-88 (Cat. No. 475904) is from Calbiochem. The tabletop 5415D centrifuge is from Eppendorf. Spectrophotometer Ultrospec 3000Pro is from Amersham Pharmacia Biotech. The axiovert 135 microscope is from Zeiss. Image acquisition is performed with a RTE/CDD-1300 camera by Princeton Scientific Instruments. Image analysis and processing are done with Metamorph software from Universal Imaging Corporation.

III. PROCEDURES

A. Measurement of Bacterial Invasion: The Gentamicin Survival Assay

This procedure is a classical method that can be adapted to study the intracellular survival kinetics of different intracellular pathogens.

Solutions

1. *Cell culture medium*: F-12K nutrient mixture (Kaighn's modification) without supplements. Store at 4°C. Warm at 37°C just before use.
2. *Complete cell culture medium*: F-12K nutrient mixture (Kaighn's modification) supplemented with 2 m*M* l-glutamine and 10% fetal bovine serum. To make 500 ml, add 5 ml of a 200 m*M* l-glutamine stock solution and 50 ml of fetal bovine serum to 445 ml of F-12K nutrient mixture (Kaighn's modification). Store at 4°C and use within 7 to 8 weeks. Warm at 37°C just before use.
3. *Complete cell culture medium supplemented with gentamicin*: Complete cell culture medium supplemented with 10 μg/ml gentamicin. To make 50 ml, add 50 μl of the gentamicin stock solution (at 10 mg/ml) to 50 ml of F-12K nutrient mixture. Prepare fresh and warm at 37°C just before use.
4. *1× PBS*: To make 500 ml, add 50 ml of the PBS, pH 7.4, 10× solution to 450 ml of sterile distilled water. Store at room temperature. Warm at 37°C just before use.
5. *Trypsin/EDTA*: Aliquot in 5-ml portions in sterile 15-ml tubes. Keep at −20°C. Warm at 37°C just before use.
6. *Trypan blue solution*: Keep at room temperature.
7. *Brain–heart infusion agar*: Suspend 52 g of the powder in 1 liter of water and dissolve by frequent agitation. Autoclave at 121°C for 15 min, cool down at 50°C, and distribute in petri dishes. Keep at 4°C.
8. *Brain–heart infusion broth*: Suspend 37 g of the powder in 1 liter of water and dissolve by frequent agitation. Autoclave at 121°C for 15 min. Cool down and keep at room temperature.

Steps

1. Two days before the invasion assay, wash the LoVo cells (grown to 90% confluence in

a 75-cm^2 flask) by removing the cell culture medium and adding 10 ml of PBS. Remove the PBS, add 1 ml of trypsin/EDTA, and incubate the cells for 5 min at 37°C in a 10% CO_2 atmosphere. After this incubation, carefully hit the flask to release the cells and dilute them in 10 ml of fresh complete cell culture medium. Take 10 μl of this solution and mix with 90 μl of a 0.4% trypan blue solution. With a visible light microscope, count viable cells (the ones not stained by the dye) on a Malassez cell counting chamber (100 squares = 1 μl) and dilute them to a final concentration of 10^5cells/ml with complete cell culture medium. Add 500 μl of this solution to each of three wells of a 24-well plate and incubate for 48 h at 37°C in a 10% CO_2 atmosphere.

2. The evening before the invasion assay, take a colony of *L. monocytogenes* from a brain–heart infusion agar plate and inoculate in 5 ml of brain–heart infusion broth contained in a 15-ml tube. Incubate overnight on a rocking platform at 37°C.
3. The day of the assay, dilute 400 μl of the bacterial overnight culture in 5 ml of fresh brain–heart infusion broth contained in a 15-ml tube and incubate it on a rocking platform at 37°C to an OD_{600} of 0.8 (approximately 8×10^8 bacteria/ml).
4. Centrifuge 1 ml of the bacterial culture for 2 min at 6000 rpm and resuspend the pellet in 1 ml of PBS. Repeat this step twice; the second time resuspend the bacterial pellet in 1 ml of cell culture medium. Measure the OD and dilute bacteria to a final solution of 10^7 bacteria/ml with cell culture medium.
5. Wash LoVo cells in each well once with 1 ml of cell culture medium and add 500 μl of the bacterial dilution (MOI of 50, as we assume that cells have grown to a final density of 10^5 cells/well). Incubate for 1 h at 37°C in a 10% CO_2 atmosphere.
6. Make a 10-fold dilution of the initial bacterial solution by adding 100 μl of this preparation to 900 μl of sterile distilled water contained in an Eppendorf tube. Make another 10-fold dilution by adding 100 μl of the first 10-fold dilution to 900 μl of sterile distilled water contained in a different Eppendorf tube. Repeat this operation twice. Take a brain–heart infusion agar plate, divide the surface into four equal parts (by tracing a cross on the bottom of the plate), and plate 50 μl of each 10-fold dilution on each part. Incubate the plate overnight at 37°C in order to have an exact record of the precise number of bacteria that were inoculated on LoVo cells.
7. After 1 h of bacterial inoculation, wash LoVo cells carefully three times with 1 ml of PBS and then add 1 ml of the cell culture medium supplemented with gentamicin. Incubate the cells for another hour at 37°C in a 10% CO_2 atmosphere.
8. Wash the cells twice with 1 ml of PBS and add 1 ml of sterile distilled water to each well. Disrupt the cells by pipetting up and down water several times. Make a 10-fold dilution by adding 100 μl of the disrupted cell solution contained in each well to a different Eppendorf containing 900 μl of steriled distilled water. Make another 10-fold dilution for each tube by adding 100 μl of the first 10-fold dilution to 900 μl of sterile distilled water contained in a different Eppendorf tube. Repeat this operation once for each individual series of tubes. Take three brain–heart infusion agar plates, divide their surface in four equal parts (by tracing a cross on the bottom of the plate), and for each series of tubes, plate 50 μl of each 10-fold dilution on a different part of the agar plate (also plate 50 μl of the initial disrupted cell solution contained on the wells). Incubate the plate overnight at 37°C.
9. The day after the invasion assay, count the colony-forming units (CFU) on the agar plates at a dilution that allows a clear discrimination of individual bacterial

colonies. The percentage of *Listeria* that were able to invade the LoVo cells (and which survived the gentamicin treatment) is evaluated as the ratio of bacteria recovered from the disrupted cell solution divided by the total number of bacteria added to the monolayers (results are expressed as the mean of bacterial counts obtained from the three different wells ± standard deviation).

Comments

1. Cells are grown preferentially without antibiotics.
2. *Listeria* are P2 microorganisms and should be manipulated in a microbiological hood.
3. The replication of bacteria inside cells can be analysed over time if different subsets of cells are infected initially at the same time and then each subset of cells is lysed at different time points after inoculation. This procedure is helpful in order to track the proliferation or killing of bacteria in the intracellular environment.

B. Measurement of Bacterial Invasion: Differential Immunofluorescence Labeling of Intracellular versus Extracellular Bacteria

Visualisation of bacteria allows evaluation of several parameters, such as the proportion of cells in the inoculated monolayer to which bacteria attach, as well as the proportion of adherent versus invasive bacteria.

Solutions

1. *Cell culture medium, complete cell culture medium, 1× PBS, trypsin/EDTA, trypan blue solution, brain–heart infusion agar, and brain–heart infusion broth*: Prepare as described in the Section III,A.
2. *Fixation solution*: 3% paraformaldehyde in PBS. To prepare 100 ml, heat 80 ml of PBS to 80°C and then add 3 g of paraformaldehyde while stirring. Mix until clear (add a few drops of 1*N* NaOH to help dissolution if the liquid does not become clear after several minutes of stirring). Add 100 μl of 100 m*M* $CaCl_2$ and 100 μl of 100 m*M* $MgCl_2$, cool down at room temperature, add PBS up to 100 ml, and check pH (should be about 7.4). Aliquot in 5-ml portions and store at −20°C. Use only freshly thawed solution.
3. *Quenching solution*: 50 m*M* NH_4Cl in PBS. Prepare 100 ml of a stock solution of 1 *M* NH_4Cl in PBS by adding 5.35 g of NH_4Cl to 80 ml of PBS. Mix until clear and add PBS up to 100 ml. Keep at room temperature. The quenching solution used in the experiment will be prepared freshly by diluting 500 μl of the 1 *M* NH_4Cl stock solution in 9.5 ml of PBS.
4. *Blocking solution*: 1% BSA in PBS. To prepare 200 ml, add 2 g of BSA to 200 ml of PBS. Mix until clear. Prepare fresh just before performing the experiment.
5. *Permeabilising/blocking solution*: 0.05% saponin in blocking solution (1% BSA in PBS). To prepare 100 ml, add 50 μg of saponin to 100 ml of blocking solution. Prepare fresh just before performing the experiment.
6. *First primary antibody solution*: 1/500 dilution of R11 anti-*Listeria* rabbit polyclonal serum in blocking solution. To prepare 500 μl, add 1 μl of the R11 serum in 499 μl of blocking solution (1% BSA in PBS). Prepare fresh before performing the experiment.
7. *First secondary antibody solution*: 1/500 dilution of Alexa 546 antirabbit serum in blocking solution. Prepare as in step 6.
8. *Second primary antibody solution*: 1/500 dilution of R11 anti-*Listeria* rabbit polyclonal serum in permeabilising solution. Prepare as in step 6, using permeabilising/blocking solution instead of blocking solution.

9. *Second secondary antibody solution*: 1/500 dilution of Alexa 488 antirabbit serum in permeabilising solution. Prepare as in step 7, using permeabilising/blocking solution instead of blocking solution.
10. *Mounting solution*: 2.4% Mowiol/6% glycerol in 20 m*M* Tris buffer. To prepare 100 ml, add 2.4 g of Mowiol and 6 g of glycerol to 6 ml of distilled water. Keep at room temperature overnight. Add 90 ml of 0.2 *M* Tris, pH 8.5, warm in a water bath at 50°C, and mix regularly until total dissolution. Spin 15 min at 4000 rpm and aliquot supernatant in 1-ml portions at −20°C. Once an aliquot is thawed, keep at 4°C.

Steps

1. Infect LoVo cells following steps 1 to 5 described in Section III,A, adding 12-mm-diameter glass coverslips to the wells before seeding the cells (step 1).
2. After 1 h of bacterial inoculation, wash LoVo cells carefully three times with 1 ml of PBS and then add 500 μl of the fixation solution. Incubate for 15 min (all steps are now performed at room temperature).
3. Wash LoVo cells carefully once with 1 ml of PBS and add 500 μl of the quenching solution in order to block free aldehyde groups. Incubate for 10 min.
4. Prepare a wet incubation chamber by humidifying a 10 × 10-cm piece of Whatman 3 mm chromatographic paper. Add a sheet of Parafilm over the Whatman paper and, for each coverslip that is going to be labelled, add a drop of 30 μl of the first primary antibody solution to the Parafilm sheet.
5. Take carefully each coverslip from the wells with the forceps, carefully remove the excess of liquid from the coverslip by touching gently the tip of the coverslip with absorbant paper, and label extracellular bacteria by putting the coverslip (cells downwards) over the drop of the first primary antibody solution. Cover with a plastic lid and incubate for 30 min.
6. Prepare a second wet incubation chamber as described in step 4 and add a drop of 30 μl of the first secondary antibody solution to the Parafilm sheet for each coverslip that is going to be labeled.
7. Take carefully each coverslip with the forceps and, in order to wash the first primary antibody solution, dip it gently 10 times in a 50-ml beaker containing 40 ml of the blocking solution. Remove excess liquid as described in step 5 and label the first primary antibody by putting the coverslip (cells downwards) over the drop of the first secondary antibody solution. Cover with the plastic lid and incubate for 25 min.
8. Prepare a third wet incubation chamber as described in step 4 and, for each coverslip that is going to be labeled, add a drop of 30 μl of the second primary antibody solution to the Parafilm sheet.
9. Take carefully each coverslip with the forceps and wash the first secondary antibody solution by dipping the coverslip 10 times in a 50-ml beaker containing 40 ml of the blocking solution and then dipping it again 10 times in a 50-ml beaker containing 40 ml of the permeabilising/blocking solution. Remove excess liquid from the coverslip as described in step 5 and label intracellular and extracellular bacteria by putting the coverslip (cells downwards) over the drop of the second primary antibody solution. Cover with the plastic lid and incubate for 30 min.
10. Prepare a fourth wet incubation chamber as described in step 5 and add a drop of 30 μl of the second secondary antibody solution to the Parafilm sheet for every coverslip that is going to be labeled.
11. Take carefully each coverslip with the forceps and wash the second primary antibody solution by dipping the coverslip

10 times in a 50-ml beaker containing 40 ml of the permeabilising/blocking solution. Remove excess liquid as described in step 5 and label the second primary antibody by putting the coverslip (cells downwards) over the drop of the second secondary antibody solution. Cover with the plastic lid and incubate for 25 min.

12. Take carefully each coverslip and wash the second secondary antibody solution by dipping it 10 times in a 50-ml beaker containing 40 ml of the permeabilising/blocking solution, 10 times in a different 50-ml beaker containing 40 ml of PBS, and 10 times in a third 50-ml beaker containing 40 ml of distilled water. Remove excess water and put the coverslip carefully over a 10-μl drop of mounting medium on top of a glass slide.
13. Visualise the preparation with a fluorescence microscope (Fig. 23.2). Extracellular bacteria, labeled with the first and second primary antibodies and, accordingly, also labeled with the first and second secondary antibodies, will appear yellow (due to the superposition of the green signal of the Alexa 488 fluorochrome and of the red signal of the Alexa 546 fluorochrome). Intracellular bacteria, however, labeled only with the second primary antibody and, consequently, labelled only with the second secondary antibody, will appear green. Count extracellular and intracellular *Listeria* in at least 100 LoVo cells and express the results in a histogram standardising the total number of bacteria to 100% and expressing accordingly the corresponding proportions of extracellular and intracellular *Listeria*.

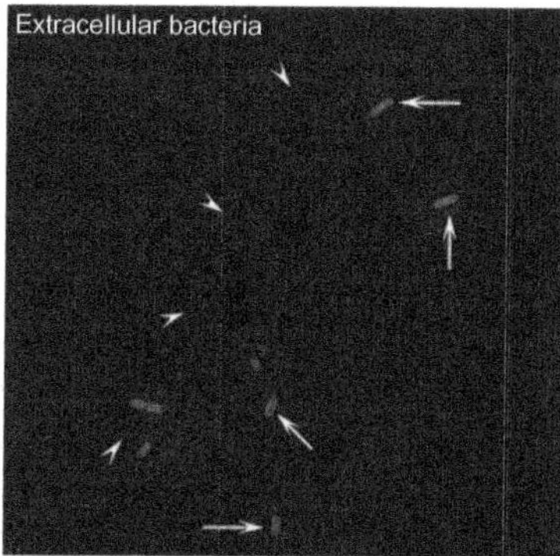

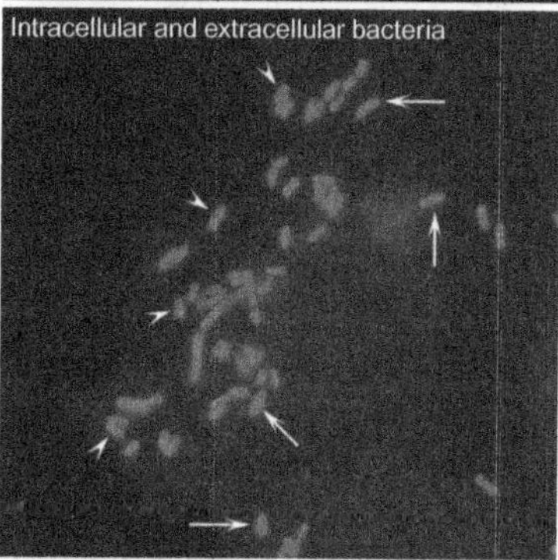

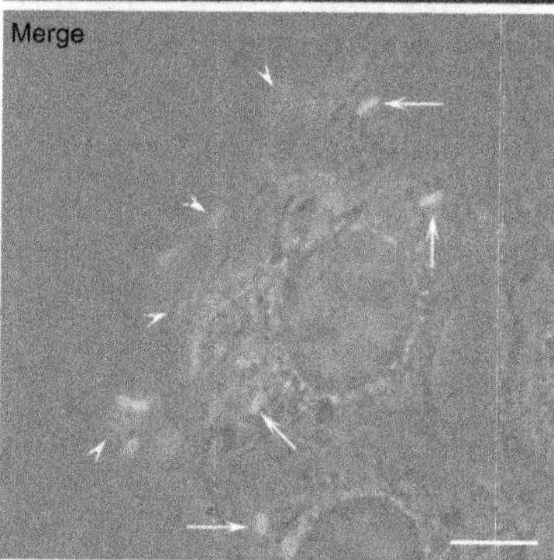

FIGURE 23.2 **Differential immunolabeling of intracellular versus extracellular bacteria.** LoVo cells were infected with *L. monocytogenes* as described in the text, and extracellular bacteria were distinguished from intracellular microorganisms by differential immunolabeling. (Top) Extracellular bacteria were labeled with an anti-*Listeria* serum without permeabilising LoVo cells. (Middle) The total population of bacteria (both extracellular and intracellular) was labeled with the same anti-*Listeria* serum after permeabilising LoVo cells. Intracellular bacteria are identified as those bacteria that are labeled only after the permeabilisation treatment. (Bottom) Phase image of LoVo cells and *Listeria*. Bar: 5 μm. Arrowheads, intracellular bacteria; arrows; extracellular bacteria.

Comments

1. Never allow cells to dry.
2. As fluorochromes are sensitive to light, perform incubations in the dark by covering with a plastic lid enveloped in aluminum foil.

3. Antibody dilutions are suggested accordingly to their use in our own laboratory. They could need finer adjustment depending on the batch used.

C. Visualisation of the Interaction of *Listeria* with Its Receptors E-Cadherin and c-Met

Two bacterial proteins, InlA (internalin) and InlB, are known to mediate the internalisation process of *Listeria* into target cells (Gaillard *et al.*, 1991; Dramsi *et al.*, 1995; Cossart *et al.*, 2003). InlA promotes entry into a subset of epithelial cells that express its cellular receptor, the adhesion molecule E-cadherin (Mengaud *et al.*, 1996; Lecuit *et al.*, 1999), a transmembrane protein that is normally involved in homophilic cell/cell interactions. Two different cellular receptors have been identified for InlB: the receptor for the globular part of the complement component C1q (gC1q-R) (Braun *et al.*, 2000) and c-Met, the receptor for the hepatocyte growth factor (Shen *et al.*, 2000). Interaction of InlA and InlB with their respective cellular receptors induces the recruitment to the site of bacterial entry of intracellular molecules that mediate cytoskeletal rearrangements necessary to induce bacterial invasion (Ireton *et al.*, 1996, 1999; Lecuit *et al.*, 2000; Bierne *et al.*, 2001). By immunofluorescence it is possible to visualise the recruitment of E-cadherin and c-Met at the site of bacterial invasion.

Solutions

1. *Cell culture medium, complete cell culture medium, 1× PBS, trypsin/EDTA, trypan blue solution, brain–heart infusion agar, and brain–heart infusion broth*: Prepare as described in Section III,A.
2. *Fixation solution, quenching solution, permeabilising/blocking solution, and mounting medium*: Prepare as described in Section III,B.
3. *Primary antibody solution*: 1/100 dilution of both the anti-E-cadherin mouse monoclonal serum BTA-1 and the anti-C-Met polyclonal serum C-28. To prepare 500 μl, add 5 μl of the anti-E-cadherin BTA-1 serum and 5 μl of the anti-C-Met C-28 serum to 490 μl of permeabilising/blocking solution. Prepare fresh before performing the experiment.
4. *Secondary antibody solution*: 1/500 dilution of both Alexa 488 antimouse serum and Alexa 546 antirabbit serum in permeabilising/blocking solution. To prepare 500 μl, add 1 μl of each secondary antibody to 498 μl of permeabilising/blocking solution freshly before use.

Steps

1. Infect LoVo cells following step 1 in Section III,B.
2. Fix and quench LoVo cells following steps 2 and 3 in Section III,B.
3. Label LoVo cells following steps 4 to 7 in Section III,B, taking into account that in the present experiment there is only one primary antibody solution (containing actually a mix of two different primary antibodies) and only one secondary antibody solution (containing a mix of two different secondary antibodies).
4. Wash and mount the cells following step 12 in Section III,B.
5. Visualise the preparation with a fluorescence microscope (Fig. 23.3). c-Met will be recognised by the red signal (from the Alexa 546 fluorochrome) and will be observed at different locations of the plasma membrane of LoVo cells. E-cadherin will be detected by the green signal (from the Alexa 488 fluorochrome) and will be found essentially at adherens junctions located in the borderline region between two different adjacent cells. *Listeria* can be visualised by phase contrast microscopy. Both receptors can also be detected at the site of entry of invading *Listeria*.

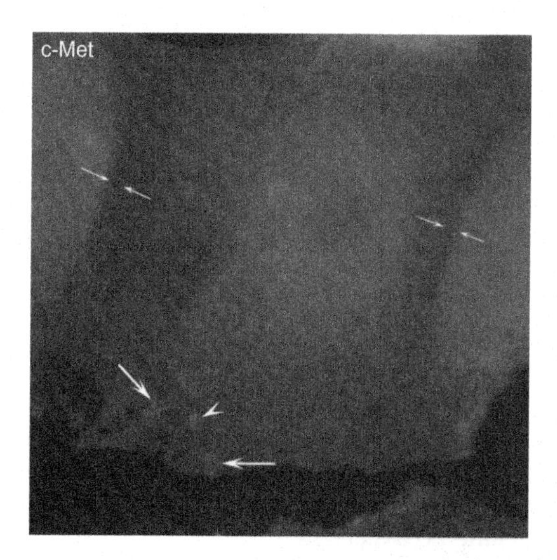

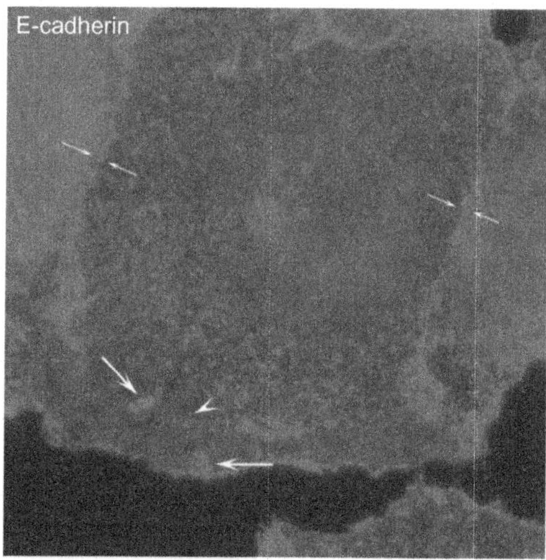

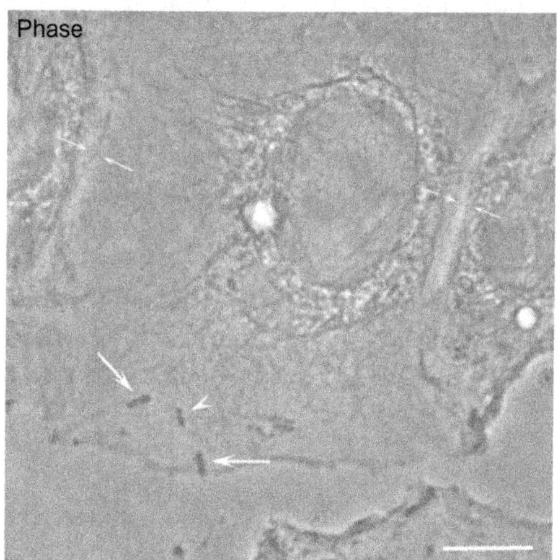

FIGURE 23.3 Interaction of *Listeria* with its receptors E-cadherin and c-Met. LoVo cells were infected with *L. monocytogenes* as described in the text, and its receptors E-cadherin and c-Met were immunolabeled. (Top left) c-Met shows a diffuse distribution in LoVo cells and is clearly absent at sites of adherens junctions formation, but is clearly recruited at sites of bacterial invasion. While some of the invading *Listeria* colocalise with both E-cadherin and c-Met, one bacterium is shown that only colocalises with c-Met and not with E-cadherin (arrowhead), suggesting that the interaction of *Listeria* with each one of their receptors can be uncoupled, as observed in cells lines such as Caco-2 (in which invasion is mediated mainly by the InlA/E-cadherin interaction) or Vero (in which invasion is driven exclusively by the InlB/c-Met interaction). (Top right) E-cadherin distribution at the plasma membrane of LoVo cells can be clearly seen at sites of adherens junction formation as well as at sites of bacterial entry. (Bottom) Phase image of LoVo cells and *Listeria*. Bar: 5 μm. Opposing arrows, adherens junction; large arrows, bacteria that recruit E-cadherin and c-Met at their site of entry; arrowheads, bacterium that only recruits c-Met at its site of entry.

D. Visualisation of *Listeria*-Induced Actin Comet Tails and Protrusions

A hallmark of *Listeria* infection is the polymerisation of cellular actin: bacteria that have escaped their internalisation vacuoles take advantage of this actin-based motility mechanism to reach and infect neighboring cells without leaving the intracellular environment. A single bacterial product, the protein ActA, is responsible for this phenomenon by recruiting to the bacterial rear end two cellular key players of the actin nucleation process: VASP and the Arp2/3 complex (Cossart, 2000). Polymerised actin can be visualised as a structure that extends from the rear end of the moving *Listeria* to the cell cytoplasm, which has been named the actin comet tail. Bacteria that have reached the plasma membrane of a primary infected cell form protrusions that extend into the cytoplasm of neighboring cells (or that extend simply to the extracellular space in the case of cells that do not have adjacent neighbors). These structures can be visualised by labeling the cellular actin with the fungal drug phalloidin in cells that have been infected with *Listeria* for at least 5 h.

Solutions

1. *Cell culture medium, complete cell culture medium, 1× PBS, trypsin/EDTA, trypan blue solution, brain–heart infusion agar, and brain–heart infusion broth*: Prepare as described in Section III,A.
2. *Fixation solution, quenching solution, permeabilising/blocking solution, and mounting medium*: Prepare as described in Section III,B.
3. *Primary antibody solution*: 1/500 dilution of R11 anti-*Listeria* rabbit polyclonal serum in permeabilising/blocking solution. To prepare 500 μl, add 1 μl of the R11 serum in 499 μl of permeabilising/blocking solution. Prepare fresh before performing the experiment.
4. *Secondary antibody and probe solution*: 1/500 dilution of Alexa 546 antirabbit serum in permeabilising/blocking solution and 1/100 dilution of phalloidin Alexa 488. To prepare 500 μl, add 1 μl of the Alexa 546 serum and 5 μl of the phalloidin Alexa 488 probe to 496 μl of permeabilising/blocking solution. Prepare freshly before use.

Steps

1. Infect LoVo cells following steps 1 to 5 and step 7 in Section III,A, adding 12-mm-diameter glass coverslips to the wells before seeding the cells (step 1) and incubating the LoVo cells in the presence of cell culture medium supplemented with gentamicin for 5 h insted of 1 h (step 7).
2. Fix and quench LoVo cells following steps 2 and 3 in Section III,B.
3. Label LoVo cells following steps 4 to 7 in Section III,B, taking into account that in the present experiment there is only one primary antibody solution (containing one primary antibody) and only one secondary antibody solution (containing a mix of one secondary antibody and one cytoskeletal probe).
4. Wash and mount the cells following step 12 in Section III,B.
5. Visualise the preparation using a fluorescence microscope (Fig. 23.4). Actin will be detected by the green signal (from the Alexa 488 fluorochrome) and will be observed at different cellular locations, such as stress fibres or cortical filaments near the plasma membrane. *Listeria* will appear as red (due to the signal of the Alexa 546 fluorochrome). Bacteria that have just lysed their phagosomal membrane and that have already started to polymerise actin will also be surrounded uniformly with the Alexa 488 fluorochrome. Those *Listeria* that broke their vacuole early will be associated with comet tails, which can be visualised starting at

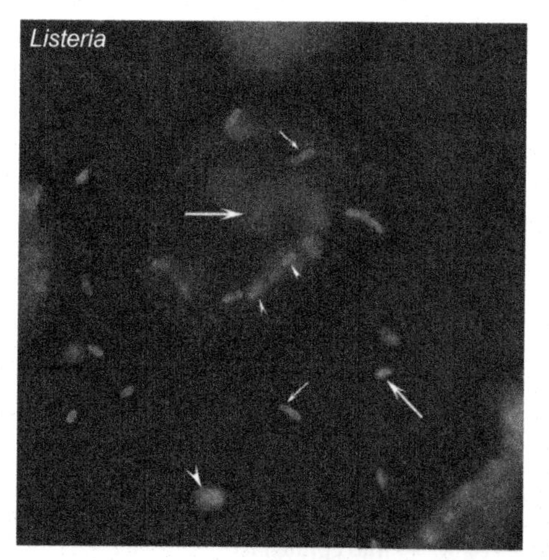

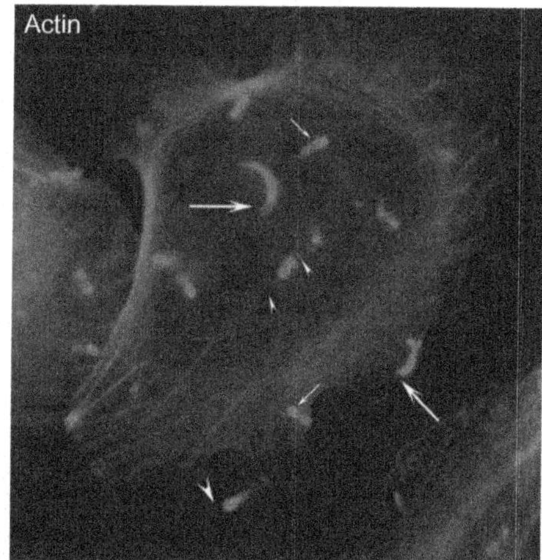

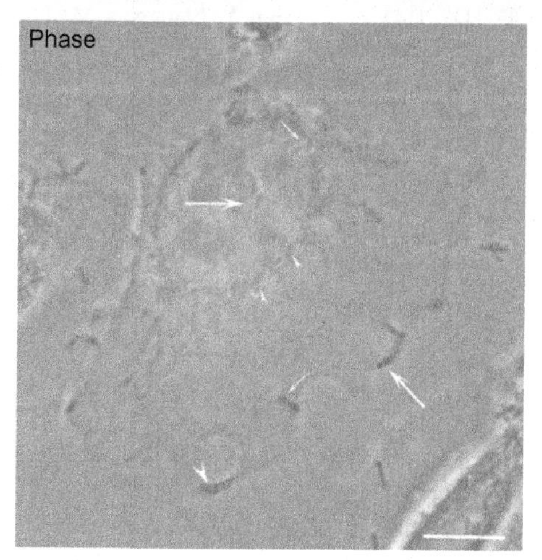

FIGURE 23.4 Actin polymerisation and protrusion formation by *Listeria*. LoVo cells were infected with *Listeria* as described in the text. Both bacteria and the actin cytoskeleton were labeled for immunofluorescence analysis. (Top left) Labeling of the total population of bacteria associated to LoVo cells. (Top right) Actin labeling with the fungal drug phalloidin. While some bacteria are not associated with actin (small arrowheads), other bacteria are detected at early phases of actin polymerisation (small arrows). Bacteria at later stages of their intracellular cycle can be observed at the tip of actin comet tails (big arrows). *Listeria* that have reached the plasma membrane can form protrusions that extend beyond the infected cell (big arrowhead). (Bottom) Phase image of LoVo cells and *Listeria*. Bar: 5 μm.

the rear pole of red bacteria and projecting into the cell cytoplasm. In the case of those *Listeria* that have reached the plasma membrane of an infected cell, a protrusion can be seen as a protruding structure that contains a red bacteria at its tip and a green actin comet tail following behind.

References

Bierne, H., Gouin, E., Roux, P., Caroui, P., Yin, H.L., and Cossart, P. (2001). A role for cofilin and LIM Kinase in Listeria induced phagocytosis. *J. Cell Biol.* **155**, 101–112.

Braun, L., Gebrehiwet, B., and Cossart, P. (2000). gC1q-R/p32, a C1q binding protein, is a novel receptor for *Listeria monocytogenes*. *EMBO J.* **19**, 1458–1466.

Cossart, P. (2000). Actin-based motility of pathogens: The Arp2/3 complex is a central player. *Cell. Microbiol.* **2**, 195–205.

Cossart, P., Pizarro-Cerda, J., and Lecuit, M. (2003). Invasion of mammalian cells by *Listeria monocytogenes:* Functional mimicry to subvert cellular functions. *Trends Cell Biol.* **13**, 23–31.

Dramsi, S., Biswas, I., Maguin, E., Braun, L., Mastroeni, P., and Cossart, P. (1995). Entry of *L. monocytogenes* into hepatocytes requires expression of InlB, a surface protein of the internalin multigene family. *Mol. Microbiol.* **16**, 251–261.

Dussurget, O., Pizarro-Cerda, J., and Cossart, P. (2004). Molecular determinauts of *Listeria monocytogenes* virulence. *Annu. Rev. Microbiol.* **58**, 587–610.

Gaillard, J.-L., Berche, P., Frehel, C., Gouin, E., and Cossart, P. (1991). Entry of *L. monocytogenes* into cells is mediated by internalin, a repeat protein reminiscent of surface antigens from Gram-positive cocci. *Cell* **65**, 1127–1141.

Ireton, K., Payrastre, B., and Cossart, P. (1999). The *Listeria monocytogenes* protein InlB is an agouist of mammalian phosphoinositide 3-kinase. *J. Biol. Chem.* **274**, 17025–17032.

Ireton, K., Payrastre, B., Chap, H., Ogawa, W., Sakaue, H., Kasuga, M., and Cossart, P. (1996). A role for phosphoinositide 3-kinase in bacterial invasion. *Science* **274**, 780–782.

Kocks, C., Gouin, E., Tabouret, M., Berche, P., Ohayon, H., and Cossart, P. (1992). *L. monocytogenes*-induced actin assembly requires the actA gene product, a surface protein. *Cell* **68**, 521–531.

Lecuit, M., and Cossart, P. (2002). Genetically-modified animal models for human infections: The *Listeria* paradigm. *Trends Mol. Med.* **8**, 537–542.

Lecuit, M., Dramsi, S., Gottardi, C., Fedor-Chaiken, M., Gumbiner, B., and Cossart, P. (1999). A single amino-acid in E-cadherin is responsible for host specificity towards the human pathogen *Listeria monocytogenes*. *EMBO J.* **18**, 3956–3963.

Lecuit, M., Hurme, R., Pizaro-Cerda, J., Ohayon, H., Geiger, B., and Cossart, P. (2000). A role for alpha- and beta-catenins in bacterial uptake. *Proc. Natl. Acad. Sci. USA* **97**, 10008–10013.

Mengaud, J., Ohayon, H., Gounon, P., Mège, J.-M., and Cossart, P. (1996). E-cadherin is the receptor for internalin, a surface protein required for entry of *Listeria monocytogenes* into epithelial cells. *Cell* **84**, 923–932.

Shen, Y., Naujokas, M., Park, M., and Ireton, K. (2000). InlB-dependent internalization of *Listeria* is mediated by the met receptor tyrosine kinase. *Cell* **103**, 501–510.

Vazquez-Boland, J.A., Kuhn, M., Berche, P., Chakraborty, T., Dominguez-Bernal, G., Goebel, W., Gonzalez-Zorn, B., Wehland, J., and Kreft, J. (2001). *Listeria* pathogenesis and molecular virulence determinants. *Clin. Microbiol. Rev.* **14**, 584–640.

CHAPTER

24

Measurement of Cellular Contractile Forces Using Patterned Elastomer

Nathalie Q. Balaban, Ulrich S. Schwarz, and Benjamin Geiger

OUTLINE

I. INTRODUCTION

The adhesive interaction of cells with their neighbors and with the extracellular matrix is a characteristic feature of all metazoan organisms. Cell adhesion is essential for cell migration, tissue assembly, and the direct communication of cells with their immediate environment. These interactions are mediated via specific cell surface receptors that specifically interact with the external surface, and link it, across the membrane, with the actin cytoskeleton (Geiger *et al.*, 2001). These multimolecular complexes are subjected to constant mechanical perturbation, generated either by the cellular contractile system or by changes in the neighborhood of the cell. To list just a few examples, in the course of cell migration, focal adhesions (FA) are formed, where contractile microfilament bundles, consisting of actin and myosin, are anchored. Their pulling on the substrate is involved in regulating the adhesion itself, as well as in coordinating the persistent forward movement of the cell. Mechanical perturbations are generated by diverse external factors, such as muscle contraction, blood flow, gravitational forces, and acoustic waves. Both internal and external forces apparently act at adhesion sites and modulate their organization and signaling activity (Geiger and Bershadsky, 2001; Riveline *et al.*, 2001; Galbraith *et al.*, 2002). In view of the major physiological significance of these cellular forces, it appears important to develop approaches for the accurate measurement of cellular forces with an appropriate sensitivity and spatial resolution (Beningo and Wang, 2002; Roy *et al.*, 2002). This article describes an approach for measuring such forces using the patterned polydimethylsiloxane (PDMS) elastomer as an adhesive substrate and cells expressing fluorescent focal adhesion molecules (Balaban *et al.*, 2001). These experiments demonstrated that cells maintain a constant stress at focal adhesion sites, of the order of $\sim 5\,nN/\mu m^2$, and that changes in cellular contractility lead rapidly to changes in FA organization. This article describes the technique used for the preparation of the patterned elastomeric substrate and for measuring the cellular forces.

II. MATERIALS AND EXPERIMENTATION

A. Lithography

Silicone wafer
Photoresist: Microposit S1805 (Shipley, Marlborough, MA)
Developer: Microposit MF-319 (Shipley)
HMDS or tridecafluorooctyltrichlorosilane (UCT, Bristol, PA)
Acetone, methanol
Spinner (Headway Research, Inc., Garland TX)
Mask aligner (Karl Suss, MJB3, Germany)
Chrome mask with desired pattern, or transparency
Digital hot plates
Oven
Scriber

B. Elastomer

PDMS: Sylgard 184, Dow Corning
No. 1 glass coverslips (diameter 25 mm)
Oven

C. Microscopy

The DeltaVision microscopy system (Applied Precision Inc., Issaquah, WA) is used in these experiments. Similar microscopy systems could be used for such purpose, provided that they are capable of accurately acquiring high-resolution images (512×512 pixels or better) at a high sensitivity (allowing for low-dose recording) and high dynamic range (12 bit or better).

D. Calibration

Bulk calibration: weights, clamps
Calibrated micropipettes with elastic constants 10–50 $nN\,\mu m^{-1}$

E. Image Analysis

Image analysis software: Priism (Applied Precision Inc.)
Data analysis software: Matlab (The Mathworks, Natick, MA) (This software can be purchase with image analysis toolbox and used to perform the image analysis as well)

F. Calculation of Forces

A detailed description of this calculation is provided in a web site established by Dr. Ulrich Schwartz (see p. 278 for additional notes/comments on this).

III. PROCEDURES

A. Lithography

This step provides molds consisting of a pattern of photoresist on silicon wafer that will be used for the patterning of elastomer substrates, a technique termed "soft lithography" (Whitesides *et al.*, 2001). It relies on access to a clean room with basic optical lithography. It also assumes the availability of the chrome mask with the pattern of interest. For high-resolution patterns (features below 1 μm), such masks are produced by electron beam lithography (also available in many clean room facilities). For low-resolution patterns (5 μm features and more), transparencies with appropriate resolution (5080 dpi) can be used.

1. Clean silicon wafer in acetone and then immediately wash with methanol and blow dry.
2. (Optional) Bake for 20 min in a 120°C oven.
3. Dispense HMDS in the middle of the wafer until full coverage and spin at 6000 rpm for 1 min.
4. Dispense S1805 photoresist in the middle of the wafer covering approximately one-third of the wafer and spin at 5000 rpm for 30 s.
5. Soft bake on hot plate at 80°C for 5 min.
6. Make sure the wafer and the mask are in contact in the mask aligner.
7. Expose for 10 s. This number is only an estimate; exposure and development times have to be adjusted for the specific exposure system. Typically an exposure of 100–140 mJ/cm^2 is recommended.
8. Develop for 10 s (see aforementioned comment).
9. Hard bake for 5 min in a 120°C oven.
10. In order to prevent sticking between the mold and the PDMS, an additional spin of HMDS can be performed at this stage. Alternatively, overnight exposure to vapors of tridecafluorooctyltrichlorosilane can be done.
11. Dice the wafer with a scriber in small pieces of about 5 × 5 mm. Each of these pieces can be reused as a mold many times for patterning of the PDMS substrates.

B. Elastomer Substrates

1. Pour about 30 ml of part A of the Sylgard 184 kit. Add 1 part of B (cross-linker) for 50 parts of A (in weight). Mix thoroughly.
2. Cover and let stand on bench until bubbles are scarce (typically 30 min).
3. Coat glass coverslips with the PDMS mixture. For large patterns that can be seen with low-resolution, long working distance objectives, put a few drops of the mixture in the middle of the coverslip on a flat surface covered with aluminium foil and let flow until the whole coverslip is covered. For short working distance objectives, a thinner layer is needed. This can be achieved by spinning No.1 coverslips at 1000 rpm on a spinner equipped with a 0.05-in. rotating head or "chuck." Alternatively, coverslips coated with PDMS can be put vertically to allow the flow of excess PDMS and attain a thinner thickness (typically 0.05 mm).
4. Place the coated coverslips on aluminium foil and bake in an oven at 65°C for 25 min until the top layer of the PDMS has started to solidify.
5. Place the rest of the PDMS mixture in the oven for bulk calibration purposes (see Section III,D).
6. Place a piece of the mold in the center of each coverslip (the patterned side of the mold should be in contact with the PDMS) and put back in oven overnight.
7. Delicately separate the mold from the patterned coverslip.
8. Glue each patterned coverslip to the botton of a 35-mm tissue culture plate with a 15-mm hole. This step can be done with melted paraffin.

9. Wash the coverslip and dish extensively with phosphate-buffered saline (PBS).
10. Incubate at 4°C overnight with a solution of 10 μg/ml of fibronectin in PBS.
11. Before plating the cells, wash the fibronectin solution with fresh plating medium twice and incubate the substrates with medium at 37°C for 1 h.
12. Plate cells at appropriate dilution and incubate immediately until observation.

C. Microscopy

Images are recorded, using phase-contrast or fluorescence optical mode, with a DeltaVision system digital microscopy system. This system is based on a Zeiss inverted microscope equipped with filter sets for multiple color microscopy and a high-resolution, scientific-grade CCD camera. Images are acquired using high numerical aperture oil objectives for the high-resolution pattern and focal adhesion detection (Fig. 24.3); long distance objectives are used for semiquantitative estimates of force using large patterns (Fig. 24.2). Image processing is conducted primarily using the Prism software of the DeltaVision system.

D. Calibration of Elastic Properties of the Elastomer

The aim of this step is to determine whether the cured PDMS is an elastomer and to obtain the two parameters that characterize its elasticity, namely the Young modulus (Y) and the Poisson ratio (σ) (Feynman, 1964).

1. *Bulk Calibration*

Cut strips of cured elastomer (see Section III,B, step 5). Typically a strip of 100 × 30 × 5 mm can be used. Hold each side with clamps wider than the strip. With a marker, draw two lines on the elastomer, away from the clips, separated by about 50 mm. The distance between the lines is marked as L in Fig. 24.1. Add increasing masses, m, to the lower clamp and measure the total increase in length, dL, until $dl > L$. Plot the relation $dl/L(m)$ with m expressed in kilograms. Verify that the relation is linear and extract the linear coefficient, α. The Young modulus (in Pa) is given by

$$Y = \frac{g}{A\alpha}$$

with $A = w^*h$, the cross section of the strip (in m^2), and $g = 9.8\,ms^{-2}$.

The Poisson ration, σ, is given by measuring the contraction in w and h:

$$\frac{dw}{w} = \frac{dh}{h} = -\sigma\frac{dl}{L}$$

The strips should be stretched overnight to verify that they relax to their original length, L, once the weights are removed.

2. *In Situ Calibration*

In order to verify that the plating of cells does not modify the elastic characteristics of

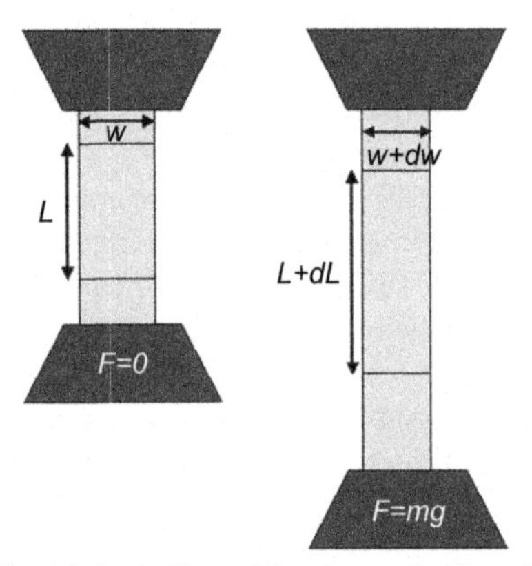

FIGURE 24.1 Bulk calibration of the elastomer. A stripe of PDMS prepared in parallel to preparation of elastomer substrates is checked for its elastic properties. The procedure consists of subjecting the stripe to increasing masses while measuring the force–extension response.

the surface, *in situ* calibration of the elasticity of the patterned substrate is performed under a microscope.

Mount the calibrated micropipette on a micromanipulator. Approach the surface slowly, keeping the micropipette parallel to the patterned elastomer surface. Once the tip of the micropipette is in contact with the surface, acquire an image of the patterned elastomer and of the tip of the pipette. Slowly move the stage by 2–3 μm in order to create a stress between the micropipette and the surface. Acquire a new image of the deformed pattern and of the micropipette bending. Usually, images at different magnifications are necessary in order to visualize the distortions of the pattern and of the micropipette. The force exerted by the micropipette is measured directly by quantifying the deflection of the pipette in the image. Verify that the calculation of force using the bulk Young modulus agrees with the known force exerted by the calibrated micropipette.

E. Image Analysis

Substrates with low resolution patterns such as the 30-mm grid shown in Fig. 24.2 can provide semiquantitative estimates of the mechanical perturbation due to cells and is useful for the rapid comparison between different cell types or conditions. However, for the quantitative measurement of the forces at single adhesion sites, high-resolution patterns of dots (Fig. 24.3a), as well as fluorescence tagging of adhesion sites, are needed (Fig. 24.3b). In this section, it is assumed that the following images are available for the same field of view.

a. Phase contrast of cells on a high-resolution pattern of dots ("dot_force.tif").
b. Phase contrast of the high-resolution pattern of dots, after forces have been relaxed ("dot_relaxed.tif"), namely after treatment with acto-myosin inhibitors or trypsinization. Alternatively, locations of the dots in the relaxed pattern can be generated directly as ("center_relaxed.dat") by assuming the regularity of the pattern.
c. Fluorescence image of adhesion sites ("focals.tif").

Registration of Images

In order to correct for eventual shifts during image acquisitions, all three images should be aligned using fixed reference points. Typically, corrections of a few pixels shift in x or y might be needed; angular shift are infrequent.

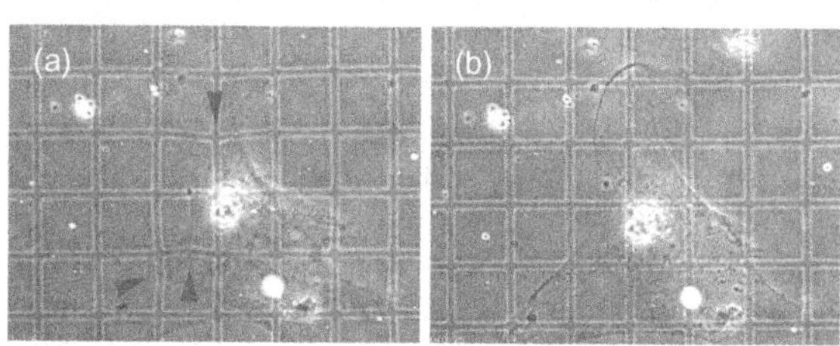

FIGURE 24.2 **Rapid semiquantitative estimate of forces (Balaban *et al.*, 2001).** A large grid-patterned elastomer is used for the easy visualization of distortions. (a) A rat cardiac fibroblast exerts forces on the elastomer substrates, which result in deformations of the grid (marked by arrows). (b) Once the forces are relaxed by adding an inhibitor of contractility, the grid returns to its original regular shape. Grid size: 30 μm.

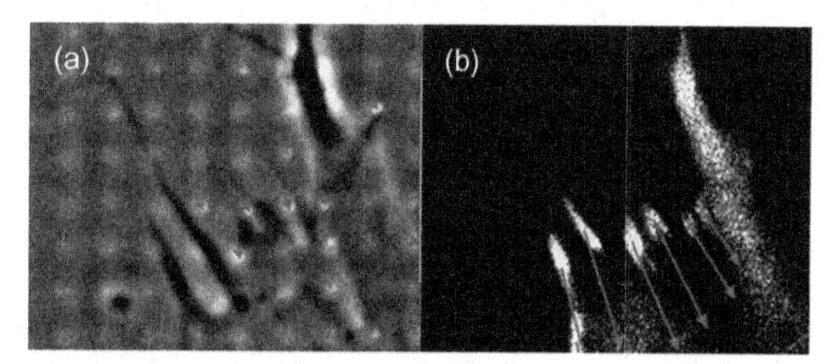

FIGURE 24.3 **From displacements to forces (Balaban *et al.*, 2001).** (a) Phase-contrast image of a small part of a fibroblast on the high-resolution dot pattern. Arrows denote displacements of the center of mass of the dots relative to the relaxed image. Pitch size: 2 μm. (b) Fluorescence image of the same field of view showing locations of the focal adhesions. The length of the arrows here indicates the forces at each focal adhesion, calculated from displacement data.

The step is performed using the registration option of the image analysis software.

Detection of the Pattern

This step automatically detects the dots in the two phase-contrast images and outputs their center-of-mass coordinates to an ASCII file. It is important to use the same parameters for both images.

a. Filter "dot_force.tif " and dot_relaxed.tif" with a large kernel in order to correct for background unevenness.
b. Depending on the image quality, use a high pass filter with a small kernel to enhance the dots contrast on both images.
c. Perform segmentation for the detection of the dots. Verify that the segmentation detected the dots properly. Correct manually for eventual defects.
d. Save the center of mass coordinates of the dots in each picture in two separate ASCII files ("center_relaxed.dat" and "center_force.dat").
e. Using Matlab, load the two ASCII files and arrange in pairs nearest neighbors according to their center of mass coordinates to identify equivalent dots between the two files. This will work for most of the image where there are no displacements. In areas where displacements are larger than the distance between dots, manual corrections might be needed (see Section III,F). Save data as an ASCII file ("lines.dat") whose first two columns are the x and y coordinates of the dots in "center_relaxed.dat" and the last two columns are the x and y coordinates of the equivalent dots from "center_force.dat".
f. Visualization of the displacements: load "lines.dat" using a procedure for drawing arrows in the image analysis software (DrawArrows in Prism) on top of the phase-contrast images. Correct for wrong or missing arrows. Save corrections.

3. *Detection of Focal Adhesions*

This step automatically detects focal adhesions in the fluorescence image of cells with GFP-tagged focal adhesion protein (Zamir *et al.*, 1999). ("focals.tif") and outputs their center-of-mass coordinates to an ASCII file ("focals.dat"). The segmentation procedure is analogous to the detection of dots in the pattern.

F. Calculation of Forces

The main advantage of flat elastic substrates is that they interfere little with cell adhesion as traditionally studied on glass or plastic surfaces. In particular, by using transparent elastomers such as PDMS, one can use the normal setups for light microscopy. The main disadvantage of flat elastic substrates is that a computational technique has to be implemented to calculate force from displacement. For the procedure described here, a package with Matlab routines can be downloaded from http://www.iwr.uni-heidelberg.de/organization/bioms/schwarz/celladhesion.html. For well-spread cells, one can assume that forces are applied to the top side of the elastic film mainly in a tangential fashion. Because the PDMS film is much thicker than typical displacements on its top side, one can moreover assume that displacements follow from forces as in the case of an elastic half-space. Because we focus on forces from single focal adhesions, where force is localized to a small region of space (namely the focal adhesion), we finally assume that each focal adhesion corresponds to one point force. Then displacement follows from force as described by the well-known Boussinesq solution for the Green function (Landau and Lifshitz, 1970). Since we only consider the case of small deformations (i.e., magnitude of strain tensor $|u_{ij}| < 1$), we deal with linear elasticity theory, and deformations resulting from different force centers can be simply added up to give overall deformation at a given point. However, because the Green function of a point force diverges for small distance r as $1/r$, we first use the Matlab program "data.m" to clear the file "lines.dat" from all displacement data points, which are closer to the positions of the focal adhesions in "focals.dat" than a distance of the order of the size of focal adhesions [this procedure can be justified with the concept of a force multipolar expansion (Schwarz *et al.*, 2002)]. This gives a new file "newlines.dat", which together with "focals.dat" is used by our main Matlab program, "inverse.m". Since we are interested in force as a function of displacement, but the Boussinesq solution from linear elasticity theory describes deformation as a function of force, we now have to solve an inverse problem. Because the elastic kernel effectively acts as a smoothing operation that removes high-frequency data, this inversion is ill posed in presence of noise in displacement data: small differences in displacement data can result in large differences in the force pattern. Due to the noise problem, we only can estimate force through a χ^2 minimization. Moreover, the target function has to be extended by a so-called regularization term, which prevents the program from reproducing every detail of displacement data through a force estimate with unrealistically large forces. Therefore, we add a quadratic term in force to the target function, which guarantees that forces do not become exceedingly large. The regularization procedure introduces an additional variable, the regularization parameter λ, whose value is fixed by the program "inverse.m" with the help of the so-called discrepancy principle, which states that the difference between measured displacement and displacement following from the force estimate should attain a certain nonvanishing value. Because the target function remains quadratic for the simple regularization term used, it can be inverted easily by singular value decomposition. For this purpose, the Matlab program "inverse.m" uses the freely available package of Matlab routines *Regularization Tools* by P. C. Hansen. Finally the force estimate is saved in a new ASCII file "forces.dat", which for each entry in "focals.dat" gives the corresponding force estimate. Force calibration assumes the value $\sigma = 0.5$ for the Poisson ratio and requires values for Young modulus Y and microscope resolution. The resolution of this procedure depends on the details of the force pattern under consideration, but can be estimated by data simulation. Even under the most favorable conditions, which can be achieved experimentally in regard to quality of displacement data, the original force cannot be reproduced completely and regularization is required to arrive at a reasonable estimate. Spatial and force resolutions have been found to be better than 4 μm and 4 nN, respectively. Therefore the calculated point force should be correlated with other properties of a focal adhesion only when no other focal adhesions are closer than the spatial resolution of our method.

IV. PITFALLS

1. The mixing of the PDMS should be done in many strokes (typically for a few minutes) and preferably in plastic dishes with lids.
2. Because of variations between batches, the 50:1 ratio of the Sylgard kit might need adjustment.
3. The first bake of the PDMS layer should be fine tuned: too much baking before putting the mold will result in a poor pattern, whereas insufficient baking will result in difficulties separating the mold from the PDMS. Baking times can be reduced by increasing the temperature.
4. Before dicing the patterned wafer, make sure the scriber is good by practicing on a similar unpatterned wafer.
5. The choice of the cells to use is important: immortalized cell lines often exert forces that are below the force resolution of this method.

6. Focal adhesions formed on PDMS may differ, at the molecular level, from those formed on other substrates and their composition (e.g., type of integrin) should be examined.

Acknowledgments

BG is the E. Neter Prof. for cell and tumor biology.

NQB is supported by the Bikura program of the Israel Science foundation and is a fellow of the Center for Complexity; The Horowitz Foundation.

References

Balaban, N. Q., Schwarz, U. S., *et al.* (2001). Force and focal adhesion assembly: A close relationship studied using elastic micropatterned substrates. *Nature Cell Biol.* **3**(5), 466–472.

Beningo, K. A., and Wang, Y. L. (2002). Flexible substrata for the detection of cellular traction forces. *Trends Cell Biol.* **12**(2), 79–84.

Feynman, R. P. (1964). *The Feynman Lecture on Physics*. Addison & Wesley.

Galbraith, C. G., Yamada, K. M., *et al.* (2002). The relationship between force and focal complex development. *J. Cell Biol.* **159**(4), 695–705.

Geiger, B., and Bershadsky, A. (2001). Assembly and mechanosensory function of focal contacts. *Curr. Opin. Cell Biol.* **13**(5), 584–592.

Geiger, B., Bershadsky, A., *et al.* (2001). Transmembrane crosstalk between the extracellular matrix–cytoskeleton crosstalk. *Nature Rev. Mol. Cell Biol.* **2**(11), 793–805.

Landau, L. D., and Lifshitz, E. M. (1970). "*Theory of Elasticity.*" Pergamon Press, Oxford.

Riveline, D., Zamir, E., *et al.* (2001). Focal contacts as mechanosensors: Externally applied local mechanical force induces growth of focal contacts by an mDia1-dependent and ROCK-independent mechanism. *J. Cell Biol.* **153**(6), 1175–1186.

Roy, P., Rajfur, Z., *et al.* (2002). Microscope-based techniques to study cell adhesion and migration. *Nature Cell Biol.* **4**(4), E91–E96.

Schwarz, U. S., Balaban, N. Q., *et al.* (2002). Calculation of forces at focal adhesions from elastic substrate data: The effect of localized force and the need for regularization. *Biophys. J.* **83**(3), 1380–1394.

Whitesides, G. M., Ostuni, E., *et al.* (2001). Soft lithography in biology and biochemistry. *Annu. Rev. Biomed. Eng.* **3**, 335–373.

Zamir, E., Katz, B. Z., *et al.* (1999). Molecular diversity of cell-matrix adhesions. *J. Cell Sci.* **112**(Pt. 11), 1655–1669.

Index

Printed in the United States
By Bookmasters